高等学校教材

General Chemistry

普通化学

王欣　耿旺昌　等编

化学工业出版社

·北京·

内容简介

General Chemistry（《普通化学》）根据我国高等院校工科专业学生对化学学习的需求及特点编写而成。书中从物质结构出发，阐明组成物质的原子及分子结构及其作用力，结合热力学、动力学、溶液、电化学、配位化学等基本化学原理，探讨当今迅速发展的材料、能源、环境、信息、国防等工程领域中的一些化学应用问题。本书内容精而不简，通用性强，注重可靠性、规范性的同时体现先进性和创新性。

本书可作为高等院校各工科专业的全英文或双语化学基础课教学用书，也可供自学者、工程技术人员参考使用。

图书在版编目（CIP）数据

普通化学 = General Chemistry : 英文/ 王欣等编.
—北京：化学工业出版社，2022.5 （2025. 5重印）
高等学校教材
ISBN 978-7-122-41365-9

Ⅰ.①普… Ⅱ.①王… Ⅲ.普通化学-高等学校-教材-英文 Ⅳ.①O6

中国版本图书馆 CIP 数据核字（2022）第 074683 号

责任编辑：成荣霞　　文字编辑：曹　敏
责任校对：张茜越　　装帧设计：王晓宇

出版发行：化学工业出版社（北京市东城区青年湖南街 13 号　邮政编码 100011）
印　　装：北京科印技术咨询服务有限公司数码印刷分部
787mm×1092mm　1/16　印张 20　字数 444 千字　　2025 年 5 月北京第 1 版第 4 次印刷

购书咨询：010-64518888　　售后服务：010-64518899
网　　址：http://www.cip.com.cn
凡购买本书，如有缺损质量问题，本社销售中心负责调换。

定　　价：88.00 元

前　言

近年来，随着国家“一带一路”倡议的制定和教育全球化进程的加快，在提升国际竞争力和加快高等教育国际化的驱动下，EMI（全英语教学）在母语为非英语的国家和地区迅速发展。当前，我国 EMI 教学面临教材选择和课程建设等多方面的问题与挑战。国外优秀的基础课教材价格高昂令人“望洋兴叹”，国内引入的部分影印版又因“年久失修”，不适应当前高等教育教学先进性和创新性的要求。*General Chemistry* 是针对我国高等院校化学化工类专业本科生化学基础课 EMI 教学需求而编写的一部通用型基础化学全英文教材。编者在积累了多年的教学体会和实际教学案例的基础上，从认识物质的基本规律出发，紧密联系当前迅速发展的材料、信息、能源、环境、国防等工程实例，深入浅出地介绍物质的组成及结构、化学的基本原理及在现代高科技中的应用等。本教材旨在向学生传授工科化学基础知识，健全学生的知识结构体系，训练学生掌握专业工程实际应用中涉及的基础化学知识和实验操作的技能，使学生能够应用化学的理论、思维去审视当今社会关注的如环境污染、能源及资源危机、工程材料的选择等热点问题。

本书突出工程应用，弱化理论推导，设置的例题习题等尽量与工程实践相结合，强调工程应用中的化学思维。教材中还融入了化学史教育。我国化学家傅鹰曾说：“化学给人以知识，化学史给人以智慧。”通过学习化学史了解知识背后的化学思维过程，了解化学概念变迁的来龙去脉，进而学习化学家的辩证思维方法，这比学到化学知识本身更为重要。通过化学史的教育将知识与智慧（方法）结合起来，对于提高现代教育质量有着非常重要的意义。因此，本教材将化学史的教育穿插于各个章节之中，努力做到将智慧与知识融为一体。

本书共分 10 章，由王欣提出编写大纲及要求，耿旺昌、张新丽、颜静等西北工业大学基础化学教研组教师分工协作，共同完成。王欣编写了第 1 章、第 4 章、第 5 章、第 10 章和第 3 章的部分内容（3.1）及附录；耿旺昌编写了第 7 章、第 8 章和第 9 章；张新丽编写了第 2 章和第 6 章；颜静编写了第 3 章的其余内容（3.2，3.3）。欧植泽、高云燕、王景霞、岳红、尹德忠等分工进行了校对及修订。全书由王欣、耿旺昌负责统稿。

西北工业大学化学与化工学院基础化学教研组其他成员对本书的编写提出了很多建设性的意见及建议，在此表示衷心的感谢！另外，在本书的编写过程中，参考了许多文献资料、国内外同类教材的部分内容，对原作者表示衷心的感谢！

由于编者水平有限，加之时间较为仓促，书中疏漏之处在所难免，抛砖引玉，恳请读者多提宝贵意见，以便改进，为国际化人才培养体系建设贡献微薄之力。

编者

Contents

Chapter 1 Introduction

Teaching contents	Learning requirements
Research object	Describe the research object of chemistry
History of chemistry	Introduce the history of chemistry
Modern chemistry and high technology	Describe some important chemical applications in modern science and technology engineering
Development prospect of chemistry	Describe some development trends of chemistry

1.1 The Purpose of Chemistry Research and Its Brief History

1.1.1 What Is Chemistry?

We are surrounded by chemistry in everyday life. Sometimes it is easy to see, as your science teacher conducts a big experiment in class. At other times, it can be extremely hard to see everyday chemistry at work, but nearly everything you touch or use has some elements of chemistry in it. What exactly is chemistry? People have different understandings of chemistry in different historical development periods, which leads to the definition of chemistry and the research objects of chemistry also in the process of dynamic development. Until the 2nd century BC, human activities related to chemistry were mainly ceramics making, copper smelting, wine making, iron making, *etc*. At that time, people did not yet have the concept of chemistry, just starting from actual needs and using the experience accumulated through long-term practice to obtain new living or production supplies. Chemistry in this period can be defined as a kind of craft (Brock W H, 2016). The main object of chemistry research is the practical transformation or processing of natural minerals, animals and plants. From the 2nd century BC to the 15th century, alchemy gradually formed pushing by the fanatical pursuit of gold by the feudal ruling class. The alchemist tried to achieve success through mysterious theories or magical substances, thereby making the chemical research objects of this period have a certain mystery. Therefore, the chemistry of this period can be defined as a secret technique. The object of chemistry research in this period was 'exploring the chemical transformation process of substances', which the alchemists did not realize.

Until the mid-17th century, the British scientist Robert Boyle proposed in his book *The Sceptical Chymist* that chemistry should no longer be regarded as a vassal of medicine or alchemy. Only after the research did it mark the emergence of modern chemistry and chemistry as a science. The object of chemistry research during this period was 'to find the composition of substances and split the compounds into elements.' Robert Boyle's widely misunderstood book elevated the status of chemistry (Principe L, 2011) and the development of chemistry entered its heyday after more than 300years. For the purpose of the following short introduction to the history of chemistry, we could use the definition of chemistry proposed by Linus Pauling and the object of chemical research has evolved into what we know as "the process of studying the composition, structure and properties of substances at the molecular and atomic level and creating new substances" (Pauling L,1970).

When Christian missionaries first began to translate western textbooks into Chinese in the 1870s, they needed a term to stand for chemistry. With the help of some Chinese scholars, such as Li Shanlan, they coined the phrase 'hua-xue', literally meaning "the study of change". This was an eminently sensible neologism, for chemistry is about change, and has also affected Japan and some other Asia countries.

1.1.2 A Brief History of Chemistry

The history of chemistry refers to a description and explanation of the formation, development and evolution of chemical science. The history of chemistry contains two connotations. The first is an objective description of the fact that researchers develop chemistry knowledge, and the second is the analysis and explanation of the dominant factors (including the chemist's thoughts, perseverance, *etc.*) behind the development of knowledge.

The Emperor Taizong of the Tang dynasty, Li Shimin said: "With history as a mirror, we can know the rise and fall of a dynasty." The Chinese chemist Fu Ying said: "Chemistry gives people knowledge, and the history of chemistry gives people wisdom." By studying the history of chemistry, you can understand the chemical thinking process behind the knowledge, understand the ins and outs of changes in chemical concepts, and then learn the dialectical thinking methods of chemists. This is more important than learning chemistry itself. Chemistry history education is the objective demand of modern chemistry education. Combining knowledge and wisdom (methods) through education in the history of chemistry is very important for improving the quality of modern education. Therefore, in this textbook, we intersect the history of chemistry in each chapter, and strive to integrate wisdom and knowledge.

The history of chemistry can be roughly divided into three periods.

(1) Ancient chemistry

From the germination of chemistry to the mid-17th century was the period of ancient chemistry. Yuanmou Ren about 1million years ago and Beijing Ren about 500,000years ago have known to use fire to heat and roast food, which has opened the earliest human chemical practice activities. From about 7000 BC to the 2nd century BC, mankind learned to make pottery. Smelting, dyeing, brewing and other handicrafts appeared one by one and were widely used based on the utilization of fire. Although these activities with practicability, experience, and scattered characteristics did not form systematic chemical knowledge, they laid a solid practical foundation for the birth of chemical science. Therefore, this period is called the germination period in the ancient chemical period. From the 2nd century BC to the 15th century, mankind carried out chemical practice activities, such as alchemy, and for the purpose of "substance conversion". Although practice has proved that "spotting into gold" and "eternal immortality" are not realistic, in this long-term practice stage, people have accumulated a lot of chemical knowledge and created a series of chemical operations such as evaporation, distillation, sublimation, and calcination. After entering the 16th century, Swiss pharmaceutical chemist Paracelsus (1493-1541) and others promoted the transformation of chemical activities from mysterious alchemy to more practical medicinal chemistry. The development of chemistry gradually embarked on the road of science. This period is called the period of medicine in the ancient chemistry.

(2) Modern period

From the middle of the 17th century to about 1895, chemistry is in a modern period. In 1661, the British chemist Robert Boyle (1627-1691) proposed the concept of scientific elements and published the book *The Sceptical Chymist*. The publication of this book marked the birth of the discipline of chemistry and the beginning of modern chemistry. In 1777, the French chemist Antoine-Laurent de Lavoisier (1743-1794) expounded the theory of combustion oxidation using quantitative chemistry experiments. In 1778, he published *Basics of Chemistry* and carried out preliminary classification of chemical elements. In 1803, the British chemist John Dalton (1766-1844) proposed scientific atomism based on the ancient Greek naive atomism and Newtonian particle theory, and believed that atoms were the smallest unit that could not be divided in chemical changes. In 1811, the Italian physicist Amedeo Avogadro (1776-1856) proposed a molecular hypothesis that "the same volume of gas contains the same number of molecules at the same temperature and pressure". This hypothesis is called Avogadro's Law. Later, the German chemists Justus von Liebig (1803-1873) and Weiler Friedrich (1800-1882) developed the theory of organic structure. The Russian chemist Mendeleev (1834-1907) discovered the periodic law of the elements and compiled the Periodic Table of the Elements. The development of these chemical theories has promoted the systematization of chemical sciences, and promoted the establishment of four basic disciplines such as inorganic chemistry, organic chemistry, analytical chemistry, and physical chemistry, and then entered the heyday of the modern chemistry.

(3) Development period

From the end of the 19th century to the present, chemistry has entered a rapid development period. In this period, the content of chemistry research has become deeper and more extensive. It began to form many cross fields and formed many cross-disciplines related to chemistry, such as pharmaceutical chemistry, biochemistry, environmental chemistry, electronics information chemistry, energy chemistry, *etc*. The characteristics of the chemical sciences during this period are undergoing the process from description to inference, from qualitative to quantitative, from macro to micro, from static to dynamic. Until present, it has spread to the field of molecular design and engineering.

1.2 Modern Chemistry and High Technology

In the past two decades, a technological revolution centered on high and new technology has been launched worldwide. These high and new technologies are distributed in the fields of national defense, energy, materials, electronic information, and environment, and have a decisive influence on a country's politics, economy and military. Chemical science is inextricably linked to these high and new technologies. Next, we will briefly introduce the application of chemical science in high and new technology from the following aspects.

1.2.1 Aerospace Technology

The two key fields of aerospace technology are materials and fuel propellants, and the development of these two fields is inseparable from chemistry.

At present, the materials widely used in aerospace fuselages, aero engines, rockets, missiles and spacecraft can be classified as metals, alloy materials, inorganic non-metallic materials and resin-based composite materials according to their chemical composition. Among them, the proportion of composite materials is an important indicator of the development level of aerospace technology. For example, the amount of composite materials used for Airbus' new ultra-wide body A-350 XWB passenger aircraft that first flew on June 14, 2013 reached 52%, exceeding the amount used in Boeing B-787 aircraft (50%), marking a new stage of the aerospace composite materials development. In the next 20 to 30years, the mainstream of the composite materials development for aerospace is carbon fiber reinforced resin matrix composite materials (CFRP, carbon fiber reinforced polymer/plastic). At present, the carbon fibers used in CFRP (6-8μm in diameter) are mainly classified into polyacrylonitrile-based carbon fibers and pitch-based carbon fibers. Compared with traditional metal-based materials, CFRP has the advantages of light weight, high strength, corrosion resistance, vibration resistance, fatigue resistance and durability. For example, the humidity in the aircraft cabin using CFRP material can be constant at 10% to 15%, but due to corrosion problems in the metal fuselage, the humidity can only be maintained between 5% to 10%. Higher humidity increases occupant comfort. There are currently three main types of resin matrix in CFRP: epoxy resin matrix, bismaleimide resin matrix and polyimide resin matrix. The corresponding use temperatures of these three types of resin matrix increase sequentially.

Due to the particularity of the CFRP application field, it has strict requirements on its service life and refresh cycle. Some of the first generation of carbon fiber aircraft will reach a service period of 25 to 30years. More and more aircraft will be scrapped, generating a large amount of CFRP waste. However, CFRP mostly uses thermosetting polymers (epoxy resin, phenolic resin, *etc.*) as the matrix resin. After curing, it forms a three-dimensional network structure, which cannot be molded or processed again and is difficult to handle. Therefore, the develop- ment of low-cost, green waste carbon fiber composite material recycling and reusing technology is urgent. The chemical recycling method is currently the most suitable method for processing CFRP waste, mainly including the pyrolysis method and the solvent decomposition method. Pyrolysis is a method that uses high temperature to decompose the resin in the composite material into small organic molecules to recover carbon fiber. According to the different reaction atmosphere, reactor and heating method, it is divided into thermal cracking, fluidized bed, vacuum cracking and microwave cracking. The solvent method refers to the use of the combined action of the solvent and heat to break the cross-link bonds in the polymer and decompose into low molecular weight polymers or small organic molecules dissolved in the solvent, thereby separating the resin matrix and the reinforcement. According to different reaction conditions and reagents used, the solvent method can be divided into nitric acid decomposition method, hydrogenation decomposition method, super / subcritical fluid decomposition method, atmospheric pressure solvent decomposition method and molten salt method.

The propellants used in aerospace are divided into liquid and solid. Conventional liquid rocket propellants include nitro oxidants such as nitrous oxide (N_2O_4) and red smoke nitric acid (HNO-27S), and hydrazine fuels such as unsymmetrical dimethylhydrazine (UDMH), methylhydrazine (MMH) and anhydrous hydrazine (HZ), *etc*. However, liquid propellants have the following shortcomings: ①The preparation time before the launch of a missile (rocket) is long, ranging from 1hour to more than two or three hours, and it is vulnerable to blows; ②Propellants are generally toxic or highly corrosive liquid, once launching a failure, the consequences are serious; ③Intolerable storage, once fueled, launching will be difficult to stop. Solid propellant can overcome the above disadvantages. Solid propellant is a kind of energetic composite material with specific properties, which is the power source of rocket and missile engines. The performance of solid propellants directly affects the combat effectiveness and survivability of missile weapons. In the 1960s and 1970s, the main solid propellants used were carboxyl-terminated polybutadiene (CTPB), hydroxyl-terminated polybutadiene (HTPB), cross-linked double base (XLDB) and composite double base (CDB) propellants. In the 1980s, double-base and composite solid propellants were further combined to produce nitrate plasticized polyether (NEPE) high-energy propellants. In the past ten years, it has mainly been based on azide propellants and high energy density materials (HEDM) propellants.

1.2.2 Energy and Environment

The energy crisis and global warming are two serious problems facing humanity in the 21st century. Both of these problems stem from the continued consumption of non-renewable fossil fuels and the releasing of the greenhouse gas, CO_2. Therefore, how to realize the development and preparation of clean energy or capture CO_2 gas and convert it into effective energy is a major scientific hotspot. Current methods of capturing CO_2 mainly include absorption, adsorption and membrane separation methods. Among them, the chemical absorption method using organic amine compounds as the main absorbent is more economical and effective. After adsorption, how to convert it into carbon resource and raw engineering material that can be effectively used is also one of the hot issues that scientists and technicians have paid attention to in recent years. For example, TiO_2, SiO_2, ZrO_2, Al_2O_3 and other oxides are used as carriers, and metals such as Co, Ni, Ru, and Rh are used as catalysts. Through catalytic hydrogenation of CO_2, they can be converted into organic raw materials such as hydrocarbons or alcohols. Using CO_2 electrocatalytic reduction to make it a useful chemical dye is also an active research field. In addition, the utilization and storage of solar energy also play a vital role in solving future energy crisis and environmental problems. The current use of solar energy is mainly in photovoltaic materials, solar cells, and the use of sunlight to catalytically split water to produce hydrogen. Recently, researchers have also successfully realized the storage of solar energy using chemical reactions. They have developed a transparent polymer film material that can store solar energy during the day and release heat when needed. It can be used for different surfaces such as window glass of buildings or human clothes.

1.2.3 Flexible Electronic Technology

Flexible Electronics refers to an emerging electronic technology that produces electronic devices based on organic or inorganic materials on foldable and extensible substrates. It has a wide range of application in the fields of information, energy, medical care, and national defense, such as flexible electronic displays, organic light-emitting diodes, and thin-film solar panels. The concept of flexible electronic devices can be traced back to the study of organic electronics. Professor Forrest, a microelectronics scientist at Princeton University, published a paper in *Nature* in 2004 to review the research status and development direction of organic electronics, and proposed the conceptual design and manufacturing method of pen-shaped flexible rollable displays (Forrest S R, 2004). In the past ten years, many scientific research institutions focused on flexible electronic technology, and conducted a lot of research on flexible electronic materials, devices and process technology. Western developed countries have formulated major research programs for flexible electronics in the design, preparation, and mass production of flexible displays and polymer electronic materials. Recently, researchers in China have drawn inspiration from the chameleon's color-changing principles and developed photonic ink materials that are expected to be used in color electronic paper.

1.3 The Role, Status and Development Prospect of Chemistry

1.3.1 The Role and Status of Chemistry

In the future, chemistry will continue to be a central science with new ideas, concepts and methods in terms of human life quality and safety. It should be said that chemistry on the 20th century played a major role in ensuring human needs for food, clothing, housing and transportation, and improving human living and health status. The problems of food, population, environment, resources and energy faced by human being in the 21st century are more serious. Solving these difficult problems depends on various disciplines, however, it always depends on the chemistry that studies the material basis in every case.

(1) Chemistry is still one of the main disciplines to solve food problems

Chemistry will design and synthesize functional molecules and structural materials, and clarify and control the mechanism of biological processes (such as photosynthesis, animal and plant growth) at the molecular level. It has laid a foundation for new agricultural materials (such as biodegradable agricultural films), biological fertilizers and biological pesticides. Using chemical and biological methods to increase the effective components of animal and plant foods for disease prevention, provide safe food and improve food storage and processing methods to reduce unsafe factors are all important contents of chemical research.

(2) Chemistry plays a key role in the national development and efficient and safe use of energy and resources

In terms of energy and resources, future chemistry should study efficient and clean conversion technology and control the chemical reactions of low-grade fuels. New energy sources, efficient and clean chemical power sources and fuel cells will become important energy sources in the 21st century. All need to start from the basic problems of chemistry. Mineral resource is non-renewable, and chemistry should study the separation and deep processing techniques and utilization of important mineral resources (such as rare earths).

(3) Chemistry continues to promote the development of materials science

All kinds of structural materials and functional materials will always be the material basis for human survival and development. Chemistry is the "source" of new materials. Every functional material is based on functional molecules. The discovery of a new structure with a certain function has caused a major breakthrough in materials science (such as fullerene). In the future, chemistry must not only design and synthesize molecules, but also assemble and construct these molecules into materials with specific functions. From superconductors, semiconductors to catalysts, drug-controlled release carriers, nanomaterials, *etc*., the structure of materials needs to be studied from the molecular level. The research on the active center of chemical mimic enzymes in the 20th century has made progress. In the future, there will be a breakthrough in the study of mimic enzymes that can be used in production, life and medical treatment. The breakthrough is based on the construction of both active centers and guaranteed active center functions of advanced structure compounds. In the 21st century, electronic information technology will develop in the direction of faster, smaller, and more powerful. At present, new technologies are springing up and developing rapidly, such as quantum computers, biological computers, molecular devices and biochips. The arrival of "information technology" requires chemists to make greater efforts to design and synthesize various substances and materials needed.

(4) Chemistry is an effective guarantee to improve the quality of human life

After meeting the need for survival, continuously improving the quality of life and survival safety is an important symbol of human progress. Chemistry can contribute to the improvement of the quality of life from three aspects: ① By studying the chemical basis of the biological effects (positive and negative) of various substances and energies, especially to understand the nature of the two sides, find out the best utilization plan. ② Research and development of environmentally friendly chemicals and daily necessities, research of environmentally friendly production methods, these two aspects are the main content of green chemistry. ③ Research on the disadvantages of small environments (such as indoor environment) for the generation, transformation and interaction in the human body propose ways to optimize the environment and establish a better living space.

Health is an important symbol of quality of life. Disease prevention is the central task of medicine in the 21st century. Chemistry can understand the pathological process at the molecular level, put forward the detection methods of early warning biomarkers and the prevention methods.

1.3.2 The Development Prospect for Chemistry

The development of disciplines in the 21st century is characterized by the intersection of disciplines to solve practical problems. The continuous development of chemistry has accelerated its integration with related disciplines. It combines the study of basic scientific problems with the solution of practical problems.

(1) Combination of structural diversity research and functional research

Facing the increasing demands of various functional molecules and materials, the research content, goals and ideas of synthetic chemistry should be greatly changed. In the future, synthetic chemistry should be able to design and synthesize new structures according to needs (functions). Synthetic chemistry should study not only traditional molecular synthesis, but also advanced structures (above the molecular level), especially ordered structures (tectonics). Combinatorial chemistry is based on the reverse thinking opposite to the traditional synthesis, coupled with the solid-phase synthesis technology, and inspired by the large-scale parallel operation of biology. It initially shows strong vitality, and the research in this field will become a new growth point. In addition, the discovery of new synthesis methods is an eternal theme.

(2) Research on complex chemical systems

At present, mathematics, physics, biology, finance and sociology are all studying complex problems. Complexity has the characteristics of multi-component, multi reaction and multi species. The structural complexity is mainly manifested in multi-level and high orderly structure. The process complexity is mainly the process of complex systems participating in chemical reactions. The complexity of state change also reflects the complexity of the process. These characteristics exist widely in biological and abiotic systems, as well as in the fields of industrial and agricultural production, medical treatment and environment. Therefore, it is of universal significance to study the chemical processes of complex systems. In the future, chemistry should clarify the chemical basis of structure and the structural changes above the molecular level, as well as the relationship between molecular level structure, properties and functions. Inspired by the research results of nano materials, physicists put forward the concept of mesoscopic scale. It is found that when an object is divided into nano scale, the properties of particles will change suddenly. Therefore, the quantum effect is proposed. For many years, chemists have believed that nature is determined by the atomic and molecular structure. In fact, many phenomena have shown that chemical properties also have scale effect, and there is a leap between chemical properties and scale. Therefore, we must pay attention to the multi-scale problem of complex systems in the future. In addition, the chemical process in complex systems is the core in the study of complex systems, because human beings are facing the changing in biology, environment, and so on. In the future, chemistry needs to study more extensive chemical behavior, establish tracking analysis methods and develop process theory.

(3) Establishment of new experimental methods and methodological research

In the future, chemical research will first develop advanced research ideas, research methods and related technologies to study the changes of molecular structure and properties at all levels. Miniaturization (such as biochip technology) and intellectualization of analytical instruments should be paid attention to. In addition, we should take notes of the establishment of time, space dynamics, on-site and real-time tracking and monitoring technologies, as well as the establishment of methods and instruments to study chemical processes in small-scale complex systems (such as scanning microscopy).

Looking forward to the future, chemistry in the 21st century is still a central science, practical science and innovative science. It will help us solve a series of major problems facing mankind. At the same time, chemistry will be further developed. We have reason to believe that chemistry in the 21st century will be more prosperous and enter the golden age.

Chapter 2 Chemical Composition and Aggregation of Substances

Teaching contents	Learning requirements
Ideal gases	Master the ideal gas laws and Dalton's law of partial pressures
Liquids, solution	Understand the basic properties of liquids and master the expression method of solution concentration
Solids	Understand the types of crystals and grasp the basic properties of various crystals
Other states	Understanding liquid crystal state, supercritical state and plasma state

2.1 Chemical Composition of the Substance

Through the previous study, we have learned that from the microscopic point of view, substances in the chemical sense are composed of particles with volume and mass, including three structures of atoms, molecules and ions; from the macroscopic point of view, chemical substances can be divided into two types: pure substance composed of a single substance and mixture composed of two or more pure substances according to the different elements. Among them, the pure substance includes the simple substance composed of the same element and the compound composed of the different elements.

Furthermore, the chemical composition of substances can be classified into metal, non-metal and rare gases. Compounds include inorganic compounds (acids, bases, salts and oxides) and organic compounds (organic small molecule compounds, organic high polymer, complexes and life macromolecules). The relationship between the chemical composition of these substances and their structures, chemical properties and applications will be described in the following chapters.

This chapter will focus on the macro aggregation state of various chemicals under certain conditions, as well as their different properties, functions and uses. In short, the three states of matter in nature are gases, liquids and solids. Gases have large molecular spacing, fast molecular movement speed, and the system is in a highly disordered state; liquids have close molecular spacing, strong intermolecular force, and the rotation and translation of molecules are active, showing significant fluidity; while solids have shorter distance between atomic or molecular, the molecules are difficult to move or rotate, so it does not have obvious fluidity. In addition, under special conditions, some substances have other aggregation states, such as liquid crystal state, supercritical state and plasma state. When the external conditions change, the states of matter can be transformed into each other.

2.2 Gases

2.2.1 Gas Phase

The gas phase is the least dense state of matter, which means the small particles that compose it(atoms and molecules) are widely separated and exert weak attractive forces on each other. Gases have no fixed shape, can diffuse arbitrarily and have compressibility. They are the most fluid of the three physical states. They can totally fill their containers, and their volumes depend on the pressure exerted on them. Temperature, pressure, volume, quantity and mass of matter are usually used to describe the state of gases. According to their chemical composition, gases can be divided into: ① gases composed of single elements, such as oxygen, nitrogen, *etc.*; ② gases composed of compounds, such as carbon dioxide, nitric oxide, *etc.*; ③ gas mixtures composed of two or more gases, the most typical example is air. In order to understand the physical state and properties of gases, gases are usually divided into ideal gases and real gases. This section focuses on the properties of ideal gases.

2.2.2 Ideal Gases

(1) Basic properties of ideal gases

An ideal gas is a term used to describe the behavior of hypothetical gases, given a set of simplifying assumptions. Ideal gas does not exist in reality, it is a kind of hypothetical gas, and it is a physical model proposed by people when studying the properties of real gases.

A gas is considered an ideal gas if it follows the four assumptions.

① Volume of individual particles is negligible.

② Attractive forces between molecules are negligible. Elastic collision can occur between molecules, but the collision is not enough to cause chemical reaction.

③ Gas molecules are not adsorbed by the vessel wall.

④ The average kinetic energy of a collection of gas particles is directly proportional to the Kelvin temperature of the gas.

(2) The ideal gas laws

In addition to pressure, gases may be characterized by volume and temperature. How each affects the others, gives rise to the ideal gas law equation.

The ideal gas equation expresses the pressure, volume, temperature, and moles of an ideal gas. This equation is as follows:

$$pV = nRT \tag{2-1}$$

Where p is the pressure, and the unit is Pa or kPa; V is the volume, and the unit is m^3; n is the number of moles, and the unit is mol; R is the ideal gas constant, and its numerical value is equal to 8.314J/(mol·K); T is the temperature in Kelvins, and the unit is K.

According to this equation, if the number of moles and temperature are constant, the product of pressure and volume is equal to a constant ($pV = k$); similarly, when volume and temperature are fixed, the pressure is directly proportional to the amount of substance ($p \propto n$). That is to say, there are four variables (p, V, n and T) in the ideal gas equation of state. If three of them are determined, another variable can be calculated.

Example 2-1 The density of the air under the standard condition is 1.29kg/m^3, try to calculate the molar mass of the air.

Solution The ideal gas equation

$$pV = nRT \tag{a}$$

$$n = m / M \tag{b}$$

$$\rho = m / V \tag{c}$$

where p—pressure of gas;

V—volume of gas;

n—the number of moles of the gas;

R—the ideal gas constant;

m—mass of gas;

M—molar mass of gas.

Combining (a),(b) and (c), we obtain

$$M = \rho \frac{RT}{p} \tag{d}$$

Under standard condition

$$p = 101325\text{Pa} = 101325\text{N} / \text{m}^2 = 101325\text{kg} \cdot \text{m} / (\text{s} \cdot \text{m})^2$$

$$T = 273.15\text{K}$$

Using the expression (d)

$$M = \rho \frac{RT}{p} = 1.29(\text{kg/m}^3) \times \frac{8.314[\text{J} / (\text{mol} \cdot \text{K})] \times 273.15\text{K}}{101325\text{kg} \cdot \text{m} / (\text{s} \cdot \text{m})^2}$$

$$= 0.0289\text{kg} / \text{mol} = 28.9\text{g} / \text{mol}$$

It is worth noting that the intermolecular force between gas particles and molecular volume of the real gases cannot be ignored, so the ideal gas equation is not suitable for the real gases in most case; however, real gases most closely resemble ideal gases at higher temperatures and lower pressures, and the result calculated by formula (2-1) is not much different from the actual measurement value. For example, at 400℃, the actual pressure when 1mol of water vapor occupies 22.4L is 248kPa, and the result of the formula (2-1) is 249kPa, which is very close to each other.

(3) Dalton's law of partial pressures

Dalton's law of partial pressures describes the properties of ideal gases. This empirical law was observed by John Dalton in 1801, and is based on the fact that, when two gases are mixed together, the gas particles tend to act independently of each other. The result is that, for a mixture of gases, the total pressure is equal to the pressures of all of the components of the mixture at the same temperature:

$$p = p_1 + p_2 + p_3 + \cdots + p_n \tag{2-2}$$

Dalton's law of partial pressures is only applicable to ideal gas mixture, and the real gases do not strictly follow it.

Example 2-2 Mix 20m^3 of oxygen with 50kPa pressure and 30m^3 of nitrogen with 200kPa pressure at 25℃. The volume is 110m^3 after mixing. Please calculate the total pressure.

Solution

The partial pressure of oxygen after mixing is

$$p_{O_2} = 20\text{m}^3 \times 50\text{kPa}/110\text{m}^3 = 9.09\text{kPa}$$

the partial pressure of nitrogen after mixing is

$$p_{N_2} = 30\text{m}^3 \times 200\text{kPa}/110\text{m}^3 = 54.55\text{kPa}$$

the total pressure after mixing is

$$p_{\text{total}} = 9.09\text{kPa}+54.55\text{kPa}=63.64\text{kPa}$$

2.2.3 Gas Storage and Transportation

In practical engineering application, gases are usually compressed in steel cylinder for storage and transportation. Gases cylinders are marked with different font colors to distinguish different gases. Table 2-1 lists some common gas storage and transportation regulations.

When using steel cylinders to store, transport and use gases, relevant safety management regulations shall be strictly followed, and improper operation is ease to cause safety problems. For example, gas cylinders shall be stored in a special warehouse, with obvious warning signs such as "danger" and "no fireworks"; when transporting gas cylinders with combustible gas, it is not only prohibited to use fireworks, but also the fire-fighting equipment shall be transported together to prevent serious losses. When using high-pressure gas cylinder, check the reading of pressure gauge frequently and pay close attention to whether there is leakage.

Table 2-1 Marking Regulations for Common Cylinders

Gas	Cylinder color	Marking of gas type		Marking of gas pressure	
		Word	Font color	Pressure/MPa	Color
Oxygen	Sky blue	Oxygen	Black	20	A white ring
				30	Two white rings
Hydrogen	Light green	Hydrogen	Bright red	20	A pale bright red ring
				30	Two pale bright red rings
Ammonia	Light yellow	Liquid ammonia	Black	—	
Nitrogen	Black	Nitrogen	White	20	A white ring
				30	Two white rings
Air(liquid)	Black	Liquid air	White	—	
Chlorine	Dark green	Liquid chlorine	White	—	
Acetylene	White	Acetylene not accessible to fire	Bright red	—	
Liquefied carbon dioxide	Aluminium white	Liquefied carbon dioxide	Black	20	A black ring
Liquefied petroleum gas(for civil use)	Silver grey	Liquefied petroleum gas	Bright red	—	

Steel cylinder is used for gas storage and transportation, mainly for hospitals, schools, research institutes and other civil institutions; in order to store ultra-high pressure gas medium, provide high pressure gas for spacecraft propulsion system, fluid management system and test system, alloy and composite gas cylinder is one of the pressure vessels for aerospace power system. At present, titanium alloy gas cylinder has gradually replaced the traditional aluminum alloy gas cylinder, and it has been widely used in aerospace field for its long life, high reliability and high working pressure. In order to adapt to the rapid development of aerospace industry, the research and development of double layer structure ultra-high pressure gas cylinder with metal shell as inner lining and carbon fiber composite winding surface has become the research focus in this field. This kind of gas cylinder has many advantages, such as lighter weight, higher container characteristic coefficient and more flexible design.

2.3 Liquids

2.3.1 Basic Properties of Liquids

Liquid is one of the existing states of natural substances, which is composed of molecules. There are attractive forces between molecules and also have molecular gap. Therefore, the properties of liquids lie between solids and gases. Liquids and gases are the same kind of fluid, they are no definite shape, the shape depend on the shape of the container. But unlike gases, liquids have a definite volume, which changes very little with the external pressure, that is, they have no significant compressibility and expansibility, which is more similar to solids in this respect.

(1) Density of liquids

The mass contained in unit volume of liquid is called the density of the liquid, which is expressed in ρ. For a homogeneous liquid, its density is expressed as

$$\rho = m / V \tag{2-3}$$

Where, m is the mass of the liquid, and V is the volume. Therefore, the unit of density is usually kg/m^3.

With the increase or decrease of temperature, the intensity and interaction force of thermal movement of liquid molecules change accordingly, which shows that the dense and volume of liquid changes with temperature. See Table 2-2 for the change of density and volume of liquid water with temperature under standard atmospheric pressure.

Table 2-2 Density and Volume of Liquid Water at Standard Atmospheric Pressure

Temperature/℃	Density/(g/cm^3)	Volume of 1g H_2O/cm^3
0	0.9998	1.0002
4	1.0000	1.0000
10	0.9997	1.0003
20	0.9982	1.0018
50	0.9880	1.0121
75	0.9749	1.0257
100	0.9584	1.0434

(2) Surface tension of liquids

The force acting on the surface of liquids to reduce the surface of liquids is called liquids surface tension. At present, there are two different explanations for its causes. One is based on the principle of keeping the lowest energy and keeping the state stable. It is generally believed that the liquid molecules interacting with the surrounding molecules are in a lower energy state, while the independent liquid molecules are in a higher energy state. Compared with the molecules inside the liquid, the liquid molecules at the gas-liquid interface have fewer neighboring molecules, so they have higher energy. In order to keep the energy minimum, the liquids need to reduce the surface area to achieve the goal of having the least high energy molecules. Another view is that the cohesion between liquid molecules is the main cause. As shown in Fig.2-1, each liquid molecule is subject to the pull from all directions of neighboring molecules, as a result, the net force on the inner molecules of the liquid is zero. The liquid molecules at the gas-liquid interface contact with each other are under the inner pull force due to the unbalanced force, so as to keep the minimum surface areas of liquids surface.

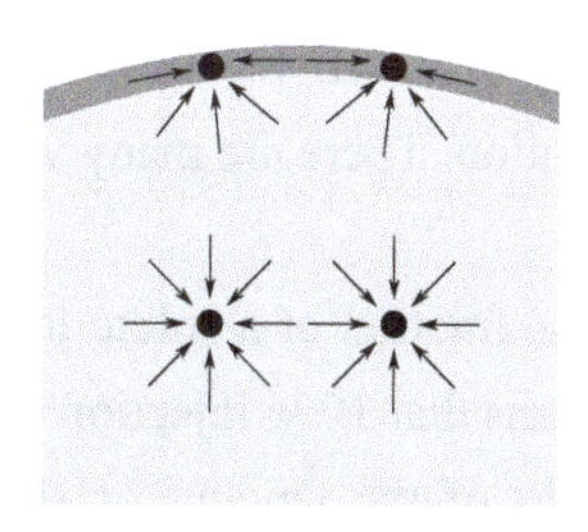

Fig.2-1 Schematic diagram of forces on liquid molecules

Because of the existence of liquids surface tension, life presents a variety of scenes, such as dew on leaves in the morning, ellipsoidal raindrops in the rain, mercury bead on smooth glass plate surface, *etc*. The principle of liquids surface tension is also applied in the development of modern science and technology. For example, the surface tension tank is a kind of pressure vessel which stores and supplies propellant for the liquid propulsion system of spacecraft. It manages the propellant according to the principle of liquids surface tension in microgravity environment, and provides the propellant without inclusions for the engine or thruster under the specified flow and acceleration conditions. At present, the surface tension tank has been widely used in space vehicles working in microgravity environment such as space shuttle, communication satellite and space station.

2.3.2 Solution

A solution is a homogeneous, stable mixture of two or more substances. Among them, the dispersed substance is called solute, and the dispersing medium is called solvent. The solvent is usually the main part of the mixture, which can be gases, liquids or solids. The solution has the same physical state as the solvent. See Table 2-3 for the classification of solutions according to the state of solute and solvent.

Table 2-3 Types of Solutions and Typical Examples

Types of solutions	State of solute	State of solvent	Typical examples
Liquid solution	Gas	Liquid	Dissolved oxygen in water, oxygen and carbon dioxide in blood, *etc.*
	Liquid	Liquid	Alcohol (ethanol in water)
	Solid	Liquid	Sugar water (sucrose in water), salt water (sodium chloride in water), body fluids contain a variety of electrolytes such as potassium and sodium
Gaseous solution	Gas	Gas	A mixture of gases, such as air
Solid solution	Gas	Solid	Hydrogen dissolved in palladium for storage
	Liquid	Solid	Deliquescence of amalgam (mercury dissolved in zinc), deliquescence of salt or sucrose
	Solid	Solid	Brass (zinc in copper), steel, polymer with plasticizer

Solution concentration is a physical variable used to qualitatively describe the relative content of each component in a solution. There are many ways to express it, and here are only several common methods.

① Mass fraction (w): the mass fraction of a solute in the total mass of the solution. For example, 30% glucose injection means that 100g injection contains 30g glucose.

② Mole fraction (x): the ratio of the amount of substance in a solution to the total amount of substance in the solution. In ethanol water system, the mole fraction of ethanol is 0.24, that is, 0.24mol ethanol / 1mol (water + ethanol).

③ Volume fraction (ϕ): At the same temperature and pressure, the ratio of the volume of a solute to the total volume of the solution. For example, 75% alcohol commonly used in medical treatment means that 75mL of alcohol is dissolved in water under normal temperature and pressure, and is prepared into 100mL solution.

④ Mass volume concentration (ρ): The mass of a solute in 1L solution, and unit is kg/m^3(kg/m^3 equals to g/L).

⑤ Quantity concentration of substance (c): The quantity of substance of a certain solute in 1L solution, and unit is mol/L.

⑥ Molarity (m_B): the quantity of a solute in 1kg solvent, and unit is mol/kg.

2.4 Solids

2.4.1 Introduction of Solids

Solid is a common form of matter. Different from gas and liquid, solid matter has a relatively fixed shape and hard texture. Generally, solids are divided into two categories: crystalline and amorphous.

A crystal is a solid in which particles (atoms, particles, molecules, *etc.*) are arranged periodically in three dimensions. Generally speaking, crystal appearance is mostly regular geometry. Relatively speaking, the particles inside the amorphous matter do not have periodic arrangement in the three-dimensional space. Although the short-range is orderly, the long-range is disordered, and its appearance is mostly irregular.

2.4.2 Crystal Characteristics

Crystal material has its inherent characteristics, which can be used to distinguish crystal and non crystal.

① Because of its long-range order, the crystal has regular geometry. For example, diamond crystal forms are octahedron, rhombus dodecahedron, tetrahedron and their aggregation. Non crystal (such as glass, rubber, *etc.*) does not have a certain geometric shape, so it is also called amorphous.

② The crystal has a fixed melting point, for example, as the most common ice. In the process of heating ice, there is no melting phenomenon below 0℃. When the temperature reaches 0℃, it starts to melt. Continue heating, the system temperature remains unchanged, and the temperature will continue to rise until all of them melt. However, because of the irregular arrangement of molecules and atoms, nanocrystal do not need to destroy their spatial lattice after absorbing heat. They are only used to improve the average kinetic energy. So when they absorb heat from the outside, they soften, and eventually become a liquid. There is no fixed melting point. For example, in the process of heating rosin, it softens at 50-70℃ and becomes liquid at above 70℃.

③ The crystal is anisotropic. Due to the periodic arrangement of particles in the crystal, the crystal shows different physical and chemical properties (hardness, thermal expansion coefficient, thermal conductivity, refractive index, *etc.*) in different directions. For example, graphite can be peeled off along the direction of its lamellar structure, or even form a single layer of graphene. At the same time, the conductivity along the lamellar structure is more than 10000 times of that in the vertical direction. Because of the disordered arrangement of particles in the amorphous, there is no anisotropy.

2.4.3 Types of Crystals

The same substance can have different crystal forms. For example, graphite and diamond are composed of carbon, but their crystal forms are different, and their physical and chemical properties are different. This phenomenon is called allomorphism. Under certain conditions, the allomorphism can change with each other. For example, under high pressure, graphite crystal can be transformed into diamond, which is also the principle of the preparation of artificial diamond.

Crystal materials can be divided into single crystal and polycrystalline. The particles in the single crystal are arranged orderly according to some rules, while the polycrystalline is formed by the aggregation of several single crystals. Because of its regular arrangement, the preparation of single crystal and its structure analysis had important applications in inorganic materials and protein chemistry. In practical operation, in the process of crystal growth, due to the change of temperature, pressure, concentration of medium components and so on, some irregularity or imperfection of particle arrangement is often caused, thus forming crystal defects. The existence of crystal defects has an obvious effect on the properties of crystals.

According to the type of matter and the attractive forces between different particles, crystals can be divided into ionic crystals, atomic crystals (also known as covalent crystal), molecular crystals and metal crystal. Table 2-4 lists the basic properties of various crystals.

Table 2-4　Basic Properties of Various Crystals

Crystal Type	Ionic crystals	Atomic crystals	Molecular crystals		Metallic crystals
Particle	Positive and negative ion	Atomic	Polar molecule	Nonpolarity molecule	Atom, positive ion
Acting force	Ionic bond	Covalent bond	Intermolecular force, hydrogen bond	Intermolecular force	Metal bond
Melting and boiling point	High	Very high	Low	Very low	—
Hardness	Hard	Very stiff	Soft	Very soft	—
Mechanical properties	Brittle		Weak	Very weak	Ductile extension
Electrical conductivity	Melting, solution conduction	Non conductor	Solid and liquid non conducting, aqueous solution conduction	Non conductor	Good conductor
Solution	Soluble in polar solvent	Insoluble	Soluble in polar solvent	Soluble in nonpolar solvent	Insoluble
Example	NaCl	Diamond	HCl	CO_2	Au

In addition to the above four types, there are also mixed crystal (containing more than two types of lattice), such as graphite, mica, black phosphorus, *etc*.

(1)Ionic crystals

Ionic crystals are formed by the combination of anions and cations through ionic bonds. The so-called ionic bond is the strong electrostatic interaction between anion and cation. The ion bond has no saturation and directionality. Most of the salts, strong bases and active metal oxides are ionic crystals, and the most typical representative is sodium chloride. The stability of ionic crystal can be reflected by its lattice energy. The lattice energy (*U*) of ionic crystal refers to the energy needed to be absorbed by removing the amount of ionic crystal per unit substance to change it into a gaseous component ion under the standard condition. The larger the lattice energy is, the stronger the ionic bond between the anion and the cation is; the higher the melting point of the ionic crystal is, the greater its hardness is, and the ionic crystal is more stable.

$$MX(s) \longrightarrow M^+(g) + X^-(g) \qquad U_{NaCl} = 786\ kJ/mol$$

Because the ionic crystals rely on the attraction and combination of anions and cations, the ions are combined with each other by ionic bond. They are arranged according to strict rules, so they have very beautiful crystal cells. Fig.2-2 shows the cell structure of several ionic crystals.

Fig.2-2 shows the cubic closest packing of the larger Cl^- ions at lattice points, while the smaller Na^+ ions fill the octahedral holes in the NaCl structure. An octahedral hole is one surrounded by six particles (six Cl^- ions in NaCl) that form the shape of an octahedron. Other ionic compounds with the same structure are AgCl, AgBr, MgO, CaO and NiO.

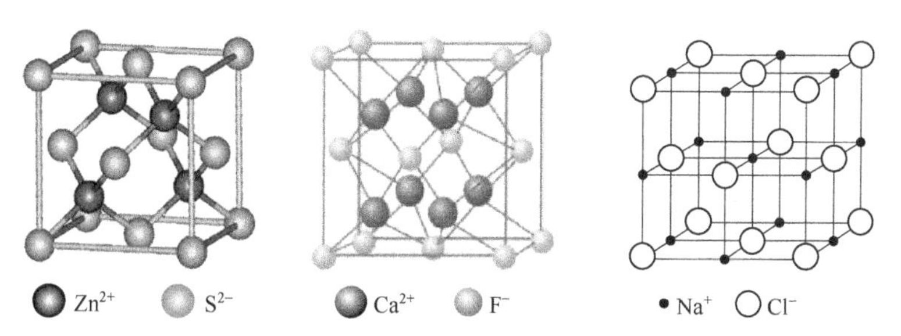

Fig.2-2 Cells structure of ZnS, CaF_2 and NaCl

(2) Network covalent crystals

Atomic crystal is a kind of crystal with the highest hardness and high melting point. Atoms in the crystal are connected by covalent bond to form a three-dimensional network structure. For example, diamond is a macromolecular solid because it has a network of C atoms each bonded tetrahedrally (four sp^3 hybrid orbitals) with the sp^3 hybrid orbitals from four other C atoms. Hence, each diamond crystal is an extremely large C_n molecule, a macromolecule, forming tetrahedral structure with strong stability in space. Therefore, diamond is the hardest natural material known. Other examples of network covalent solids include quartz (SiO_2), and silicon carbide (SiC).

(3) Molecular crystals

Molecular crystals are composed of polar or nonpolar molecules, which are connected by weak interactions such as hydrogen bond and van der Waals force. Because of the weak intermolecular forces, molecular crystals generally have lower melting point, boiling point and smaller hardness. With the electrons tightly held in the covalent bonds within the molecules, molecular solids are poor conductors of heat and electricity. Many molecular crystals are gaseous or liquid at room temperature, for example, O_2 and CO_2 are gases; ethanol and glacial acetic acid are liquids; iodine, naphthalene are highly volatile and can be sublimed directly without melting.

For crystals of the same type of molecules, the melting and boiling points increase with the increase of molecular weight, for example, the order of melting and boiling points of halogen is: $F_2 > Cl_2 > Br_2 > I_2$. With the increase of the number of carbon atoms, the melting boiling point of organic compounds increases. However, there are van der Waals force and hydrogen bond interaction among HF, H_2O, NH_3, CH_3CH_2OH, *etc.* the melting and boiling points of them are high.

(4) Metallic crystals

Metal crystal is a three-dimensional ordered structure formed by metal atoms and ions through metal bonds. Metal bond is a kind of special covalent bond, also known as metal modified covalent bond. It was put forward by Dutch scientist Lorenz in 1916 according to the theory of free electron.

Lorenz believes that atoms and ions of particles in metal crystals share the free electrons in the crystals, and free electrons can move in the whole crystals. Therefore, metal bonds can be understood as "metal atoms lose valence electrons, form a skeleton, and then soak in the ocean of electrons". Due to the existence of these delocalized free electrons in the metal crystals, the metal bond has no directivity and saturation. Free electrons can move freely in the whole crystals, so metals are good conductors of heat and electricity and have a wide range of melting points. Metal atoms can also slide with each other while keeping the metal bond unbroken, so the metal also has ductility.

2.4.4 Solid's Applications

Metal material is a kind of very important solid material, which has important application in national defense and military industry. For example, the most important aircraft manufacturing is closely related to chemical crystallography. At present, the thrust to weight ratio of advanced aeroengine reaches 12-15, and the gas temperature before turbine reaches 1800-2100℃, which requires the research and development of a new generation of high-temperature materials, such as Ti-Al metal matrix composite resistant to 816℃, Nb-Si alloys resistant to 1200-1400℃, ceramic materials resistant to 1538℃, IR base alloys resistant to 1800℃, and heat insulation coating resistant to 1371℃. Titanium alloys have the excellent performance required by aviation structure, which have been paid attention to as early as the 1950s. However, due to the poor machinability of titanium at room temperature, only pure titanium or low-strength titanium alloy parts with simple shape can be manufactured. Later, due to the development of hot forming methods and equipment, the advanced titanium alloy structure can be expanded in the application of aerospace vehicles.

2.5 Other Forms of Matter

In addition to the main forms mentioned above, there are other material forms in nature, such as liquid crystal state, supercritical state and plasma state.

2.5.1 Liquid Crystals

Liquid crystal is a kind of special material which is different from solid, liquid and gas in a certain temperature range. It has not only the unique birefringence of anisotropic crystal, but also the fluidity of liquid. Generally, it can be divided into thermotropic liquid crystal and lyotropic liquid crystal.

In 1850, the Prussian doctor Rudolf Virchow and others found that the extract of nerve fibers contained an unusual substance. In 1877, Otto Lehmann, a German physicist, first observed the phenomenon of liquid crystal with a polarizing microscope, but he did not understand the cause of this phenomenon.

On March 14, 1883, Friedrich Reinitzer, a plant physiologist, observed the abnormal behavior of cholesteryl benzoate (as shown in Fig.2-3) during hot melting when he studied cholesterol derivatives with a polarizing microscope. He found that the substance melted at 145.5℃ and produced a turbid substance with luster. When the temperature rose to 178.5℃, the luster disappeared and the liquid was transparent. When this clear liquid cools a little, turbidity reappears, and appears blue in an instant.

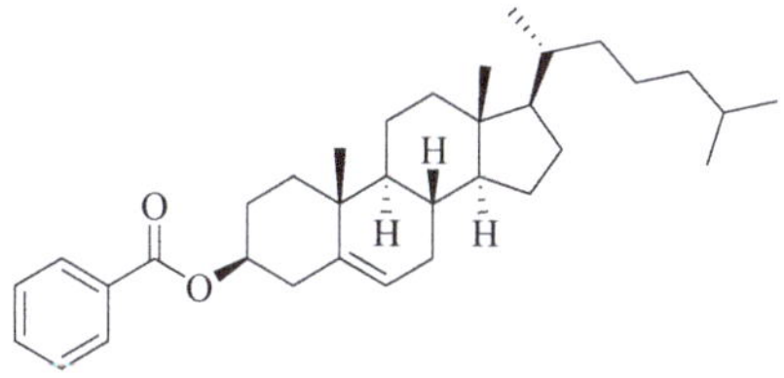

Fig.2-3 Molecular structure of cholesterol benzoate

After repeatedly confirming his findings, Reinitzer consulted Otto Lehmann. At that time, Lehmann designed and manufactured a microscope with heating function to explore the process of liquid crystal cooling and crystallization. Later, equipped with polarizer, it become an important tool for in-depth study of liquid crystal. Through the research, Lehmann thought that the polarized light property is the unique property of the substance. It is because of their contributions to the discovery and research of liquid crystals, Reinitzer and Lehmann are known as the father of liquid crystals.

In fact, the formation of liquid crystal is due to the formation of some ordered structure. Therefore, liquid crystal has a special photoelectric effect, and its interference, scattering, rotation and other phenomena are all controlled by electric field. Based on this, people have developed liquid crystal display. In short, when there is an electric field, the liquid crystal molecules are arranged in order and light can pass through, when the electric field is removed, the liquid crystal molecules are arranged in disorder and light cannot pass through. However, due to the orientation of liquid crystal, there is a problem of viewing angle in liquid crystal display, which is also the bottleneck of the development of liquid crystal display technology.

2.5.2 Supercritical State

Supercritical state is a special form of matter. In the supercritical state, the pressure and temperature of a substance exceed its critical pressure (p_c) and critical temperature (T_c) at the same time. In other words, the comparative pressure (p / p_c) and the comparative temperature (T / T_c) of a substance in the supercritical state are greater than 1 at the same time.

Supercritical state is a special fluid. Near the critical point, it has great compressibility. If the pressure is increased properly, its density will be close to the density of ordinary liquid, so it has a good performance of dissolving other substances, such as supercritical water can dissolve *n*-alkanes. On the other hand, the viscosity of supercritical state is only 1/12 to 1/4 of that of general liquid, but its diffusion coefficient is 7 to 24 times larger than that of general liquid, which is similar to that of gas.

One of the most important applications of supercritical state is supercritical extraction. The basic principle of supercritical fluid extraction is: when the gas is in supercritical state, it has a density similar to that of the liquid. Although its viscosity is higher than that of the gas, it is obviously lower than that of the liquid. Its diffusion coefficient is 10-100 times of that of the liquid. Therefore, it has better permeability and strong dissolving capacity for the material, and it can extract some components from the material. At the same time, the density and dielectric constant of supercritical fluid increase with the increase of the pressure of the closed system, and as the polarity of supercritical fluid increases, the components with different polarity can be extracted in different parts by using program boosting. After extraction, change the temperature or pressure of the system to make the supercritical fluid become a common gas to escape, the extracted components in the material can be completely or basically completely separated, so as to achieve the purpose of extraction and separation.

2.5.3 Plasma State

Plasma is an ionized gaseous substance composed of positive and negative ions. When the gas molecules are heated to a high enough temperature, the outer electrons get rid of the atomic nucleus and become free electrons, just like the students who run to the playground after class. When electrons leave the nucleus, the process is called ionization. At this time, matter becomes a uniform "paste" composed of positively charged nuclei and negatively charged electrons. Therefore, it is also called plasma. The total positive and negative charges in these plasma are equal, so it is nearly electrically neutral, so it is called plasma.

Plasma is actually a common matter in the universe. There are plasmas in the sun, stars and lightning, accounting for 99% of the universe. In the 21st century, people have mastered and used electric and magnetic fields to control plasma. The most common plasma is high-temperature ionized gas, such as arc, luminous gas in fluorescent lamp, lightning and aurora in nature.

Plasma has a wide range of applications, such as plasma sensor, plasma microscope, plasma smelting, plasma welding, plasma spraying and so on. Plasma display technology is a new display technology developed in recent years. The mixture gas is filled between the thin glass plates, and the voltage is applied to produce the ion gas, then the plasma gas is discharged to react with the phosphor in the substrate to produce the color image. Plasma color TV, also known as "wall mounted TV", is not affected by magnetic force and magnetic field. It has the advantages of thin body, light weight, large screen, bright color, clear picture, high brightness, small distortion and space saving. Compared with the narrow viewing angle (about 120 degrees) of LCD, the viewing angle of plasma display can reach 160 degrees.

2.6 Change of State

Through the above study, we have learned that there are all kinds of substances in nature, most of which exist in the form of solid, liquid and gas. In order to describe the different aggregation states of matter, "phase" is used to express the "appearance" of solid, liquid and gas. In a broad sense, the so-called phase refers to the homogeneous material part with the same physical properties in the material system. There are interfaces between different phases, which can be separated by mechanical methods. Under a certain pressure and temperature, the existing state of the substance can change correspondingly, and the change of the aggregation state of the substance can also be called phase transition. Generally, the phase transition process of pure substance is shown in Fig.2-4.

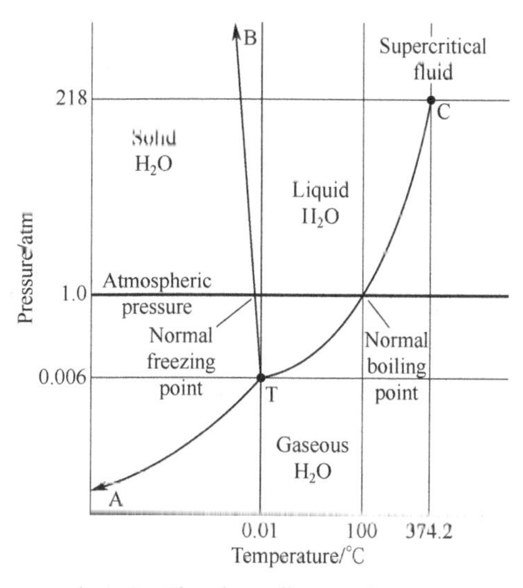

Fig.2-4 The phase diagram for water
(1atm = 101325Pa)

2.6.1 Three Phase Diagram of Pure Substance

Before we understand the process of state change, we first introduce some basic concepts related to the phase transition process.

① Phase: solid phase, liquid phase and gas phase, *i.e.* three basic states of material existence.

② Vaporization line (condensation line): the gas-liquid coexistence, indicating the change of boiling point of liquid with pressure.

③ Melting line (solidification line): the state of liquid-solid coexistence, indicating the change of solidification point of liquid with pressure.

④ Sublimation line (sublimation line): the state of gas-solid coexistence, indicating the change of gas sublimation or solid sublimation process with pressure.

⑤ Critical point: the end point of the vaporization line. Any pure substance has a unique and definite critical point, *i.e.* the critical parameter (critical temperature T_c, critical pressure p_c, critical volume V_c) is unique and definite, which is an important parameter of the actual gas properties.

⑥ Triple point: the intersection of three phase equilibrium lines, which is the coexistence of gas, liquid and solid. At a certain temperature and pressure, gas, liquid and solid coexist in a state of equilibrium. The corresponding temperature and pressure are called three-phase point temperature (T_{TP}) and three-phase point pressure (p_{TP}), respectively. Similar to the critical point, each substance has a certain three-phase point temperature and pressure, which is also one of the important parameters of the actual gas properties. For example, the exact value of the three-phase point of water was determined by the late chemist Professor Huang Ziqing in 1938, which was 0.0098℃ 0.610kPa.

From the three-phase diagram, we can judge which phase may exist under a given temperature and pressure. We can see the conditions of coexistence and mutual transformation of any two equilibriums in solid, liquid and gas phases, as well as the conditions of coexistence of three-phase equilibriums.

Using the water phase diagram as an example (see Fig.2-4), the region that represents the solid state is to the left of the line BT and above line AT. At low pressures and high temperatures, water is a vapor and the vapor region is underneath the lines AT and TC. Between the solid and vapor phase, bound by lines BT and CT, are the range of the liquid state. Line BT separates the solid and liquid regions and is thus the melting-point line. Note that line BT is nearly vertical because pressure changes have little effect on the melting point. Line TC is the vapor-pressure line that separates the liquid and vapor regions, as the pressure decreases, the boiling point of water decreases. Line TC terminates at its critical point (point C), which is 374℃ and 218atm (1atm = 101325Pa). Line AT is the vapor-pressure line, it gives the pressures and temperatures at which the solid and vapor phases are in equilibrium. The triple point conditions for water are 0.610kPa and 0.0098℃.

All matter has its own phase diagram. Knowing the phase diagram is helpful to observe and understand the changing law of the state of matter.

2.6.2 Vaporization of Liquids

The process of matter changing from liquid to gas is called vaporization. Vaporization is the most common phase change in our life. For example, steam is produced when water is heated.

There are two forms of vaporization: evaporation and boiling.

Evaporation is a process of vaporization on the surface of liquid, which can occur at any temperature. The gas in dynamic equilibrium with liquid phase is called saturated steam, and its pressure is saturated vapor pressure. It should be noted that:

① The vapor pressure does not change with the volume of the container, nor is it related to the quantity of the liquid substance.

② The vapor pressure increases with the increase of temperature. When the temperature is fixed, the vapor pressure of liquid is fixed.

③ Condensation is the reverse process of evaporation, thus heat is absorbed by evaporation and heat is released by condensation.

Example 2-3 The vapor pressure of benzene is p = 12.3kPa at 25℃. At 25℃, existing 0.120mol benzene, try to calculate the following:

① When benzene is gasified and the vessel pressure is kept equal to the vapor pressure, what is the volume (considered as ideal gas)?

② If the volume of benzene vapor is 10.9L, what is the vapor pressure of benzene?

③ If the volume of benzene vapor is 40.0L, what is the vapor pressure of benzene?

Solution ① When benzene is gasified completely, its volume can be calculated according to the ideal gas laws:

$$pV = nRT$$

$$V = \frac{nRT}{p} = \frac{0.120\text{mol} \times 8.314[\text{kPa} \cdot \text{dm}^3 / (\text{mol} \cdot \text{K})] \times 298\text{K}}{12.3\text{kPa}} = 24.17\text{dm}^3$$

② When the volume of benzene vapor is 10.9dm^3, benzene has not been completely gasified, which is a state of gas-liquid coexistence. The pressure of benzene vapor is its saturated vapor pressure, which is 12.3kPa.

③ When the volume of benzene vapor is 40.0dm^3, benzene has been completely gasified and becomes unsaturated vapor. The pressure can be calculated according to the ideal gas laws.

$$p = \frac{nRT}{V} = \frac{0.120\text{mol} \times 8.314[\text{kPa} \cdot \text{dm}^3 / (\text{mol} \cdot \text{K})] \times 298\text{K}}{40.0\text{dm}^3} = 7.43\text{kPa}$$

Boiling is a vaporization process that takes place in the whole liquid. It can only take place when the temperature reaches the boiling point. When the liquid is heated to a certain temperature above the normal boiling point, it begins to boil, and then the temperature drops to the normal boiling point, which is called overheating. Overheating is easy to cause accidents, especially when dealing with flammable liquids (such as ether, acetone, alcohol, *etc.*), when the liquid splashed with bubbles meets the flame, it will cause a fire. In chemical experiments, when heating this kind of liquid, the method of stirring or adding zeolite is usually used to prevent it.

2.6.3 Liquefaction of Gases

Gas liquefaction is a physical process in which gases are transformed into liquids. Under normal pressure, many gases only need to be cooled to enter the liquid state. For example, at

101kPa, the water vapor can be liquefied when the temperature is lower than 100℃. Some gases can also be liquefied under pressure. For example, chlorine may be liquefied under pressure at room temperature. And some gases, such as carbon dioxide, must be reduced to a certain temperature before the compressed volume can be liquefied.

Liquefying gases is usually for some scientific, industrial or commercial purposes, including: analyzing the basic properties of gases (intermolecular force), storing or transporting gases (liquefied petroleum gases) and commercializing its special properties. For example, liquid oxygen is used to supply patients with respiratory problems; liquid nitrogen can be used for medical low-temperature operation and preservation of precious samples such as stem cells. As a refrigerant, liquid carbon dioxide can be used for food storage and artificial rainfall; in the national defense industry, liquid ammonia is used to make propellants for rockets and missiles.

2.6.4 Other State Changes

In addition to the gas-liquid phase transition, there are also solid-liquid phase transition (*i.e.* melting and solidification) and gas-solid phase transition (*i.e.* sublimation and sublimation) between the three phases of matter. The process of solid matter absorbing heat and melting at a certain temperature, and becoming liquid is melting; the process of liquid matter releasing heat under the condition of temperature and pressure reduction, and becoming solid is solidification. When the temperature is raised below the pressure of three-phase point, the process of substance changing from solid state to gas state without liquid state is called sublimation; the opposite process is called condensation. Examples of such phase transitions are also common in nature. For example, water coagulates into ice, ice melts, camphor sublimates, and water vapor in the atmosphere coagulates into frost or snow.

This kind of phase change is always accompanied by significant energy change, which has become a research hotspot of energy storage materials. For example, when the temperature is lower than 0℃, the water changes from liquid to solid (freezing); when the temperature increases, the water changes from solid to liquid (melting). In the process of freezing, a lot of cold energy is absorbed and stored, while in the process of melting, a lot of heat energy is absorbed. This is the most typical example of phase change materials. At present, a variety of phase change materials have been developed for energy storage or temperature regulation, mainly including hydrated salt and waxy phase-change materials, which have been widely used in military, aerospace, construction, clothing, refrigeration equipment, communication and power system.

Exercises

1. Is there an ideal gas in nature? Can absolute ideal gas be liquefied?

2. Mix 21g nitrogen and 32g hydrogen in a container, set the total pressure of the gas as p, and try to find the partial pressure of nitrogen and hydrogen.

3. Please list several methods which were used to express the concentration of following reagents in laboratory: concentrated ammonia water, concentrated sulfuric acid and concentrated hydrochloric acid.

4. What is the essential difference between crystals and non crystals? Try to list some crystal types of common substances you know.

5. Please describe the aggregation state of water in combination with the three-phase diagram, and describe the phase transition process from the point of view of state transformation.

Chapter 3 Atomic Structure and Periodic Table of Elements

Teaching contents	Learning requirements
Atomic structure theory	Describe the development of the atomic models Understand the multi-electron atom model Wave-particle duality Bohr's theory Schrödinger equation for hydrogen atom Understand the shape of atomic function orbitals and master the rules of electron configurations
Periodic Table of Elements	Knowing the blocks of the Periodic Table of Elements State the periodic law, and discuss the contributions made by Meyer, Mendeleev, Ramsay, and Mosely in establishing the form of the Periodic Table of Elements
Atomic properties	Describe and explain the periodic variation of atomic radii, ionization energy and electron affinity energy

3.1 Atomic Structure

This chapter is devoted to a discussion of the Rutherford-Bohr model of the atom. The two giants of modern physics, Rutherford and Bohr, have not collaborated on the model. However, they both made a major contribution to it: Rutherford by introducing the concept of the atomic nucleus with electrons revolving about the nucleus in a cloud, and Bohr by introducing the idea of electron angular momentum quantization and by deriving from first principles the kinematics of the hydrogen atom and on electron atoms in general.

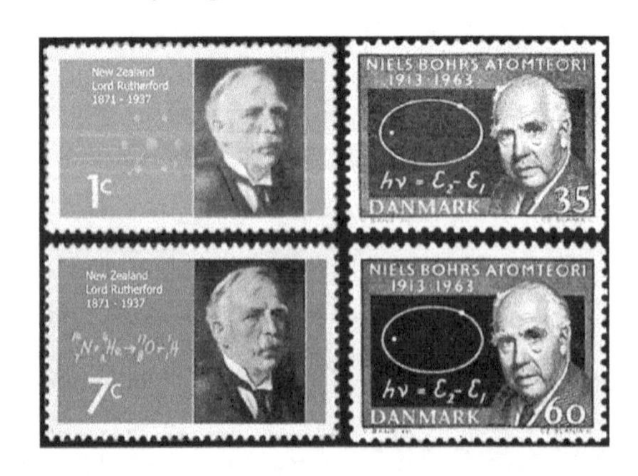

The chapter deals first with the atomic structure theory development in detail. The successes and limitations of the Rutherford-Bohr model are discussed, while the Hartree approximation for multi-electron atoms is introduced, and the experimental confirmation of the validity of the atomic model is presented. The chapter concludes with a discussion of the Schrödinger equation for the ground state of the hydrogen atom providing several sample calculations for the ground state of the hydrogen atom based on the Schrödinger equation. Next, elements and the Periodic Table is introduced.

3.1.1 Electrons in Atom

(1) Subatomic particles

The most familiar particles in atoms are protons, neutrons and electrons. According to modern atomic structure theory, an atom is made up of some subatomic particles. In the physical science, subatomic particles are particles much smaller than atoms. There are more than 30 subatomic particles were found in atom. If their vibration isomers are counted, there will be more than 300 particles. Some of particles can transformed into another, with the release and absorption of energy. They can be classified by composition as elementary particles and composite particles.

Elementary particles are not made of other particles according to current theories, such as electron, quarks, *etc*. Composite subatomic particles are bound states of two or more elementary particles, such as protons or atomic nuclei. For example, a proton is made of two up quarks and one down quark, while the atomic nucleus of helium-4 is composed of two protons and two neutrons. The neutron is made of two down quarks and one up quark.

Nobody knows if it is possible to cut an electron or quark but particle physicists keep trying. Chemistry concerns itself with how electron sharing binds atoms into structures such as crystals and molecules. Nuclear physics deals with how protons and neutrons arrange themselves in nuclei.

(2) Atomic spectra

Atomic spectra are very powerful tools for the study of atomic structures. If the source of a spectrum produces light having only a relatively small number of wavelength components, then a discontinuous spectrum is observed. For example, if the light source is an electric discharge passing through a gas, only certain colors are seen in the spectrum. They can be observed as colored lines with dark spaces between them. These discontinuous spectra are called atomic or line spectra.

The production of the line spectrum of helium is illustrated in Fig.3-1. The light source is a low-pressure helium gas lamp. When an electric discharge is passed through the lamp, helium atoms absorb energy and emit light. The light is passed through a narrow slit and then dispersed by a prism. The colored components of the light are detected and recorded on photographic film. Each wavelength component appears as an image of the slit: a thin line. In all, five lines in the spectrum of helium can be seen with the unaided eye.

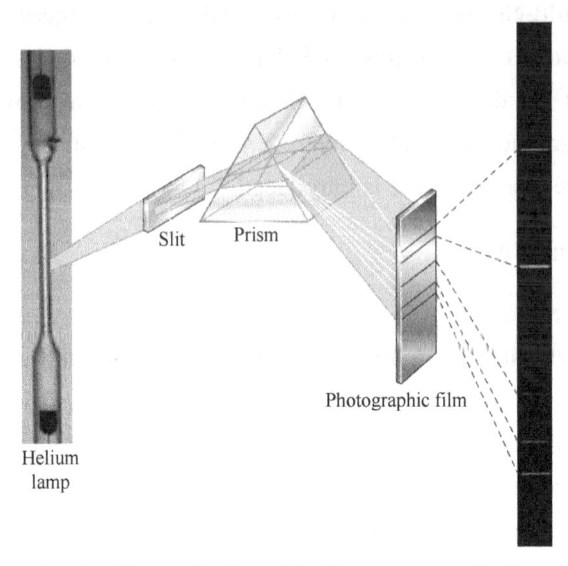

Fig.3-1 The production of the line spectrum of helium

(3) Hydrogen spectrum

The earliest extensively studied atomic spectrum is hydrogen(Fig.3-2). Light from a hydrogen lamp appears to the eye as a reddish-purple. The principal wavelength component of this light is red light of wavelength 656.3nm (nanometer). Three other lines appear in the visible spectrum: a greenish-blue line at 486.1nm, a violet line at 434.0nm, and another violet line at 410.1nm.

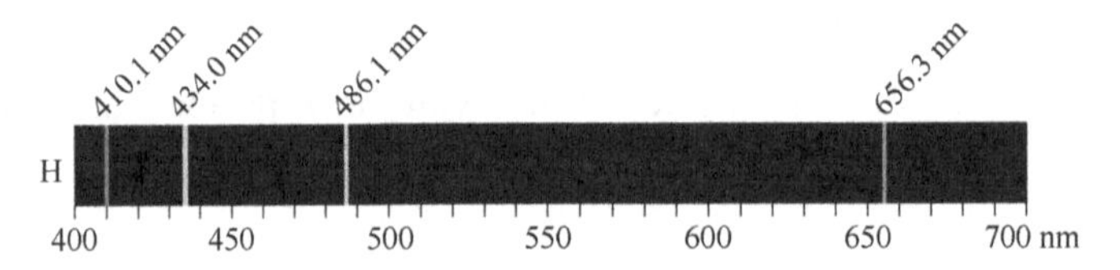

Fig.3-2 The balmer series for hydrogen atoms—a line spectrum

In 1885, Johann Balmer deduced the following formula for the wavelengths of these spectral lines.

$$\lambda = \frac{Bm^2}{m^2 - n^2} \tag{3-1}$$

In this equation (3-1), B is a constant and m and n represent integers. When n is 2 and m is 3, the wavelength of the red line is obtained. With $n = 2$ and $m = 4$, the greenish-blue line is obtained. The remaining two lines in the visible spectrum are obtained by using $n = 2$ with $m = 5$ and $m = 6$. The series of lines having $n = 2$ is now known as the Balmer series (Fig.3-3).

Balmer also speculated that if other values of n were used, then other series in the infrared and ultraviolet regions could be generated. We will see that this is indeed the case.

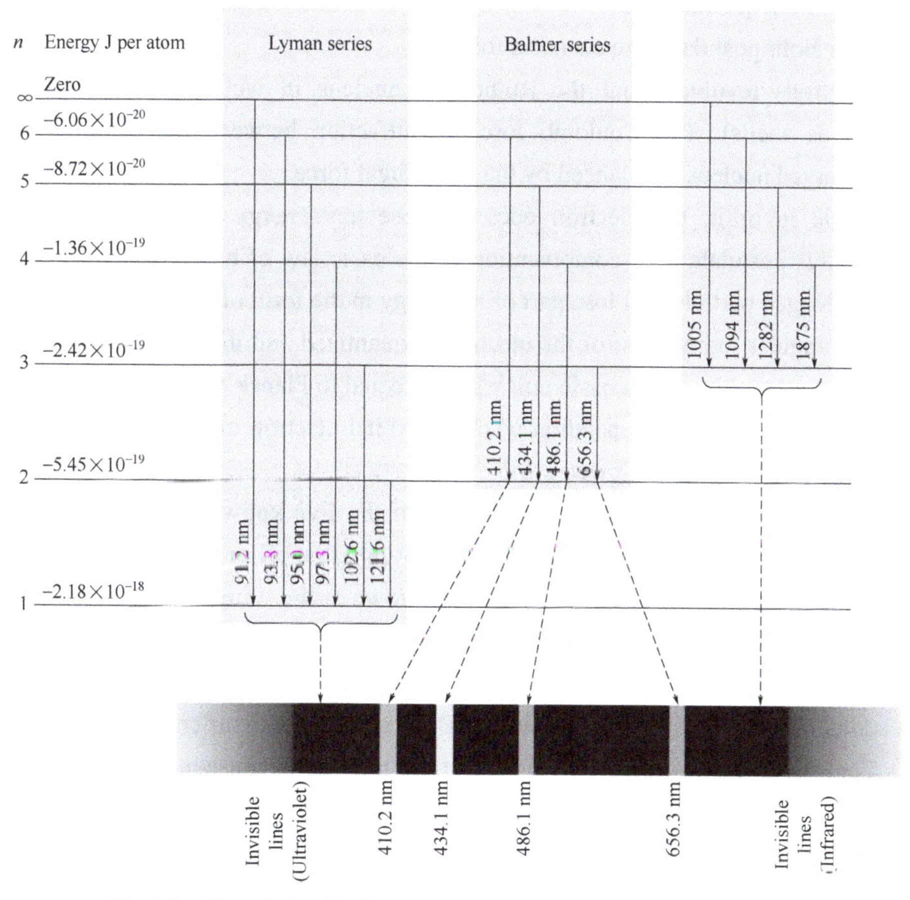

Fig.3-3 The relationship between electronic transition and spectral lines

Balmer's equation was later found to be a special case of the Rydberg formula devised in 1888. In the equation (3-2), ν is frequency, c is rate of light (the value is 2.998×10^{8}m/s), λ is length of light; R_H is the Rydberg constant for the hydrogen atom (the value is 1.09678×10^{7} /m).

$$\nu=\frac{c}{\lambda}=R_{\mathrm{H}}\left(\frac{1}{n^{2}}-\frac{1}{m^{2}}\right) \tag{3-2}$$

3.1.2 Bohr's Theory of the Hydrogen

In 1913, Niels Bohr combined Rutherford's concept of the nuclear atom with Planck's idea of the quantized nature of the radiative process and developed from first principles, an atomic model that successfully deals with one-electron structures like the hydrogen atom and one-electron ions such as singly ionized helium, doubly ionized lithium, *etc*., forming a hydrogen-like or hydrogenic structure. The model, known as the Bohr model of the atom, is based on four postulates that combine principles of classical mechanics with the concept of angular momentum quantization.

The four Bohr postulates are stated as follows:

① Electrons revolve about the Rutherford nucleus in well-defined, allowed orbits (referred to as shells). The Coulomb force of attraction between the electrons and the positively charged nucleus is balanced by the centrifugal force.

② While in orbit, the electron does not lose any energy despite being constantly accelerated (this postulate is in contravention of the basic law of nature which states that an accelerated charged particle will lose part of its energy in the form of radiation).

③ The angular momentum of the electron is quantized and the angular momentum can have only integral multiples of a basic unit which is equal to Planck constant h.

④ An atom or ion emits radiation when an orbital electron makes a transition from an initial allowed orbit to a final allowed orbit.

According to the Bohr atomic model, each of the five known-series of the hydrogen spectrum arises from a family of electronic transitions that all end at the same final state. The Lyman, Brackett and Pfund series were not known at the time when Bohr proposed his model; however, the three series were discovered soon after Bohr predicted them with his model.

In addition to its tremendous successes, the Bohr atomic model suffers several limitations:

① The orbital electron in revolving about the nucleus is constantly accelerated and by virtue of its charge should lose part of its energy in the form of photons (Larmor law) and spiral into the nucleus. With its assumption that the electron, while in an allowed orbit, emits no photons the Bohr model is in contravention of Larmor law.

② The model does not predict the relative intensities of the photon emission in characteristic orbital transitions.

③ The model fails to explain the observed fine structure of hydrogen spectral lines where each spectral line is further composed of closely spaced spectral lines.

④ The model does not work quantitatively for multi-electron atoms.

⑤ The model does not explain the splitting of a spectral line into several lines in a magnetic field (Zeeman effect).

⑥ The model does not explain the splitting of a spectral line into several lines in an electric field (Stark effect).

The idea of atomic electrons revolving about the nucleus should not be taken too literally; however, the Bohr model for the one-electron structure combined with angular momentum quantization serves as a reasonable intermediate step on the way to more elaborate and accurate theories provided with quantum mechanics and quantum electrodynamics. The Schrödinger quantum theory dispensed with the picture of electrons moving in well defined orbits but the Bohr theory is still often used to provide the first approximation to a particular problem, because it is known to provide reasonable results with mathematical procedures that are, in comparison with those employed in quantum mechanics, significantly simpler and faster.

3.1.3 Schrodinger Equation for Hydrogen

(1) Wave-particle duality

To explain the photoelectric effect, Einstein suggested that light has particle-like properties, which are displayed through photons. However, other phenomena are best understood in terms of the wave theory of light, such as the dispersion of light into a spectrum by a prism. Then, light appears to have a dual nature.

In 1924, Louis de Broglie, considering the nature of light and matter, offered a startling proposition: Small particles of matter may at times display wave-like properties. de Broglie argued that the relationship derived by Einstein for the momentum of a photon, should also apply to particles of matter.

$$\lambda = \frac{h}{mv} \tag{3-3}$$

Table 3-1 lists the de Broglie wavelengths of several particles. The de Broglie wavelength of microscopic particles is greater than the particle diameter and has significant fluctuations, such as the electrons in Table 3-1; the wavelength of macroscopic particles is so small that it cannot be measured at all. It performs as a particle and can be dealt with by classical mechanics.

Table 3-1　de Broglie Wavelength of Some Physical Particles

Particles	m/kg	v/(m/s)	λ/m	Particle diameter/m
electron	9.1×10^{-31}	1×10^{6}	7.3×10^{-10}	2.8×10^{-15}
H atom	1.6×10^{-27}	1×10^{6}	4.0×10^{-13}	7.4×10^{-11}
[illegible] atom	$2.1\times10^{[illegible]}$	1×10^{6}	3.2×10^{-15}	5.3×10^{-10}
A bullet	1×10^{-2}	1×10^{3}	6.6×10^{-35}	1×10^{-2}
A satellite	8000	7900	1.0×10^{-41}	9

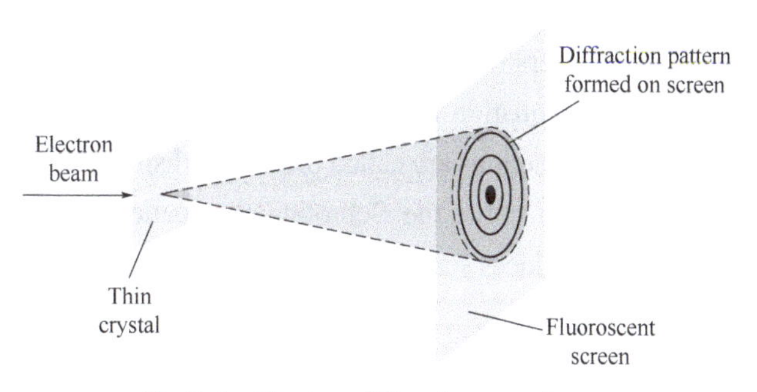

Fig.3-4　Electron diffraction experiment

In 1927, C. J. Davisson and L. H. Germer of the United States showed that a beam of slow electrons is diffracted by a crystal of nickel. In a similar experiment of the same year, G. P. Thomson of Scotland directed a beam of electrons at a thin metal foil. He obtained the same pattern for the diffraction of electrons by aluminum foil with X-rays of the same wavelength. Thomson and Davisson shared the 1937 Nobel Prize in physics for their electron diffraction experiments(Fig.3-4).

G. P. Thomson was the son of J. J. Thomson, who had won the Nobel Prize in physics in 1906 for his discovery of the electron. It is interesting to note that Thomson the father showed that the electron is a particle, and Thomson the son showed that the electron is a wave. Father and son together demonstrated the wave-particle duality of electrons.

When we talk about the moving style of a very tiny mass particle, like an electron, its wave nature will be so obvious that we cannot ignore it. That's why Bohr's theory is wrong, in which, only particle nature is considered. Schrödinger established the wave mechanics model of atom based on particle and wave nature.

(2) Schrödinger equation

In 1927, Schrödinger proposed an equation to express the motion style of electrons in atom. It's a second order partial differential equation known as the Schrödinger equation. The Schrödinger equation is accepted as a basic postulate of quantum mechanics. It cannot be derived from other equations.

$$\frac{\partial^2\psi}{\partial x^2}+\frac{\partial^2\psi}{\partial y^2}+\frac{\partial^2\psi}{\partial z^2}+\frac{8\pi^2 m}{h^2}(E-V)\psi=0 \tag{3-4}$$

Formula (3-4) is the Schrödinger equation for the hydrogen atom or hydrogen-like ion. Where ψ is the wave function of the electron, m is the mass of the electron, E is the total energy of the electron, V is the potential energy of the electron (For a hydrogen atom, the potential energy of the electron is $-e^2/r$, and e is the charge of the electron. r is the distance between the electron and the nucleus).

Wave functions are most easily analyzed in terms of the three variables required to define a point. In the usual Cartesian coordinate system, these three variables are the x, y, and z dimensions. In the spherical polar coordinate system, they are r, θ(theta) and Φ(phi). Each orbital ψ has three quantum numbers to define it since the hydrogen atom is a 3d system.

Solutions to the Schrödinger equation for the hydrogen atom give not only energy levels but also wave functions. These wave functions are called orbitals to distinguish them from the orbits of the incorrect Bohr theory. How to solve the Schrodinger equation is not the task of general chemistry. The following only introduces some important concepts related to wave function.

3.1.4 The Shapes of the Orbitals

The wave function of the hydrogen atom can be decomposed into the product of the radial and angular parts. The general formula is:

$$\psi_{n,l,m}(r,\theta,\varphi)=R_{n,l}(r)\cdot Y_{l,m}(\theta,\varphi) \tag{3-5}$$

Here, $R_{n,l}(r)$ is the radial part, which only changes with the distance r. $Y_{l,m}(\theta,\varphi)$ is the angular part of the wave function, which varies with the angle (θ,φ) changes. Table 3-2 shows the radial part and the angle part of the wave function of a hydrogen-like atom.

Table 3-2 The Angular and Radial Wave Function of A Hydrogen-like Atom

Quantum number			Radial part, $R_{n,l}(r)$	Angular part, $Y_{l,m}(\theta,\varphi)$
n	l	m		
1	0	0	$2\sqrt{\frac{1}{a_0^3}}e^{-r/a_0}$	$\sqrt{\frac{1}{4\pi}}$
2	0	0	$\sqrt{\frac{1}{8a_0^3}}\left(2-\frac{r}{a_0}\right)e^{-r/2a_0}$	$\sqrt{\frac{1}{4\pi}}$
2	1	1	$\frac{1}{2\sqrt{6}}\left(\frac{1}{a_0}\right)^{3/2}\left(\frac{r}{a_0}\right)e^{-r/2a_0}$	$\left(\frac{3}{4\pi}\right)^{1/2}\sin\theta\cos\varphi$
2	1	0		$\left(\frac{3}{4\pi}\right)^{1/2}\cos\theta$
2	1	1		$\left(\frac{3}{4\pi}\right)^{1/2}\sin\theta\sin\varphi\cos\theta$

(1) Probability density

The wave function, ψ, is only a mathematical function, and there are no observable physical properties. However, the electron probability density $|\psi|^2$ or charge density is a function of r:

$$|\psi|^2=\frac{1}{\pi a_0^3}e^{-2r/a_0} \tag{3-6}$$

Fig.3-5 is a pattern of dots represents the electron probability or charge density in a plane with the nucleus at its center. The closer the spacing between dots, the higher is the probability of finding an electron and the greater is the electron charge density.

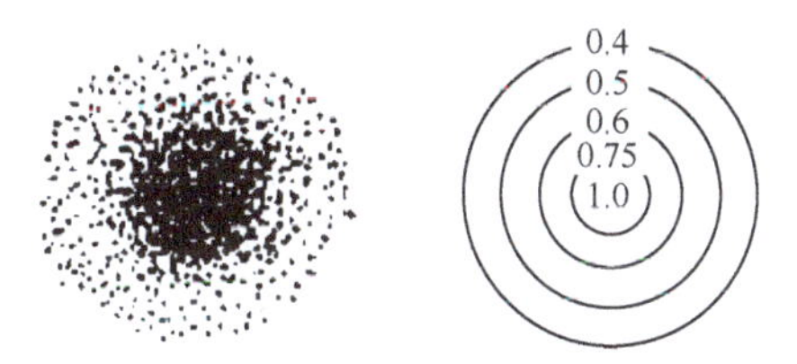

Fig.3-5 The electron probability and charge density for the ground state hydrogen atom

(2) Quantum number and electron orbitals

When solving the Schrödinger equation, many functions, $\psi1$, $\psi2$, $\psi3$, ... can be obtained.

However, due to the limitation of the physical significance of $|\psi|^2$, only some of ψ^2 are reasonable. It is determined by the values of the three parameters n, l, and m_l. A reasonable combination of (n, l, m_l) can describe a state of electron. Usually, we call this parameter a quantum number.

All orbitals with the same value of n are in the same principal electronic shell or principal level, and all orbitals with the same n and l values are in the same subshell, or sublevel.

Principal electronic shells are numbered according to the value of n, principal quantum number, a non-zero positive integer. The first principal shell consists of orbitals with n=1; the second principal shell of orbital, with n=2 and so on. The value of n relates to the energy and the most probable distance of an electron from the nucleus. The higher the value of n, the greater the election energy; and on average, the electron is the farther from the nucleus. The quantum number l determines the angular distribution, or shape, of an orbital and m_l determines the orientation of the orbital.

The number of subshells in a principal electronic shell is the same as the number of allowed values of the orbital angular momentum quantum number, l. In the first principal shell, with n=1, the only allowed value l is 0, and there is a single subshell. The second principal shell (n=2), with the allowed values l is 0 and 1, consists of two subshells; the third principal shell (n=3) has three subshells (l=0, 1 and 2); and so on. Or, to put the matter in another way, because there are n possible values of the l quantum number and that is, 0,1,2,⋯,(n−1), the number of subshells in a principal shell is equal to the principal quantum number. As a result, there is one subshell in the principal shell with n=1, two subshells in the principal shell with n=2, and so on. The name given to a subshell, regardless of the principal shell in which it is found, depends on the value of the l quantum number. The first four subshells are

s subshell	p subshell	d subshell	f subshell
$l=0$	$l=1$	$l=2$	$l=3$

The number of orbitals in a subshell is the same as the number of allowed values of m_l for the particular value of l. Recall that the allowed values of m_l are 0, ±1, ±2, ⋯, ±l, and thus the total number of orbitals in a subshell is 2l+1. The names of the orbitals are the same as the names of the subshells in which they appear.

s orbitals	p orbitals	d orbitals	f orbitals
$l=0$	$l=1$	$l=2$	$l=3$
$m_l=0$	$m_l=0,\pm1$	$m_l=0,\pm1,\pm2$	$m_l=0,\pm1,\pm2,\pm3$
one s orbital	three p orbitals	five d orbitals	seven f orbitals
in an s subshell	in a p subshell	in a d subshell	in an f subshell

The orbitals in same subshell have the same energy, which is called degenerate orbital or equivalent orbital, which have the same n and l, but m_l is different. In the presence of an external magnetic field, high-precision spectroscopy experiments can distinguish them.

Wave mechanics provides three quantum numbers with which we can develop a description of electron orbitals. However, in 1925, two scientists, George Uhlenbeck and Samuel Goudsmit, proposed that some unexplained features of the hydrogen spectrum could be understood by assuming that an electron acts as if it spins. There are two possibilities for electron spin. Thus, these two possibilities require a fourth quantum number, the electron spin quantum number, m_s. The value of electron spin quantum number is $+\frac{1}{2}$ or $-\frac{1}{2}$, that does not depend on any of the other three quantum numbers. Therefore, there can be two electrons with different spin directions on each orbital. Such two electrons are often called paired electrons.

(3) The shapes of the orbitals

The angular function of s, p, and d orbitals is shown in Fig.3-6.

① For s orbitals (l=0), the angular function has no angular dependence. The polar graph of this function is a sphere. For this reason, s orbitals have a spherical shape.

② For p orbitals (l=1), there are three angular functions. Although the mathematical forms of these functions are different, their polar graphs reveal that they are identical in shape but oriented differently in space. The three p orbitals are labeled p_x, p_y, or p_z, to signify that they are oriented along the x, y, or z axes. In contrast to what we saw for s orbitals, the angular functions for the p orbitals do not have a constant value and p orbitals do not have spherical shapes. For example, the wave function p_z has an angular maximum along the positive z axis. In three dimensions, the polar graph for each p orbital consists of two spheres tangent to the origin, as shown in Fig.3-6(b). The phase (positive or negative) is included in these graphs to indicate where Y has positive or negative values. We will see in Chapter 4 that the phase of the orbital is an important consideration when developing models for describing chemical bonding.

③ The angular functions with l=2 are more complicated. It turns out that the angular functions for d orbitals (l=2) possess two angular nodes, whereas p orbitals (l=1) possess one angular node and s orbitals (l=0) have no angular nodes. In general, the number of angular nodes is equal to the value of l. The angular function Y consists of four lobes oriented along the x and y axes. Notice that the phase is positive for two of the lobes and negative for the other two.

Cross-sections of the angular functions for the d_{xy}, d_{yz}, d_{xz} and d_{z^2} are also displayed in Fig.3-6. We observe that four of them have the same shape as $d_{x^2-y^2}$, but they are oriented differently with respect to the axes.

The angular functions for f, g, h, and so on, orbitals have rather complicated shapes because of the larger number of angular nodes. These orbitals are not often encountered, and so we will not consider their shapes at all.

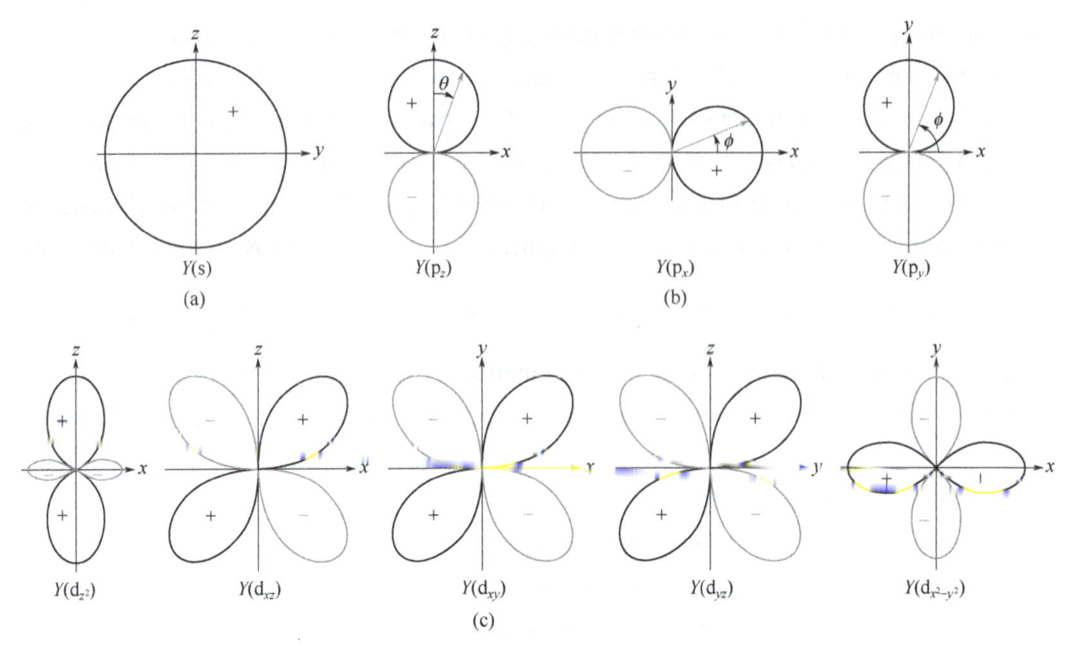

Fig.3-6 The angular function of s, p and d orbitals

[The angular functions Y for s, p and d orbitals are shown in (a), (b) and (c), respectively]

3.1.5 Electron Configurations

The electron configuration of an atom is a designation of how electrons are distributed among various orbitals in principal shells and subshells. Later, we will find that many of the physical and chemical properties of elements can be correlated with electron configurations. Here, we will see how the results of wave mechanics, expressed as a set of rules, can help us to write probable electron configurations for the elements. There are three rules for assigning electrons to orbitals.

(1) Electrons occupy orbitals in a way that minimizes the energy of the atom

The total energy of an atom depends not only on the orbital energies but also on the electronic repulsions that arise from placing electrons in particular orbitals. That is, the orbital filling order cannot be reliably predicted by consideration of orbital energies alone. The exact order of filling of orbitals has been established by experiment and it is this order based on experiment that we must follow in assigning electron configurations to the elements.

To predict the electron configuration for an atom's ground state, electrons are put into the orbitals with the lowest energy possible, placing no more than two electrons in an orbital.

The order of subshell filling is related to n, the principal quantum number, and l. In general,

- electrons fill orbitals in order of increasing $(n + l)$ and
- when two or more subshells have the same $(n + l)$ value, electrons fill the orbital with the lower n value.

These general rules result in the following orbital filling order (with only a few exceptions).

$$1s, 2s, 2p, 3s, 3p, 4s, 3d, 4p, 5s, 4d, 5p, 6s, 4f, 5d, 6p, 7s, 5f, 6d, 7p \quad (3\text{-}7)$$

It is equally important to remember that, for the reasons described above, this filling order does not represent the relative energy ordering of the orbitals. Some students find the diagram pictured in Fig.3-7 a useful way to remember this order, but the best method of establishing the order of filling of orbitals is based on the periodic table.

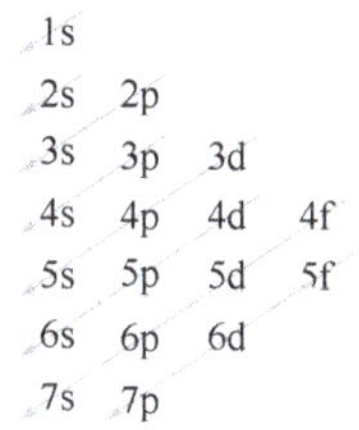

1s
2s 2p
3s 3p 3d
4s 4p 4d 4f
5s 5p 5d 5f
6s 6p 6d
7s 7p

Fig.3-7 The order of filling of electronic subshells

(2) Pauli exclusion principle: Only two electrons may occupy the same orbital, and these electrons must have opposite spins

The Pauli exclusion principle states that no two electrons within an atom can have the same set of four quantum numbers (n, l, m_l, and m_s). The limits on possible values for the four quantum numbers means that a single orbital can accommodate no more than two electrons, and when an orbital contains two electrons, those electrons must have opposite spins.

Because of this limit of two electrons per orbital, the capacity of a subshell for electrons can be obtained by doubling the number of orbitals in the subshell. Thus, the s subshell consists of one orbital with a capacity of two electrons; the p subshell consists of three orbitals with a total capacity of six electrons, and so on.

(3) Hund's rule: When orbitals of identical energy (degenerate orbitals) are available, electrons initially occupy these orbitals singly and with parallel spins

Carbon has six electrons, four in the 1s and 2s orbitals and two in the 2p orbitals. When electrons occupy a subshell with multiple orbitals such as 2p, Hund's rule of maximum multiplicity applies.

This rule means that we must place electrons singly in each orbital with parallel spins before pairing them to ensure that Hund's rule is followed. A simplified statement of Hund's rule is that, for a given configuration, the arrangement having the maximum number of parallel spins is lower in energy than any other arrangement arising from the same configuration.

This behavior can be rationalized as follows. Because electrons all carry the same electric charge, if the available orbitals all have the same energy, then by placing them in different orbitals the electrons are spatially as far apart as possible.

① Electron configurations for elements in short periods 1-3

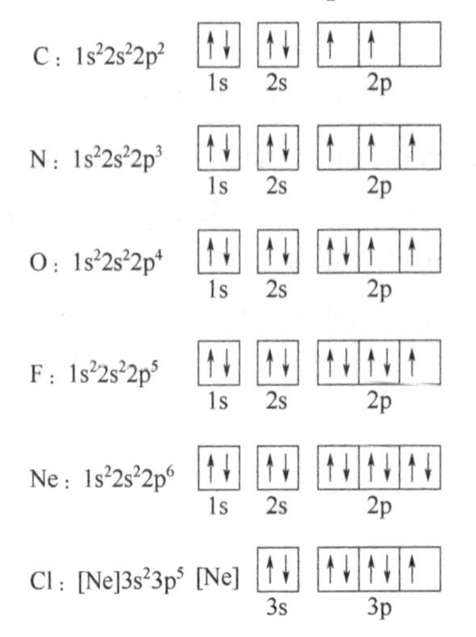

The electron configurations of nitrogen, oxygen, fluorine, and neon are shown here. The highest-energy orbital for all second-row elements is in the second energy level (2s or 2p). For multielectron species, when electron configurations are written, noble gas notation is often used to represent filled shells. These filled shells are called core electrons. In noble gas notation, the symbol for a noble gas is written within square brackets in front of the spdf orbital, noncore electrons. For example, the electron configuration for chlorine is written using noble gas notation as shown here.

The electrons beyond the core electrons are the valence electrons for an element. The valence electrons are the highest-energy electrons and are the electrons least strongly attracted to the nucleus. these electrons are involved in chemical reactions and the formation of chemical bonds. The fact that elements within a group have similar electron configurations and the same number of valence electrons suggests that elements within a group have similar properties.

② Electron configurations for elements in periods 4-7

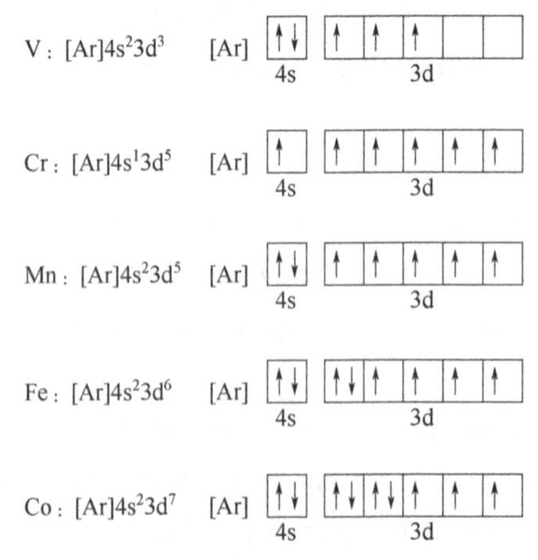

Ni: $[Ar]4s^23d^8$ [Ar] [↑↓] [↑↓|↑↓|↑↓|↑|↑]
4s 3d

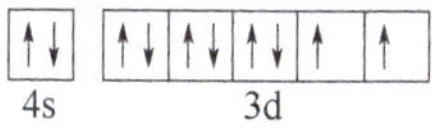

Cu: $[Ar]4s^13d^{10}$ [Ar] [↑] [↑↓|↑↓|↑↓|↑↓|↑↓]
4s 3d

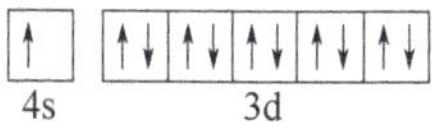

Zn: $[Ar]4s^23d^{10}$ [Ar] [↑↓] [↑↓|↑↓|↑↓|↑↓|↑↓]
4s 3d

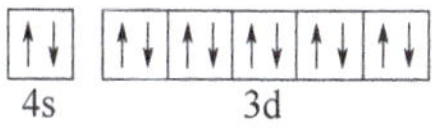

Both chromium and copper do not follow the general filling order. The reason is considered complicated and related to the similar energies of the 4s and 3d orbitals in multielectron atoms. The observed ground-state configurations for both Cr and Cu involve half-filled subshells or full-filled subshells. Thus, the supposed "special stability" of half-filled and full-filled subshells is sometimes used as an explanation for why Cr and Cu have the observed configurations.

Such an explanation raises the question, "What is the origin of this special stability?" If this special stability exists, then all the atoms below Cr in group 6 and below Cu in group 11 should also have half-filled or full-filled subshells. However, experiment reveals that this is not always the case. Most notably, for tungsten W, the ground-state configuration is the predicted one.

W $[Xe]\,4f^{14}5d^46s^2$

3.2 Chemical Elements and the Periodic Table

All atoms of a particular element have the same atomic number, *Z*, and, conversely, all atoms with the same number of protons are atoms of the same element. The chemical properties of atoms and periodized functions of the atomic number *Z* are governed mainly by electrons with the lowest binding energy, *i.e.*, by outer shell electrons commonly referred to as valence electrons. The periodicity of chemical and physical properties of elements (periodic law) was first noticed by Dimitri Mendeleev, who in 1869 produced a periodic table of the then-known elements.

Since Mendeleev's time the periodic table of elements has undergone several modifications as the knowledge of the underlying physics and chemistry expanded and new elements were discovered and added to the pool. However, the basic principles elucidated by Mendeleev are still valid today.

In a modern periodic table of elements each element is represented by its chemical symbol and its atomic number. The periodicity of properties of elements is caused by the periodicity in electronic structure. Although electron configurations may seem rather abstract, they actually lead us to a better understanding of the periodic table. Around 1920, Niels Bohr began to promote the connection between the periodic table and quantum theory. The chief link, he pointed out, is electron configurations. Elements in the same group of the table have similar electron configurations.

The periodic table of elements is now commonly arranged in the form of seven horizontal rows or *periods* and eighteen vertical columns or groups. Elements with similar chemical and physical properties are listed in the same column.

The periods in the periodic table are of increasing length as follows:

① Period 1 has two elements: hydrogen and helium.

② Periods 2 and 3 have eight elements each.

③ Periods 4 and 5 have 18 elements each.

④ Period 6 has 32 elements but lists only 18 entries with entry under lanthanum at Z = 57 actually representing the lanthanon series of 15 elements with atomic numbers from 57 (lanthanum) to 71 listed separately. Synonyms for lanthanon are lanthanide, lanthanoid, rare earth, and rare-earth element.

⑤ Period 7 has 32 elements as well (ranging in atomic number Z from 87 to 118), but lists only 18 entries with entry under actinium at Z = 89 actually representing the actinide series of 15 elements with atomic numbers from 89 to 103 listed separately. Synonyms for actinon are actinide and actinoid.

Ground state electron configurations for all elements is tabulated in Appendix Ⅶ. We have taken three groups of elements from the periodic table and written their electron configurations. The similarity in electron configuration within each group is readily apparent. If the shell of the highest principal quantum number—the outermost, or valence, shell—is labeled n, then

① The group 1 atoms (alkali metals) have one outer-shell (valence) electron in an s orbital, that is, ns^1.

② The group 17 atoms (halogens) have seven outer-shell (valence) electrons, in the configuration ns^2np^5 .

③ The group 18 atoms (noble gases)—with the exception of helium, which has only two electrons—have outermost shells with eight electrons, in the configuration ns^2np^6.

Fig.3-8 shows that the periodic table can be divided into four blocks. The blocks are classified according to the types of subshell that is filled with the highest-energy electrons for the elements in it.

① s block. The s orbital of highest principal quantum number (n) fills. The s block consists of groups 1 and 2 (plus He in group 18).

② p block. The p orbitals of highest quantum number (n) fill. The p block consists of groups 13, 14, 15, 16, 17, and 18 (except He).

③ d block. The d orbitals of the electronic shell (the next to outer most) fill. The d block includes groups 3, 4, 5, 6, 7, 8, 9, 10, 11, and 12.

④ f block. The f orbitals of the electronic shell fill. The f block elements are the lanthanides and the actinides.

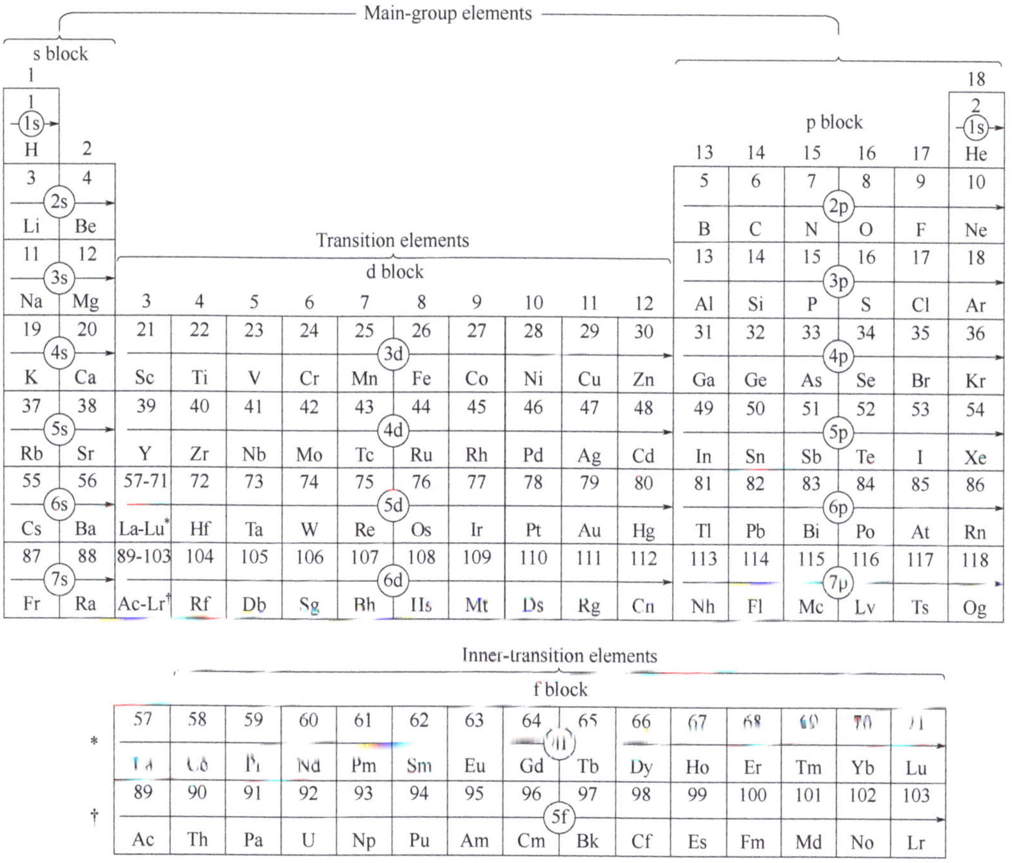

Main-group elements

s block | p block | Transition elements (d block)

1	2	3	4	5	6	7	8	9	10	11	12	13	14	15	16	17	18
1 H (1s)																	2 He (1s)
3 Li	4 Be (2s)											5 B	6 C	7 N (2p)	8 O	9 F	10 Ne
11 Na	12 Mg (3s)											13 Al	14 Si	15 P (3p)	16 S	17 Cl	18 Ar
19 K	20 Ca (4s)	21 Sc	22 Ti	23 V	24 Cr	25 Mn (3d)	26 Fe	27 Co	28 Ni	29 Cu	30 Zn	31 Ga	32 Ge	33 As (4p)	34 Se	35 Br	36 Kr
37 Rb	38 Sr (5s)	39 Y	40 Zr	41 Nb	42 Mo	43 Tc (4d)	44 Ru	45 Rh	46 Pd	47 Ag	48 Cd	49 In	50 Sn	51 Sb (5p)	52 Te	53 I	54 Xe
55 Cs	56 Ba (6s)	57-71 La-Lu*	72 Hf	73 Ta	74 W	75 Re (5d)	76 Os	77 Ir	78 Pt	79 Au	80 Hg	81 Tl	82 Pb	83 Bi (6p)	84 Po	85 At	86 Rn
87 Fr	88 Ra (7s)	89-103 Ac-Lr†	104 Rf	105 Db	106 Sg	107 Bh (6d)	108 Hs	109 Mt	110 Ds	111 Rg	112 Cn	113 Nh	114 Fl	115 Mc (7p)	116 Lv	117 Ts	118 Og

Inner-transition elements (f block)

*	57 La	58 Ce	59 Pr	60 Nd	61 Pm	62 Sm	63 Eu	64 Gd (4f)	65 Tb	66 Dy	67 Ho	68 Er	69 Tm	70 Yb	71 Lu
†	89 Ac	90 Th	91 Pa	92 U	93 Np	94 Pu	95 Am	96 Cm (5f)	97 Bk	98 Cf	99 Es	100 Fm	101 Md	102 No	103 Lr

Fig.3-8 Blocks of elements and the periodic table

Year 2019 is the 150th anniversary of Mendeleev's discovery of the periodic table and the 100th anniversary of the IUPAC. The periodic table of the chemical elements is one of the most important and influential achievements in modern science, reflecting not only the nature of chemistry, but also the nature of physics, biology and other basic science.

3.3 Atomic Chemical Properties

3.3.1 Atomic Radius

Unfortunately, atomic radius is hard to define. We have seen that atomic orbitals extend, in principle, to infinity. Although the probability of finding an electron decreases with increasing distance from the nucleus, there is always nonzero probability of finding an electron at very large distances from the nucleus. Thus, an atom has no precise outer boundary. We might describe an effective atomic radius as, say, the distance from the nucleus within which 95% of the electron charge density is found, but this distance cannot be measured experimentally.

From an experimental standpoint, we cannot make a measurement of the radius of a single, isolated atom. We can, however, obtain a measure of the size (radius) of an atom when it is combined with other atoms. For this reason, we define atomic radius in terms of internuclear distance.

We will emphasize an atomic radius based on the distance between the nuclei of two atoms joined by a chemical bond. The covalent radius is one-half the distance between the nuclei of two identical atoms joined by a single covalent bond. The ionic radius is based on the distance between the nuclei of ions joined by an ionic bond. Because the ions are not identical in size, this distance must be properly apportioned between the cation and anion. One way to apportion the electron density between the ions is to define the radius of one ion and then infer the radius of the other ion. The convention we have chosen to use is to assign O^{2-} an ionic radius of 140 pm. An alternative apportioning scheme is to use F^- as the reference ionic radius. When using ionic radii data, carefully note which convention is used and do not mix radii from the different conventions. Starting with a radius of 140 pm for O^{2-}, the radius of Mg^{2+} can be obtained from the internuclear distance in MgO, the radius of Cl^- from the internuclear distance in $MgCl_2$, and the radius of Na^+ from the internuclear distance in NaCl. For metals, we define a metallic radius as one-half the distance between the nuclei of two atoms in contact in the crystalline solid metal. Similarly, in a solid sample of a noble gas the distance between the centers of neighboring atoms is called the van der Waals radius. There is much debate about the values of the atomic radii of noble gases because the experimental determination of the van der Waals radii is difficult; consequently, the atomic radii of noble gases are left out of the discussion of trends in atomic radii.

The angstrom unit, Å (1Å = 0.1nm), has long been used for atomic dimensions. The angstrom, however, is not a recognized SI unit. The SI units are the nanometer (nm) and picometer (pm). Fig.3-9 illustrates the definitions of covalent, ionic, and metallic radii by comparing these three radii for sodium.

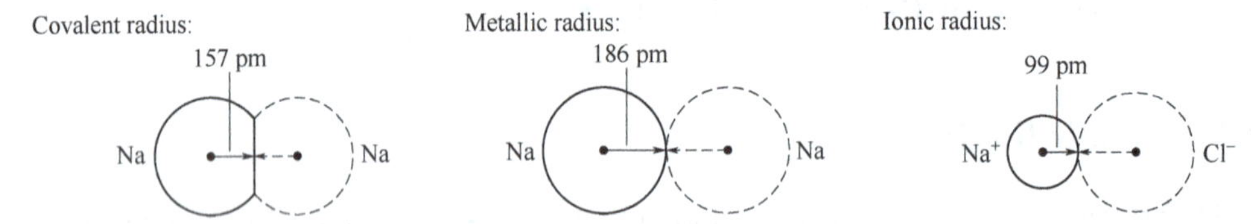

Fig.3-9　Covalent, metallic, and ionic radii compared atomic radii are represented by the solid arrows

[The covalent radius is based on the diatomic molecule $Na_2(g)$, found only in gaseous sodium. The metallic radius is based on adjacent atoms in solid sodium, Na(s). The value of the ionic radius of Na^+ is obtained by the comparative method described in the text]

Fig.3-10 is a plot of atomic radius against atomic number for a large number of elements. In this plot metallic radii are used for metals and covalent radii for nonmetals. Fig.3-10 suggests certain trends in atomic radii, for example, large radii for group 1, decreasing across the periods to smaller radii for group 17. Atomic radius decreases from left to right through a period of elements and increases from top to bottom through a group.

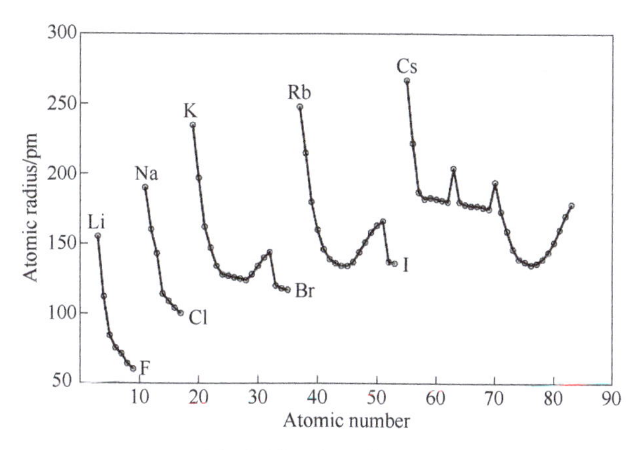

Fig.3-10 Atomic radii

[The values plotted are metallic radii for metals and covalent radii for nonmetals. Data for the noble gases are not included because of the difficulty of measuring covalent radii for these elements (only Kr and Xe compounds are known). The explanations usually given for the several small peaks in the middle of some periods are beyond the scope of this discussion]

3.3.2 Ionization Energy

When discussing metals, we talked about metal atoms losing electrons and thereby altering their electron configurations. But atoms do not eject electrons spontaneously. Electrons are attracted to the positive charge on the nucleus of an atom, and energy is needed to overcome that attraction. The more easily its electrons are lost, the more metallic an atom is considered to be. The ionization energy, *I*, is the quantity of energy a gaseous atom must absorb to be able to expel an electron. The electron that is lost is the one that is highest in energy, and therefore, is most loosely held. Ionization energies are usually measured through experiments in which gaseous atoms at low pressures are bombarded with photons of sufficient energy to eject an electron from the atom. Ionization energies are sometimes expressed in the unit electron-volt (eV). One electron-volt is the energy acquired by an electron as it falls through an electric potential difference of 1 volt. It is a very small energy unit, especially suited to describing processes involving individual atoms. When ionization is based on a mole of atoms, kJ/mol is the preferred unit(1eV per atom = 96.49kJ/mol). Sometimes the term ionization potential is used instead of ionization energy.

Here are two typical values.

$$Mg(g) \longrightarrow Mg^{+}(g) + e^{-} \qquad I(Mg) = 738\text{kJ/mol}$$
$$Mg^{+}(g) \longrightarrow Mg^{2+}(g) + e^{-} \qquad I(Mg^{+}) = 1451\text{kJ/mol}$$

① The symbol *I*(Mg) represents the first ionization energy of Mg—the energy required to strip one electron from a neutral gaseous atom.

② $I(Mg^+)$ represents the ionization energy of and thus, the second ionization energy of Mg.

③ Further ionization energies are denoted by $I(Mg^{2+})$, $I(Mg^{3+})$, and so on. Each succeeding ionization energy is invariably larger than the preceding one due to the progressively increasing Z_{eff}.

In the case of magnesium, for example in the second ionization the electron once freed has to move away from an ion with a charge of +2 (Mg^{2+}). More energy must be invested than for a freed electron to move away from an ion with a charge of +1 (Mg^+). This is a direct consequence of Coulomb's law, which states, in part, that the force of attraction between oppositely charged particles is directly proportional to the magnitudes of the charges.

First ionization energies (I) for many atoms are plotted in Fig.3-11.

With relatively few exceptions, ionization energies increase from left to right across a period and decrease from top to bottom within a group. The noble gases that contain the most stable electronic configurations and the highest ionization potentials are identified, as are the alkali elements that contain the least stable electronic configurations and the lowest ionization potentials with only one valence electron in the outer shell.

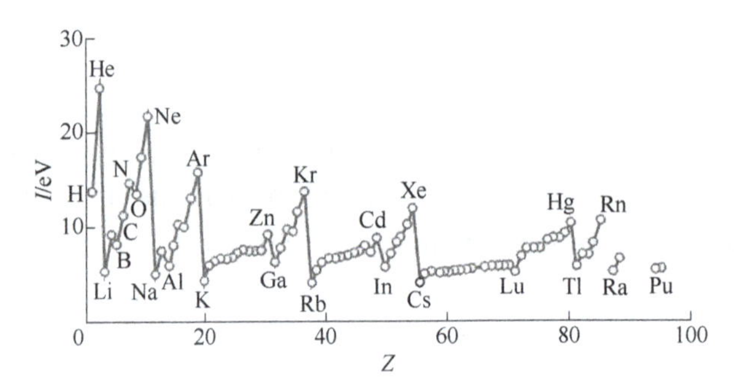

Fig.3-11 Ionization energy of atoms against atomic number Z

The variation of ionization energies across a period or within a group is essentially the opposite of that observed for atomic radii. This observation leads us to the following conclusion: Ionization energies decrease as atomic radii increase.

The statement above is easily rationalized: the farther an electron is from the nucleus, the more easily it can be removed. The decrease in ionization energy that accompanies an increase in atomic radius is evident when we compare the ionization energies and atomic radii of the alkali metal atoms (Table 3-3).

Table 3-3 Atomic Radii and First Ionization Energies of the Alkali Metal (Group 1) Elements

Items	Li	Na	K	Rb	Cs
Atomic radius/pm	155	190	235	248	267
Ionization energy/(kJ/mol)	520.2	495.8	418.8	403.0	375.7

Table 3-4 lists ionization energies for the third-period elements. With minor exceptions, the trend in moving across a period (follow the colored stripe) is that atomic radii decrease, ionization energies increase, and the elements become less metallic, or more nonmetallic, in character. Table 3-4 lists stepwise ionization energies. Note particularly the large breaks that occur along the zigzag diagonal line.

Table 3-4 Ionization Energies of the Third-period Elements (in kJ/mol)

Location	Na	Mg	Al	Si	P	S	Cl	Ar
First	495.8	737.7	577.6	786.5	1012	999.6	1251.1	1520.5
Second	4562	1451	1817	1577	1903	2251	2297	2666
Third		7733	2745	3232	2912	3361	3822	3931
Fourth			11580	4356	4957	4564	5158	5771
Fifth				16090	6274	7013	6542	7238
Sixth					21270	8496	9362	8781
Seventh						27110	11020	12000

Consider magnesium as an example. The removal of two electrons from the shell of Mg gives an ion with the configuration [He]$2s^2 2p^6$. To remove a third electron requires taking an electron from the lower energy shell. Consequently, the third ionization energy of Mg is much larger than the second ionization energy—so much larger that Mg^{3+} is not produced in ordinary chemical processes. Similarly, we do not encounter Na^{2+} or Al^{4+} in ordinary chemical processes.

Now let us turn our attention to the exceptions to the regular trend in first ionization energies. For example, why is the first ionization energy of Al smaller than that of Mg and the first ionization energy of S smaller than that of P?

Because Al is immediately to the right of Mg, we might expect the first ionization energy of Al to be larger than that of Mg, yet it is not. To understand why, we must consider the particular electrons lost. As illustrated by the orbital diagrams in Fig.3-12, Mg loses a 3s electron whereas Al loses a 3p electron. Although Z_{eff} for the 3p electron of Al is greater than that of the 3s electrons in Mg, the 3p electron of Al is less penetrating, of higher energy, and more easily removed.

Why is the first ionization energy of S lower than that of P?

As suggested by Fig.3-13, not only is it lower, but the first ionization energies of S, Cl, and Ar are all lower than expected and considerably lower than the values predicted by extrapolating the first ionization energies of the elements immediately preceding them. The explanations focus on the fact that the ionization of S, Cl, or Ar involves the removal of a paired electron whereas the ionization of Al, Si, or P involves the removal of an unpaired electron.

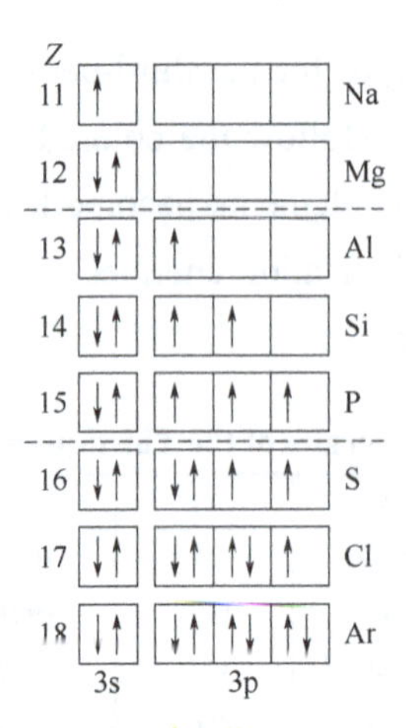

Fig.3-12 Orbital diagrams showing the valence electron configurations for atoms of the third row elements

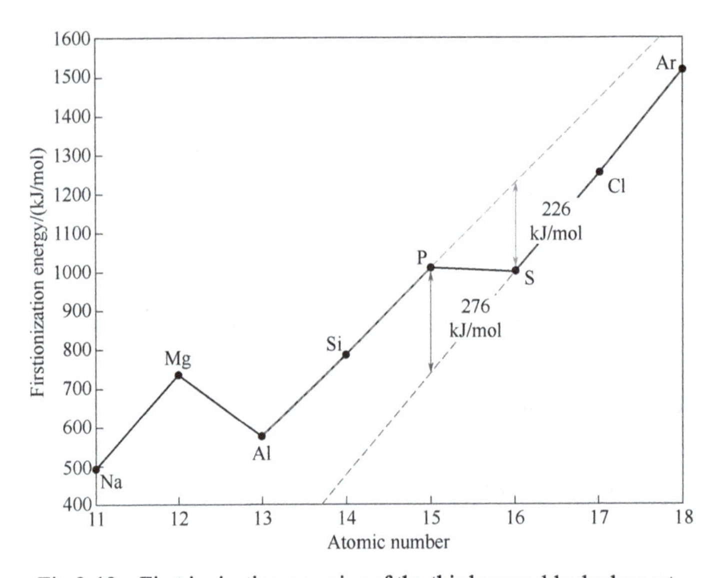

Fig.3-13 First ionization energies of the third row p-block elements

According to the first explanation, electron-electron repulsions are the key consideration. Paired electrons occupy the same orbital and are, on average, closer together than electrons in separate orbitals. Thus, they experience extra repulsion and are more easily removed. Although the electron-electron repulsions increase with the number of electrons in the 3p orbitals, there is a significant increase in the electron repulsions, and a corresponding decrease in the first ionization energy (approximately 226kJ/mol), once orbital sharing begins, that is, as we proceed from P([Ne]$3s^2 3p^3$) to S([Ne]$3s^2 3p^4$). The second explanation focuses instead on the extent of screening and the strength of the electron-nucleus attractions. Unpaired electrons with parallel spins tend to avoid each other more, screen each other less, experience a higher effective nuclear charge, interact more strongly with the nucleus, and are harder to remove. According to this line of reasoning, the extra electron-nuclear attraction causes the first ionization energy of P([Ne]$3s^2 3p^3$) to be greater by approximately 276kJ/mol than the value expected by the backward-extrapolation of the first ionization energies of S, Cl, and Ar. Conversely, paired electrons screen each other to a greater extent, interact less strongly with the nucleus, and are more easily removed.

Is the observed dip in the first ionization energy that occurs as we move from group 15 to 16 caused by increased electron-electron repulsions or by decreased electron-nucleus attractions? It is difficult to answer this question complete certainly, but it is clear that dip occurs once orbital sharing begins. The difficulty of providing an unambiguous explanation for the observed dip is not totally unexpected. As we pointed out previously, the energy of an atom is a delicate balance of electron-electron repulsions and electron-nucleus attractions. The rationalization of ionization energies is further complicated by the fact that, for example, the first ionization energy of P depends on the energies of both P([Ne]$3s^2 3p^3$) and P^+([Ne]$3s^2 3p^2$). Similarly, the first ionization energy of S depends on the energies of both S([Ne]$3s^2 3p^4$) and S^+([Ne]$3s^2 3p^3$). Thus, the decrease in ionization energy that occurs as we move from P to S depends on a delicate balance of electron-electron repulsions and electron-nucleus attractions in four different species.

3.3.3 Electron Affinity

Ionization energy is the energy change for the removal of an electron. Let's consider the energy change associated with the addition of an electron. The thermochemical equation for the addition of an electron to a fluorine atom is

$$F(g) + e \longrightarrow F^-(g) \qquad \Delta_{ea}H = -328\text{kJ/mol}$$

Notice that the process above is exothermic, meaning that energy is given off when an F atom gains an electron. Electron affinity (E_{ea}) can be defined as the enthalpy change ($\Delta_{ea}H$) that occurs when an atom in the gas phase gains an electron. According to this definition, the electron affinity of fluorine is a negative quantity.

Some representative values of $\Delta_{ea}H$ are plotted in Fig.3-14. To interpret these values, we have to consider the type of orbital in which the incoming electron has to be accommodated and the effect of the incoming electron on the electron-electron repulsions and electron-nucleus attractions. For many atoms, negative is an indication that there is generally net attraction between an atom and an incoming electron. This net attraction arises because the electrons of the neutral atom do not completely shield an incoming electron from the nuclear charge.

Not all atoms have a measurable electron affinity. Despite this, the trends in electron affinities generally follow those of the other periodic properties.

Elements lower in a group generally have less negative electron affinity values than those higher in a group.

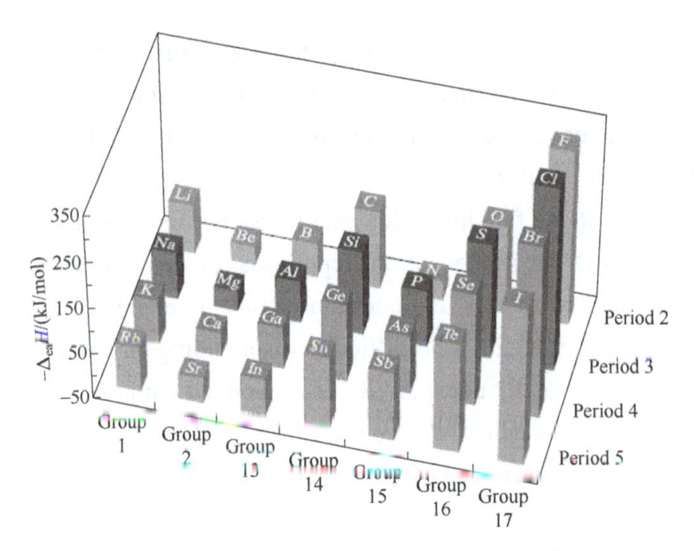

Fig.3-14 Electron affinities of some main group elements

Electron affinity values generally are more negative in elements farther to the right across a period. There are some notable exceptions, however, in the Group 2 and 15 elements.

Exercises

1. How would the Balmer equation have to be modified to predict lines in the infrared spectrum of hydrogen?

2. What is ΔE for the transition of an electron from n=5 to n=2 in a hydrogen atom? What is the frequency of the spectral line produced?

3. Calculate the increase in energy, in joules, when an electron in the hydrogen atom is excited from the first to the third energy level.

4. Without doing detailed calculations, indicate which of the following electron transitions in the hydrogen atom results in the emission of light of the longest wavelength. (a) n=4 to n=3; (b) n=1 to n=2; (c) n=1 to n=6; (d) n=3 to n=2.

5. What electron transition in a hydrogen atom, starting from n = 7 will produce light of wavelength 410nm?

6. Write an acceptable value for each of the missing quantum numbers.

(a) $n = 3, l = ?, m_l = 2, m_s = +\frac{1}{2}$

(b) $n = ?, l = 2, m_l = 1, m_s = -\frac{1}{2}$

(c) $n = 4, l = 2, m_l = 0, m_s = ?$

(d) $n = ?, l = 0, m_l = ?, m_s = ?$

7. On the basis of the periodic table and rules for electron configurations, indicate the

number of (a) 2p electrons in N; (b) 4s electrons in Rb; (c) 5s electrons in As; (d) 4f electrons in Au; (e) unpaired electrons in Pb; (f) elements in group 14 of the periodic table; (g) elements in the 6 period of the periodic table.

8. Based on the relationship between electron configurations and the periodic table, give the number of (a) outer-shell electrons in an atom of Sb; (b) electrons in the fourth principal electronic shell of Pt; (c) elements whose atoms have six outer-shell electrons; (d) unpaired electrons in an atom of Te; (e) transition elements in the sixth period.

9. To what neutral atom do the following valence-shell configurations correspond? Indicate whether the configuration corresponds to the ground state or an excited state.

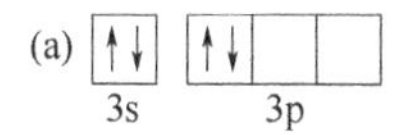

(b) [↑↓] [↑ | ↑ | ↓]
3s 3p

(c) [↑↓] [↑↓ | ↑ | ↑]
3s 3p

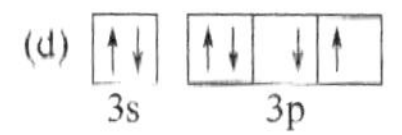

Chapter 4 Chemical Bonding and Molecular Structure

Teaching contents	Learning requirements
Lewis theory	Describe the use of the octet rule for writing a Lewis structure and the use of formal charges for determining the plausibility of a given Lewis structure
Valence bond method	Use valence bond theory to describe bond formation in terms of the overlap of atomic orbitals. Distinguish between sigma and pi bonds. Use bond-dissociation energies to estimate the enthalpy change for a gas phase reaction
Valence-shell electron-pair repulsion (VSEPR) theory	Use VSEPR theory to determine the shapes of molecules
Hybrid orbital theory	Discuss the concept of orbital hybridization and use appropriate hybridization schemes to describe bonding in molecules
Molecular polarization and intermolecular forces	Discuss the concept of molecular polarity, dipole moment, molecular polarization, and polarizability. Discuss the different types of intermolecular forces, such as London dispersion force, dipole-dipole force, and hydrogen bonding force. Describe how the properties of liquids are related to intermolecular forces

4.1 Chemical Bonding

In this chapter, we will describe the interactions between atoms called chemical bonds. Most of the discussion centers on the Lewis theory, which provides one of the simplest methods of representing chemical bonding. We will also explore another relatively simple theory for predicting probable molecular shapes.

4.1.1 Lewis Theory

In 1916, the American chemist G. N. Lewis proposed the covalent bond theory of sharing electron pairs between atoms. He believes that each atom in the molecule should have a stable electron layer structure similar to rare gas atoms, which is formed by a shared electron pair between atoms. Lewis theory provides one of the simplest methods of representing chemical bonding.

① Electrons, especially those of the outmost (valence) electronic shell, play a fundamental role in chemical bonding.

② In some cases, electrons are transferred from one atom to another. Positive and negative ions are formed and attract each other through electrostatic forces called ionic bond.

③ In other cases, one or more pairs of electrons are shared between atoms. A bond formed by the sharing of electrons between atoms is called covalent bond.

④ Electrons are transferred or shared in such a way that each atom acquires an especially stable electron configuration. Usually this is a noble gas configuration with eight outermost electrons.

It can be summarized as follows:

One or more pairs of electrons are shared between atoms. A bond formed by the sharing of electrons between atoms is called a covalent bond. Sharing a pair of electrons is a single covalent bond. Two pairs of electrons is a double covalent bond. Three pairs of electrons means a triple covalent bond. Double and triple covalent bonds are known as multiple covalent bonds.

The Lewis theory of bonding describes a covalent bond as the sharing of a pair of electrons, but this does not necessarily mean that each atom contributes an electron to the bond. A covalent bond in which a single atom contributes both of the electrons to a shared pair is called a coordinate covalent bond.

$$\left[\begin{matrix} & H & \\ H: & N & :H \\ & H & \end{matrix}\right]^+ \qquad \left[\begin{matrix} H\cdot & \ddot{O} & \cdot H \\ & H & \end{matrix}\right]^+$$

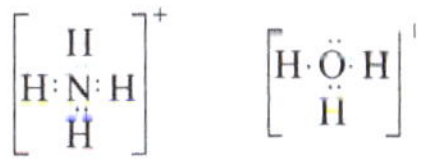

The bond formed between the N atom of NH_3 and the H^+ ion is a coordinate covalent bond. It is important to note, however, that once the bond has formed, it is impossible to say which of the four N—H bonds is the coordinate covalent bond. Thus, a coordinate covalent bond is indistinguishable from a regular covalent bond.

Another example of coordinate covalent bonding is found in the familiar hydronium ion.

Lewis developed a special set of symbols for his theory. A Lewis symbol consists of a chemical symbol to represent the nucleus and core (inner-shell) electrons of an atom, together with dots placed around the symbol to represent the valence (outer-shell) electrons. We will place single dots on the sides of the symbol, up to a maximum of four. Then we will pair up dots until we reach an octet. Lewis symbols are commonly written for main-group elements but much less often for transition elements.

A Lewis structure is a combination of Lewis symbols that represents either the transfer or the sharing of electrons in a chemical bond. Some Lewis structures are written in Fig.4-1.

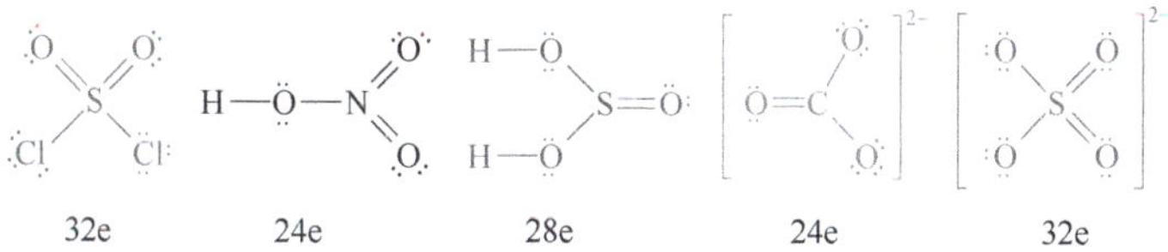

32e (a) SO_2Cl_2 24e (b) HNO_3 28e (c) H_2SO_3 24e (d) CO_3^{2-} 32e (e) SO_4^{2-}

Fig.4-1 Lewis structures

Lewis's theory of covalent bonds preliminarily explained the nature of covalent bonds different from ionic bonds, and advanced the understanding of molecular structure. However, because this theory is based on classic theory and regards electrons as static negative charges, there are certain limitations. For example: it cannot explain why the outermost electrons of the central atom of some molecules are less than 8 (such as BF_3, *etc.*) or more than 8 (such as PCl_5, *etc.*), but these molecules still exist. Nor can it explain why two atoms can combine together when there is charge repulsion between two electrons in a common electron pair. Until 1927 German chemists W. Heitler and F. London applied quantum mechanics to study the structure of hydrogen molecules and began to really understand the covalent bonds. This is the beginning of modern covalent bonds theory. Later, chemists Pauling, Mulliken, Hund and others have successively researched and developed this theory, and established a modern valence bond theory (VB Method), hybrid orbital theory, valence shell electron pair repulsion theory (VSEPR) and molecular orbital theory (MO method).

4.1.2 The Covalent Bond: A Quantum Mechanical Concept

Let's take the formation of the H—H covalent bond as an example.

Heitler and London applied quantum mechanics to study the process of forming hydrogen molecules from hydrogen atoms, and obtained the relationship between the energy E of the H_2 molecule and the nuclear distance R, as shown in Fig.4-2.

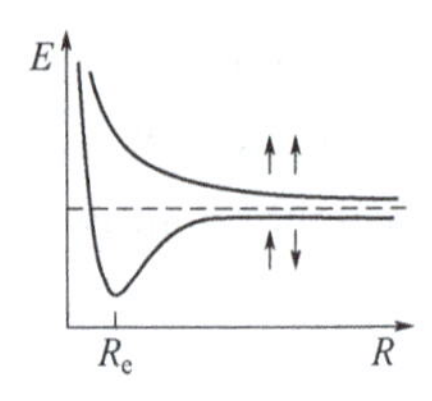

Fig.4-2 The relationship between the energy of hydrogen molecules and the nuclear distance

① Two atoms spinning in same direction (red): as the distance of the two atoms shortening, the potential energy increase and no covalent bond will form in whichever condition.

② Two atoms spinning in opposite direction (green): as the distance of two atoms shortening, the potential energy has a lowest point at R_e, which indicates the formation of a covalent bond.

The formation of covalent bonds is essentially a quantum mechanical effect. When atoms are close to each other, their atomic orbitals overlap with the same sign, resulting in an enhanced interference effect. It destroys the original balance in the atom and reduces the energy of the atom.

Generally speaking, the covalent bond is due to the overlap of orbitals when atoms are close to each other, and the energy of atoms is reduced by sharing spin-opposite electron pairs.

For H_2, the formation of covalent bonds increases the electron cloud density between the two nuclei and the electron cloud between the nuclei is attracted by the two nuclei. This is how H_2 is formed.

4.1.3 Introduction to the Valence Bond Method

Recall the region of high electron probability in a H atom that we described in Chapter 3 through the mathematical function called a 1s orbital. As the two H atoms pictured in Fig.4-2 approach each other, these regions begin to interpenetrate. We say that the two orbitals overlap. When the two atoms are close enough that their atomic orbitals overlap, a covalent bond can be formed. Bond formation is imagined to occur through a redistribution of electron probability density. It involves an increase in electron probability density between the two positively charged nuclei. In the process, the energy of the system is lowered.

A description of covalent bond formation in terms of atomic orbital overlap is called the valence bond method. The creation of a covalent bond in the valence bond method is normally based on the overlap of half-filled orbitals, but sometimes such an overlap involves a filled orbital on one atom and an empty orbital on another. The valence bond method gives a localized electron model of bonding: Core electrons and lone-pair valence electrons retain the same orbital locations as in the separated atoms, but the bonding electrons do not. Instead, they are described by an electron probability density that includes the region of orbital overlap and both nuclei.

4.1.4 Characteristics of Covalent Bond

(1) Saturability

The orbitals in an atom provided to form covalent bonds are limited. Thus, the number of covalent bonds cannot be greater than the number of unpaired electrons of an atom.

For example $H-Cl$ $H-O-H$ $N\equiv N$

In H_2O molecule, oxygen atom can only form two single bonds with two hydrogen atoms rather than form double bonds with one hydrogen atom. In fact, hydrogen atoms can only form single bond, since it has only one single electron. Nitrogen atoms in N_2 molecule can form triple bond because they contain more unpaired electrons in atomic orbitals.

(2) Orientation

Each orbital for the formation of bond provided by an atom has its own shape and orientation. Their shape and orientations determine the way that orbitals overlap. Enhanced interference occurs in the same sign part of the wave function. The electron probability density of the p orbital is highest at the petal head. According to the maximum overlap principle, the overlaps between s+s, s+p, and p+p orbitals have four basic types (shown in Fig.4-3): three "head-to-head" ways and one "side-by-side" way. They are called effective overlaps. Fig.4-4 shows the forbidden overlaps. They are not effective overlaps.

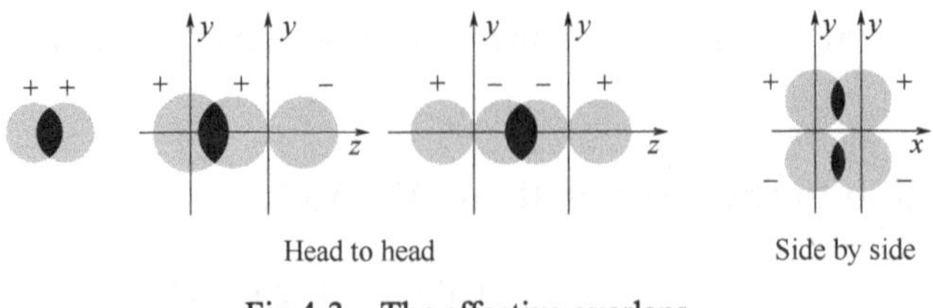

Fig.4-3 The effective overlaps

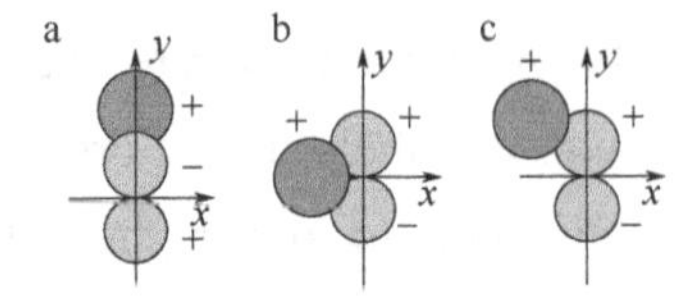
Fig.4-4 The forbidden overlaps

4.1.5 Types of Covalent Bond

Different overlap modes determine different types of covalent bond. “Head-to-head” overlap fashion produce a sigma bond, designated σ bond. And “side-by-side” overlap of two parallel orbitals is called a pi bond, designated π bond.

(1) σ bond

The covalent bonds in H_2, HCl and Cl_2 molecular are typical σ bonds. The overlap region is located on the line between two nuclei of the two atoms.

Fig.4-5 is the overlap of atomic orbitals in the formation of hydrogen-to-fluorine bond. 1s of H and 2p of F overlaps head-to-head. Two unpaired electrons pair with opposite spin.

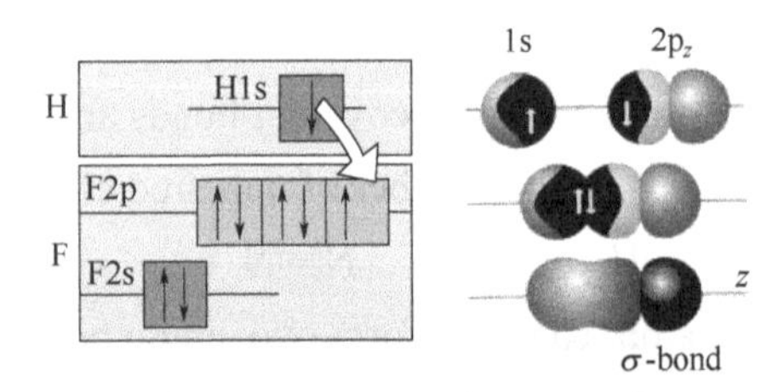

Fig.4-5 The orbital overlap in HF

N_2: the overlap of nitrogen orbitals in formation of N—N bond are shown in Fig.4-6. Each nitrogen atom contains three 2p orbitals containing an unpaired electron. But because they are perpendicular to each other, only one of them can overlap head to head to form a σ bond.

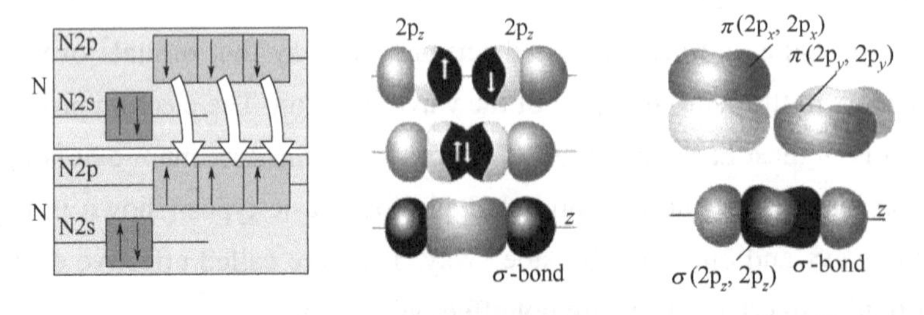

Fig.4-6 The orbital overlaps in N_2

(2) π bond

Still in the nitrogen molecule, once a σ bond is formed in the z-axis direction, the left two 2p orbitals can overlap side-by-side and form two π bonds. Therefore, there are one σ bond and two vertical π bonds in the nitrogen molecule. Note that only one σ bond can be formed, since the 2p orbitals vertical to each other. Keep in mind that there is only one σ bond in double and triple bonds. The π bonds are always generated with σ bond.

4.1.6 Bond Parameters

(1) Bond energy

Bond energy, symbolized by D, is the quantity of energy required to break one mole of covalent bonds in a gaseous species. Energy is released when isolated atoms join to form a covalent bond, and energy must be absorbed to break apart covalently bonded atoms.

It is not hard to picture the meaning of bond energy for a diatomic molecule, because there is only one bond in the molecule, *e.g.* in HCl.

$$E(\text{H—Cl}) = D(\text{H—Cl}) = 431\text{kJ/mol}$$

With a polyatomic molecule, such as H_2O, the situation is different. The energy needed to dissociate one mole of H atoms by breaking one O—H bond per H_2O molecule. It's different from the energy required to dissociate one mole of H atoms by breaking the bonds in OH (g). The two O—H bonds in H_2O are identical; therefore, they should have identical energies. This energy, which we can call the O—H bond energy in H_2O, is the average of the two values listed.

$$E(\text{O—H}) = \frac{D_1 + D_2}{2} = \frac{502 + 426}{2} = 464\,(\text{kJ / mol})$$

An average bond energy is the average of bond-dissociation energies for a number of different species containing the particular bond. Understandably, average bond energies cannot be stated as precisely as specific bond-dissociation energies.

Table 4-1 lists the average bond energy of some covalent bonds. As you can see in the table, double bond and triple bond have higher bond energies than single bonds between the same atoms.

The enthalpy change (approximate value) of the reaction can be calculated using bond energy. For example, the reaction of ammonia synthesis

$$\begin{array}{lccccc} & N_2 & + & 3H_2 & = & 2NH_3 \\ D/(\text{kJ/mol}) & 946 & & 436 & & 389 \end{array}$$

In the reaction, one N≡N bond (in N_2) and three H—H bonds (in $3H_2$) are broken to generate six N—H bonds (in $2NH_3$). According to thermodynamic principles, the following reaction steps can be imagined

Table 4-1　The Average Bond Energy and Length of Some Covalent Bonds

Bond	*D*/(kJ/mol)	Length/pm	Bond	*D*/(kJ/mol)	Length/pm
H—H	436	74	C—H	414	109
C—C	347	154	N—H	389	101
C═C	611	134	O—H	464	96
C≡C	837	120	S—H	368	136
N—N	159	145	C—N	305	147
O—O	142	145	C—O	360	143
S—S	264	205	C═O	736	121
F—F	158	128	C—Cl	326	177
Cl—Cl	244	199	N—Cl	134	—
Br—Br	192	228	Cl—H	431	—
I—I	150	267	N≡N	946	110

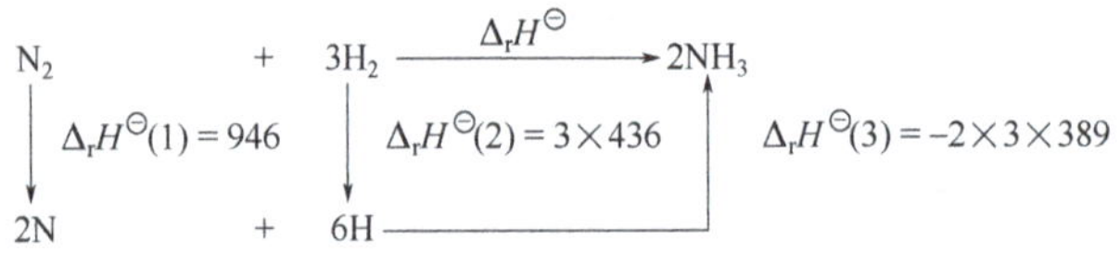

$$N_2 + 3H_2 \xrightarrow{\Delta_r H^\ominus} 2NH_3$$

$$N_2 \xrightarrow{\Delta_r H^\ominus(1) = 946} 2N \qquad 3H_2 \xrightarrow{\Delta_r H^\ominus(2) = 3\times 436} 6H \qquad 2N + 6H \xrightarrow{\Delta_r H^\ominus(3) = -2\times 3\times 389} 2NH_3$$

According to the definition of bond energy and Hess's Law,

$$\Delta_r H^\ominus = \Delta_r H^\ominus(1) + \Delta_r H^\ominus(2) + \Delta_r H^\ominus(3) = 946 + 3\times 436 - 2\times 3\times 389 = -80(\text{kJ/mol})$$

Using the bond energy to calculate the enthalpy change of the reaction can be summarized as the following formula,

$$\Delta_r H^\ominus \approx \Sigma D_{\text{reactions}} - \Sigma D_{\text{products}}$$

(2) Bond length

The internuclear distance of the bonding atoms in the molecule is called the bond length. The bond length is generally equal to the sum of the covalent radii of the bonding atoms, and the common unit is picometer (pm). Generally speaking, smaller atomic radius and greater bond energy correspond to shorter bond length. Table 4-1 also lists some covalent bond length data.

(3) Bond angle

In molecules, the angle between bonds is called the bond angle. The bond angle is a reflection of the direction of the covalent bond, and it is one of the important data that determines the geometric configuration of the molecule. The bond length and bond angle together can determine the geometry of the molecule. Table 4-2 lists the bond angle and geometric configuration of some molecules.

Table 4-2 The Bond Angle and Molecular Geometry of Some Molecules

Molecule	Bond angle	Electron-group geometry	Molecular geometry
$HgCl_2$	180°	Linear	Linear
H_2O	104.5°	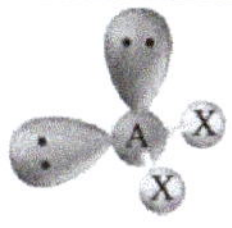 Tetrahedral	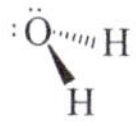 Bent
SO_2	119.3°	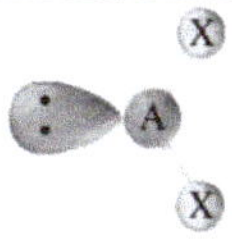 Trigonal Planar	Bent
BF_3	120°	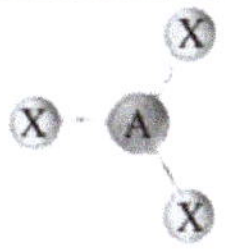 Trigonal Planar	Trigonal planar
SO_3	120°		
NH_3	107.3°	 Tetrahedral	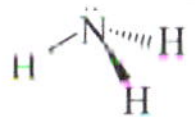 Trigonal pyramidal
SO_3^{2-}	106°		Trigonal pyramidal
CH_4	109.5°	 Tetrahedral	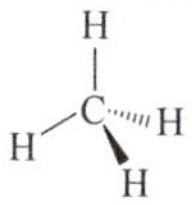 Tetrahedral
SO_4^{2-}	109.5°		Tetrahedral
PCl_5	90°, 120°	 Trigonal bipyramidal	Trigonal bipyramidal
SF_4	90°, 120°	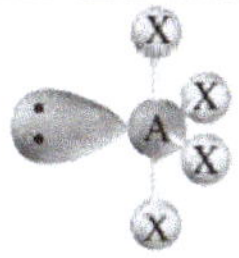 Trigonal bipyramidal	Seesaw geometry
ClF_3	90°	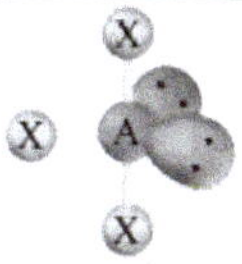 Trigonal bipyramidal	T-shape
XeF_2	180°	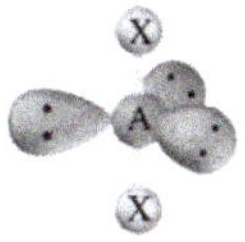 Trigonal bipyramidal	Linear

(Continued)

Molecule	Bond angle	Electron-group geometry	Molecular geometry
SF_6	90°	Octahedral	Octahedral
BrF_5	90°	Octahedral	Square pyramidal
XeF_4	90°	Octahedral	Square planar

4.2 Valence-Shell Electron-Pair Repulsion (VSEPR) Theory

The shape of a molecule is established by experiment or by a quantum mechanical calculation confirmed by experiment. The results of these experiments and calculations are generally in good agreement with the valence-shell electron-pair repulsion theory (VSEPR theory). H. Sidgwich and H. Powell first proposed related concepts in 1940, and later they developed it into modern theory. The theory has achieved surprising success in predicting the shape of polyatomic molecules, but it's a simple extension of Lewis theory.

VSEPR theory proposes that electron pairs repel each other, whether they are in chemical bonds (bond pairs) or unshared (lone pairs). Electron pairs assume orientations about an atom to minimize repulsions. This results in particular geometric shapes for molecules.

VSEPR theory predicts the distribution of electron groups, and in these molecules electron groups are arranged around the central atom in a manner that minimizes repulsion. The shape of a molecule, however, is determined by the location of the atomic nuclei. To avoid confusion, we will call the geometric distribution of electron groups the electron-group geometry and the geometric arrangement of the atomic nuclei—the actual determinant of the molecular shape—the molecular geometry.

The geometric shapes of CH_4, NH_3 and H_2O are summarized in Table 4-2. All three molecules have a tetrahedral arrangement of electron groups around the central atom.

However, molecular shapes (or molecular geometries) are established by focusing only on the positions of the atoms bonded to a common center. Pretend that lone pairs are invisible when visualizing the molecular shapes. In CH_4, there are no lone pairs and the C atom sits at the center of a tetrahedron; the molecular shape is tetrahedral. Pretending the lone pair on the N atom in NH_3 is invisible, the N atom is the vertex in a pyramid having a triangular base; the molecular shape is called trigonal pyramidal. In H_2O, the O and H atoms form a V-shape and the molecular shape is V-shaped or bent.

For tetrahedral electron-group geometry, we expect bond angles of 109.5°, known as the tetrahedral bond angle. In the CH_4 molecule, the measured bond angles are, in fact, 109.5°. The bond angles in NH_3 and H_2O are slightly smaller: 107° for the H—N—H bond angle and 104.5° for the H—O—H bond angle. We can explain these less-than-tetrahedral bond angles by the fact that the charge cloud of the lone-pair electrons spreads out. This forces the bond-pair electrons closer together and reduces the bond angles.

VSEPR theory works best for second-period elements. The predicted bond angle for H_2O is close to the measured angle of 104.5°. For H_2S, however, the predicted value of 109.5° is not in good agreement with the observed value 92°. Even though VSEPR theory does not give an accurate prediction for the angle in H_2S, it does provide an indication that the molecule is bent.

Possibilities for Electron-Group Distributions

The most common situations are those in which central atoms have two, three, four, five, or six electron groups distributed around them. The cases for five- and six-electron groups are typified by molecules with expanded valence shells.

Electron-group geometries

① two electron groups: linear

② three electron groups: trigonal planar

③ four electron groups: tetrahedral

④ five electron groups: trigonal bipyramidal

⑤ six electron groups: octahedral

The relationship between electron-group geometry and molecular geometry is summarized in Table 4-2. To understand all the cases in Table 4-2, we need two more ideas.

① The closer two groups of electrons are forced, the stronger the repulsion between them. The repulsion between two electron groups is much stronger at an angle of 90° than at 120° or 180°.

② Lone-pair electrons spread outer than bond-pair electrons. As a result, the repulsion of one lone pair of electrons for another lone pair is greater than, that between two bond pairs. The order of repulsive forces, from strongest to weakest, is

$$\text{lone pair-lone pair} > \text{lone pair-bond pair} > \text{bond pair-bond pair}$$

4.3 Hybrid Orbital Theory and Molecular Structure

4.3.1 Hybrid Orbital Theory

How to use the covalent bond theory to explain the regular tetrahedral space configuration of methane molecules? The result is not ideal. In 1931, Pauling and Slater established the hybrid orbital theory by using the concept of "hybridization". Hybrid orbital theory was on the basis of valence bond theory. Although it still belongs to modern valence bond theory in essence, it enriches and develops modern valence bond theory in terms of bonding ability and molecular spatial configuration.

(1) Ground state and excited state

For example, for carbon atoms, a paired electron on 2s can be excited to an empty 2p orbital, forming four single-electron orbitals. Where does the excitation energy come from? Pauling believes that this energy can be obtained from the hybridization process. Excitation and hybridization possibly act synergistically in the formation of bond.

(2) Hybridization and hybrid orbitals

The mathematical process of replacing pure atomic orbitals with reformulated atomic orbitals for bonded atoms is called hybridization, and the new orbitals are called hybrid orbitals.

The number of hybrid orbitals is equal to the number of atomic orbitals that are combined.

Highlights of the theory

① The orbital compositions have changed.

② The energy of the orbitals has changed.

③ The shape of the orbitals has changed.

④ The direction in which the orbitals extend has changed.

All these changes are more conducive to the formation of covalent bonds.

4.3.2 Hybridization Scheme

(1) The sp hybrid orbitals

Let's take beryllium fluoride as an example. Be atoms has one 2s orbital with a pair of electrons. It also has three empty 2p orbitals. One electron in the 2s orbital can be excited to an empty 2p orbital, forming two orbitals containing a single electron. These two orbitals are hybridized to form two sp hybrid orbitals containing a single electron. Each hybrid orbital consists of half s orbital and half p orbital. These two hybrid orbitals are linearly distributed in space because of the least electrostatic repulsion. If the sp hybrid orbitals overlap with the s or p orbitals from other atoms, a linear molecular structure forms.

Keep in mind that when this hybridization scheme is used to describe bonds formed by atoms other than beryllium, there are also two unhybridized p orbitals (not shown in Fig.4-7) that are oriented along the *y* and *z* axis.

Besides beryllium fluoride, examples of sp hybridization include beryllium dichloride, mercury dichloride, cyanic acid, acetylene and CO_2, *etc*. It needs to be mentioned that in cyanic acid molecule, the carbon atom is sp hybridized. Among the two hybrid orbitals, one overlaps with the s orbital of hydrogen atom to form a σ bond, and another hybrid orbital overlaps head to head with a p orbital of N atom to form the second σ bond. The C atom still has two unhybridized orbitals left. They can overlap side by side with the two p orbitals of the N atom to form two π bonds. Therefore, the bond between carbon and nitrogen is triple bond in cyanic acid molecule.

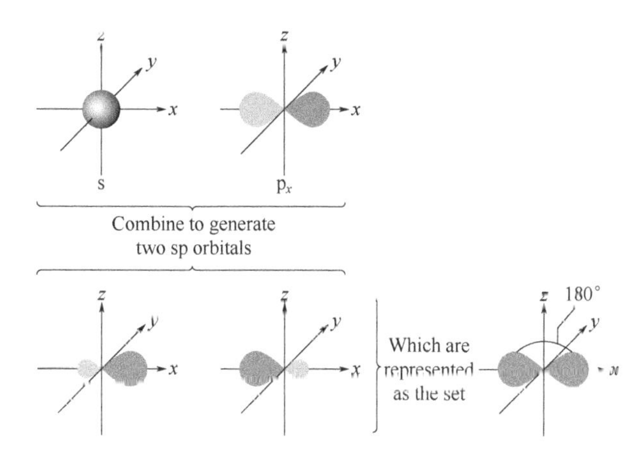

Fig.4-7 The sp hybridization scheme

The sp hybrid orbitals are appropriate for describing bonding in a linear arrangement of atoms.

(2) The sp^2 hybrid orbitals

In BF_3 and BCl_3, the boron has four orbitals but only three electrons in its valence shell. For most boron compounds, the appropriate hybridization scheme combines the 2s and two 2p orbitals into three sp^2 hybrid orbitals and leaves one p orbital unhybridized. Valence-shell orbital diagrams for this hybridization scheme for boron are shown here,

The scheme is outlined in Fig.4-8. The sp^2 hybridization scheme corresponds to trigonal-planar electron group geometry and 120° bond, as in BF_3. Note again that in the hybridization schemes of valence bond theory, the number of orbitals is conserved, that is, in an sp^2 hybridized atom there are still four orbitals: three sp^2 hybrids and an unhybridized p orbital.

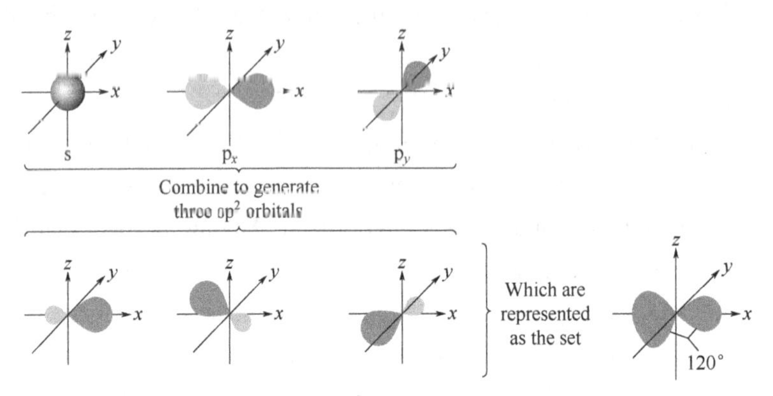

Fig.4-8 The sp^2 hybridization scheme

(3) The sp^3 hybrid orbitals

Because of the least electrostatic repulsion, the four hybrid orbitals are tetrahedral distributed in space. Each hybrid orbital consists of one quarter s orbital and three fourths p orbital(Fig.4-9).

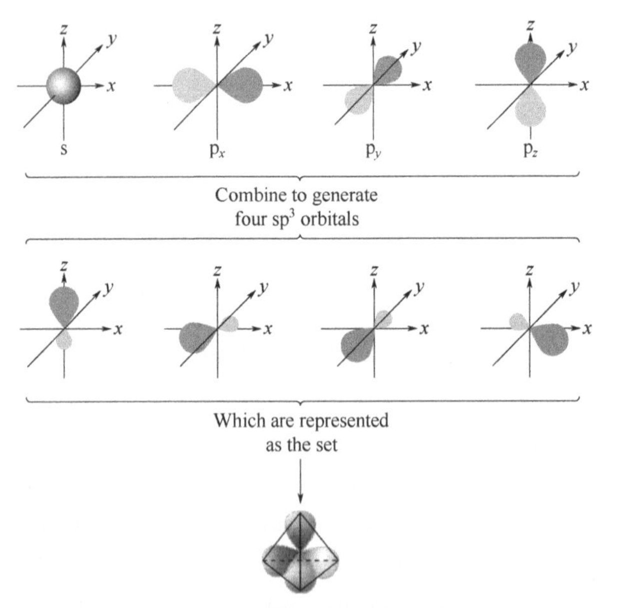

Fig.4-9 The sp^3 hybridization scheme

The C atom excites a 2s electron to 2p subshell. Four single-electron orbitals are generated. C can adopt sp^3 hybridization and forms a tetrahedron structure. These four hybrid orbitals bond with four H atoms head to head, and a tetrahedral molecule—methane with H—C—H 109.5° bond angles is obtained.

Bonding in H_2O and NH_3

The sp^3 hybridization can also be used to explain the spatial configuration of water and ammonia molecules. In these molecules, lone-pair electrons occupy some hybrid orbitals and affect the spatial configuration of molecules, *e.g.* bent structure of water, dihedral angle of hydrogen peroxide and trigonal pyramidal of ammonia.

The repulsive force of the lone electron pair is greater than that of the covalent bond, so the bond angle is less than 109.5°. These angles are in reasonably good agreement with the experimentally observed bond angles of 104.5° in water and 107° in NH_3.

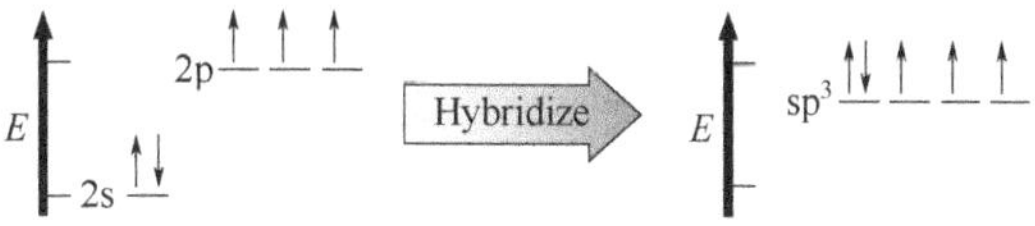

E: 2s ↑↓, 2p ↑ ↑ ↑ —Hybridize→ E: sp^3 ↑↓ ↑ ↑ ↑

Even though the sp^3 hybridization scheme seems to work quite well for H_2O and NH_3, both theoretical and experimental (spectroscopic) evidences favor a description based on unhybridized p orbitals of the central atoms. The H—O—H and H—N—H bond angles expected for 1s and 2p atomic orbital overlaps are 90°, which does not conform to the observed bond angles. One possible explanation is that because O—H and N—H bonds have considerable ionic character, repulsions between the positive partial charges associated with the H atoms force the H—O—H and H—N—H bonds to "open up" to values greater than 90°. The issue of how best to describe the bonding orbitals in H_2O and NH_3 is still unsettled and underscores the occasional difficulty of finding a single theory that is consistent with all the available evidence.

4.4 Molecular Polarity and Polarization

4.4.1 Polar Covalent Bond and Nonpolar Covalent Bond

Electronegativity (Abbreviated as EN) describes an atom's ability to compete for electrons with other atoms to which it is bonded. In general, the lower its EN, the more metallic an element is; and the higher the EN, the more nonmetallic it is. EN increases from left to right in a period and decreases from top to bottom in a group of the Periodic Table.

A covalent bond in which electrons are not shared equally between two atoms is a polar covalent bond. In such a bond, electrons are displaced towards the more nonmetallic element. For example, in HCl, Cl attracts electrons more strongly than H does; the electron charge density is greater near the Cl atom than the H atom. The center of negative charge lies closer to Cl nucleus than the center of positive charge and this produces a separation of the centers of positive and negative charges. H—Cl bond is called polar. In molecule with two same atoms such as H_2, each H atom has the same electron affinity, and the electron charge density between two H atoms is evenly distributed. The centers of positive and negative charge coincide at a point midway between the atom nuclei. H—H bond is called nonpolar covalent bond.

The electronegativity differences, ΔEN, as the absolute value of the difference in EN values of the bonded atoms describes the amount of polar character in a covalent bond. And if ΔEN for two atoms is very small, the bond between them is essentially covalent; if ΔEN is large, the bond is essentially ionic. For intermediate values of ΔEN, the bond is described as polar covalent. A useful rough relationship between ΔEN and percent ionic character of a bond is presented in Fig.4-10.

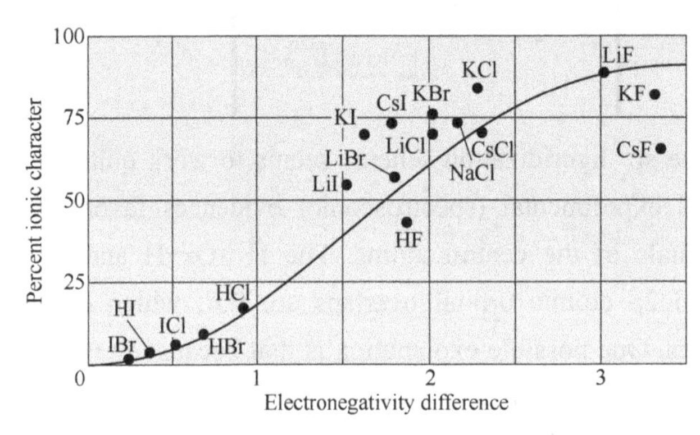

Fig.4-10 Percent ionic character of a chemical bond as a function of electronegativity difference {The curve represents the equation: ionic character = $100\% \times [1-\exp(-0.25\Delta EN^2)]$}

Large EN differences are found between the more metallic and the more nonmetallic elements. Combinations of these elements are expected to produce bonds that are essentially ionic. Small EN differences are expected for two nonmetal atoms, and the bond between them should be essentially covalent. Thus, even without a compilation of EN values at hand, you should be able to predict the essential character of a bond between two atoms. Simply assess the metallic/nonmetallic characters of the bonded elements from the Periodic Table.

4.4.2 Dipole Moment and Molecular Polarity

In a molecule, because the amount of positive charge carried by the nucleus is equal to the amount of negative charge carried by electrons, the entire molecule is electrically neutral. However, within the molecule, the molecules can be divided into polar molecules and non-polar molecules according to the distribution of these two charges. Just as any object has a center of gravity, it can be assumed that either the positive or the negative charge in the molecule has a "charge center". Molecules whose positive and negative charge centers coincide are called nonpolar molecules, and molecules whose positive and negative charge centers do not coincide are called polar molecules.

The polarity of molecules can be expressed by dipole moment. If the amount of charge carried by the positive and negative charge centers in the molecule is q, and the distance between the two centers (dipole length) is l, the product is called the dipole moment, represented by the symbol μ, and the unit of the dipole moment is Debye (10^{-30}C·m), that is

$$\mu = ql$$

The value of the dipole moment can be determined by experiment, and they can be used to measure the strength of the molecular polarity. Generally, the large dipole moment value means the strong polarity of the molecule. A molecule with zero dipole moment is a non-polar molecule. Table 4-3 lists the dipole moments of some molecules. For diatomic molecules, the molecular polarity is consistent with the polarity of covalent bond. For polyatomic molecules, the polarity of the molecule cannot be judged simply by the polarity of the covalent bond forming the molecule. For example, in NH_3 molecule and BF_3 molecule, the N—H and B—F bonds are polar covalent bonds, but from the dipole moment values in Table 4-3, NH_3 is a polar molecule, and BF_3 is a non-polar molecule. This shows that in polyatomic molecules, in addition to the polarity of the bond, there is another factor that affects the polarity of the molecule: the geometry of the molecules. The molecular geometry of the BF_3 is trigonal planar, the angle between the three B—F bonds are 120° and the three bonds are highly symmetrical. So the positive and negative charge centers of the entire molecule overlap and the dipole moment is zero. The geometry of NH_3 molecule is a triangular pyramid, and the polarity of the three N—H bonds cannot be offset, resulting in the molecular polarity.

Table 4-3 Dipole Moments of Some Molecules and Their Geometries

Formula	$\mu/10^{-30}$C·m	Geometry	Formula	$\mu/10^{-30}$C·m	Geometry
HF	6.39	Linear	HCN	9.84	Linear
HCl	3.50	Linear	H_2O	6.14	Bent
HBr	2.50	Linear	SO_2	5.37	Bent
HI	1.27	Linear	H_2S	3.14	Bent
CO	0.40	Linear	CS_2	0	Linear
N_2	0	Linear	CO_2	0	Linear
H_2	0	Linear			
NH_3	5.00	Trigonal pyramidal	$CHCl_3$	3.44	Tetrahedral
BF_3	0	Trigonal planar	CH_4	0	Regular tetrahedral
			CCl_4	0	Regular tetrahedral

4.4.3 Molecular Polarization

In addition to the molecular properties, the molecular dipole moment is also affected by the external electric field. When the non-polar molecule is induced by the applied electric field, the positive and negative charge centers will move in the opposite direction and produce molecular polarity. The electrons of the atoms in the molecule are affected by the field, and the electron cloud deforms and produces induced dipole. This phenomenon is called polarization. Not only nonpolar molecules, polar molecules can also polarize to produce induced dipole. Dipoles of polar molecules are aligned by field. The induced dipole will strengthen the permanent dipole of polar molecules. The total dipole moment is the sum of the inherent dipole moment of the polar molecule (dipole moment when there is no external electric field) and the induced dipole moment.

The degree of polarization of the molecule (the value of the induced dipole moment) is related to the degree of deformation of the electron cloud and the strength of the external electric field (E).

$$\mu_{\text{induced}} = \alpha E$$

In the formula, α is the polarizability, which is equal to the induced dipole moment per unit electric field strength, and the unit is $C \cdot m^2/V$, which suggests that the molecular polarizability is related to the diffusion degree of the electron cloud. The larger the α value, the more easily the molecule is deformed and polarized. So polarizability increases as the number of electrons in a molecule increases, and therefore, it increases with increasing molar mass and molecular size.

4.5 Intermolecular Forces

Whether it is an ionic bond, a metal bond, or the covalent bond introduce in this chapter, it is a relatively strong effect between atoms, and its bond energy is generally between 150 and 800kJ/mol. In addition to these strong forces, there are some types of intermolecular forces (IMFs) known collectively as van der Waals forces. This kind of binding energy is generally in the range of tens kJ/mol. Gas molecules can condense into liquids and solids, mainly due to this force. Intermolecular forces have a greater influence on the properties of substances, especially physical properties. The fundamental cause of IMFs is the polarity and deformability of molecules.

4.5.1 van der Waals Forces

(1) Dipole-dipole intermolecular forces

Dipole-dipole intermolecular forces are the attractive forces that occur between two polar molecules. The polar molecules have a permanent dipole; there is a partial positive charge (δ^+) at one end of the molecule and a negative partial charge (δ^-) at the opposite end. When two polar molecules such as hydrogen chloride approach each other, an electrostatic force of attraction forms between the positive and negative partial charges on the adjacent molecules (Fig.4-11).

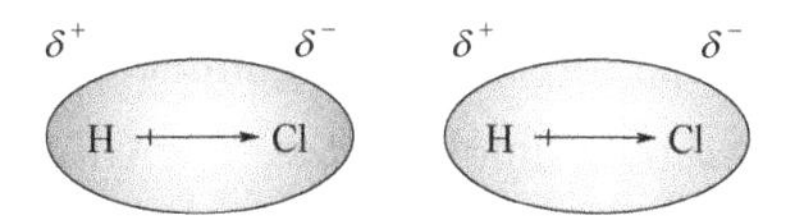

Fig.4-11 Dipole-dipole IMFs between two HCl molecules

The polar molecules tend to line up with the positive end of one dipole pointed toward the negative end of a neighboring dipole. Ordinarily, thermal motion upsets this orderly array. This tendency for dipoles to align themselves can affect physical properties, such as the melting points of solids and the boiling points of liquids. For example, the dipole-dipole IMFs between HCl molecules allow gaseous HCl to condense to form a liquid at low temperatures. Liquid HCl consists of HCl molecules held together by a network of dipole-dipole IMFs (Fig.4-12). Dipole-dipole IMFs are generally weaker than the ionic and covalent forces in ionic solids and covalently bonded compounds, respectively.

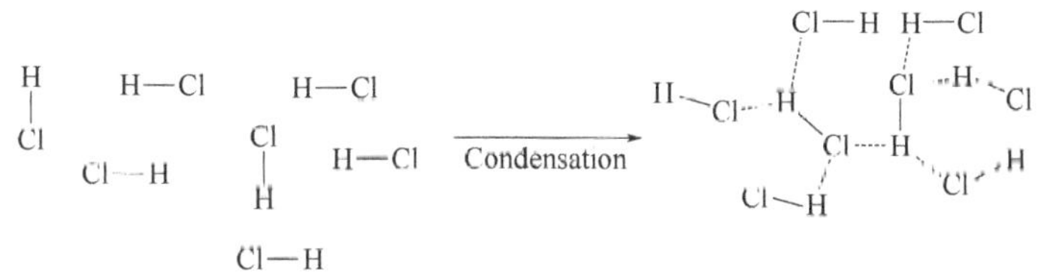

Condensation

Fig.4-12 Condensation of HCl molecules

(2) Dipole-induced dipole forces

Dipole-induced dipole intermolecular forces are attractive forces that occurs between a polar molecule (with a permanent dipole) and a nonpolar molecule (with a temporary or induced dipole). Dipole-induced dipole interactions result from the fact that all molecules are polarizable. Nonpolar molecules do not have permanent dipoles, but it is possible to create a temporary dipole, called an induced dipole, in a nonpolar molecule. Consider the interaction between water, a polar molecule, and oxygen, a nonpolar molecule. When the negative end of an H_2O molecule approaches an O_2 molecule, the negative partial charge repels the O_2 electron cloud, distorting it and creating a temporary dipole on the O_2 molecule (Fig.4-13). Note that the positive end of a water molecule can also induce a dipole in an O_2 molecule by attracting the electron cloud instead of repelling it. Induced dipoles are temporary; when the H_2O and O_2 molecules separate, the oxygen electron cloud is no longer distorted.

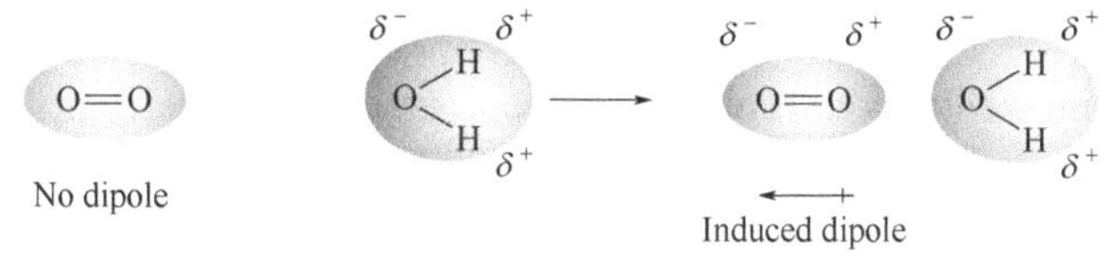

Fig.4-13 Water induces a temporary dipole in O_2

The electronic charge cloud of a molecule (polar or nonpolar) will always be distorted to some extent by the approach of a polar molecule. Although dipole-induced dipole IMFs are generally weaker than dipole-dipole forces, the magnitude of the induced dipole can result in these forces being stronger than dipole-dipole forces. The CF_4 and CHF_3 molecules have very similar polarizabilities, but CHF_3 is polar and CF_4 is not. The difference in the dipole moments manifests itself as a dramatic difference in boiling point. The intermolecular attractions between CHF_3 molecules include additional contributions—namely, dipole-dipole and dipole-induced dipole interactions—that are not present between pairs of CF_4 molecules.

(3) Induced dipole-induced dipole forces

Although the intermolecular forces described in the previous section are important for understanding the interactions between polar molecules, these two types are usually not the main component of IMFs. For example, even nonpolar substances will liquefy if the temperature is lowered sufficiently.

Induced dipole-induced dipole intermolecular forces are attractive forces that occur between nonpolar molecules. As you saw in the previous discussion, a polar molecule can induce a temporary dipole in a nonpolar molecule. It is also possible for two nonpolar molecules to induce temporary dipoles in each other. Consider iodine, which consists of nonpolar I_2 molecules. When two I_2 molecules approach each other, the electron clouds repel and temporary dipoles form in each I_2 molecule (Fig.4-14).

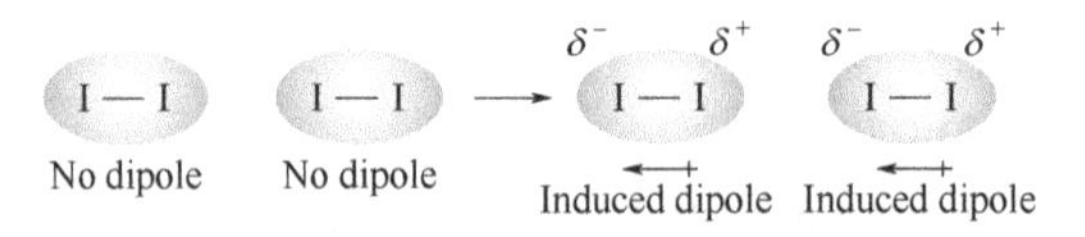

Fig.4-14 Induced dipoles formed by I_2 molecules

The attractive forces between these temporary dipoles, induced dipole-induced dipole IMFs, are also called London dispersion forces (or simply dispersion forces or London forces), the latter in honor of Fritz London who, in 1928, offered a theoretical explanation of these forces.

Dispersion forces are generally weak, but they can be stronger than dipole-dipole forces when they occur between highly polarizable molecules. They are important in all liquids, even those consisting of polar molecules, because every molecule's electron cloud can be distorted by the partial charges in surrounding molecules.

For dispersion force, we can summarize as follows:

① In the normal condition, a nonpolar molecule has a symmetrical charge distribution.

② In the instantaneous condition, a displacement of the electronic charge produces an instantaneous dipole with a charge separation represented as δ^+ and δ^-.

③ In an induced dipole, the instantaneous dipole on the left induces a charge separation in the molecule on the right. The result is an instantaneous dipole-induced dipole attraction.

(4) Effect of polarizability on physical properties

Nonpolar species such as hydrocarbons and diatomic elements have weak dispersion (induced dipole-induced dipole) IMFs, so many are gases at room temperature. However, with the increase of the molar mass, the dipole is more easily induced and the polarizability of compound increases. As shown by the data in Table 4-4, as the number of electrons increases in a series of nonpolar compounds, so does the enthalpy of vaporization and boiling point.

Table 4-4 Molar Mass, $\Delta H_{vap}^{\ominus}$, and Boiling Point for Some Nonpolar Species

Compound	Molar mass/(g/mol)	$\Delta H_{vap}^{\ominus}$ /(kJ/mol)	Boiling point/℃
He	4.0	0.08	2268.9
Ne	20.3	1.7	2246.1
N_2	28.0	5.6	2195.8
O_2	32.0	6.8	2183.0
Ar	39.9	6.4	2185.9
Cl_2	70.9	20.4	234.0
Br_2	159.8	30.0	58.8

The effect of dispersion forces on physical properties can also be seen in the physical properties of a series of straight-chain hydrocarbons (Table 4-5). Each hydrocarbon is nonpolar, and each has a similar molecular structure, differing only in molar mass and the length of the carbon chain. As shown by the data in the table, as molecular size increases, so does the strength of the IMFs which indicated by increasing enthalpy of vaporization and boiling point.

Table 4-5 Selected Physical Properties of Some Hydrocarbons Molecular

Formula	Name	$\Delta H_{vap}^{\ominus}$ /(kJ/mol)	Boiling point/℃
CH_4	Methane	8.2	2161
C_2H_6	Ethane	14.7	288
C_3H_8	Propane	19.0	242
C_6H_{14}	Hexane	28.9	69
C_8H_{18}	Octane	34.4	126
$C_{10}H_{22}$	Decane	38.8	174
$C_{18}H_{38}$	Octadecane	54.5	317

In general, there are only dispersing forces between non-polar molecules. Table 4-6 lists the data of three forces in some molecules. In many cases, dispersion force is the force that accounts for the largest proportion of several intermolecular forces, except those strong polar molecules, such as H_2O and NH_3, *etc.*

Table 4-6 Intermolecular Forces Data in Some Molecules

Formula	Dipole-dipole forces /(kJ/mol)	Dipole-induced dipole forces/(kJ/mol)	Dispersing forces /(kJ/mol)	Total forces /(kJ/mol)
Ar	0	0	8.493	8.493
H_2	0	0	1.674	1.674
CH_4	0	0	11.297	11.297

(Continued)

Formula	Dipole-dipole forces /(kJ/mol)	Dipole-induced dipole forces/(kJ/mol)	Dispersing forces /(kJ/mol)	Total forces /(kJ/mol)
HI	0.025	0.113	25.857	25.995
HBr	0.685	0.502	21.924	23.112
HCl	3.305	1.004	16.820	21.12
CO	0.00293	0.00837	8.745	8.756
NH_3	13.305	1.548	14.937	29.790
H_2O	36.358	1.925	8.996	47.279

When assessing the importance of van der Waals forces, consider the following statements:

① Dispersion (London) forces exist between all molecules. They involve displacements of all the electrons in molecules, and they increase in strength with increasing molecular size. The forces also depend on molecular shapes.

② Forces associated with permanent dipoles involve displacements of electron pairs in bonds rather than in molecules as a whole. These forces are found only in substances with resultant dipole moments (polar molecules). Their existence adds to the effect of dispersion forces also present.

③ When comparing substances of roughly comparable molecular sizes, dipole forces can produce significant differences in properties such as melting point, boiling point, and enthalpy of vaporization.

④ When comparing substances of widely different molecular sizes, dispersion forces are usually more significant than dipole forces.

4.5.2 Hydrogen Bonding

(1) Conditions for forming hydrogen bonding

Hydrogen bonding is an unusually strong type of dipole-dipole IMF that occurs between molecules with H—N, H—O, or H—F bonds. Hydrogen bonding is especially strong for two reasons:

① The elements bonded to hydrogen are very electronegative, and hydrogen has a relatively low electronegativity, resulting in very polar H—N, H—O, and H—F bonds. The highly polar bonds result in large partial charges and therefore a strong force of attraction between molecules.

② Hydrogen has a very small size, allowing hydrogen-bonding molecules to approach closely. This decrease in distance between polar molecules results in a strong force of attraction.

Although hydrogen bonding is simply an electrostatic attraction and not an actual chemical bond like a covalent bond. In a hydrogen bond a H atom is covalently bonded to a highly electronegative atom, which attracts electron density away from the H nucleus. This in turn allows the H nucleus, a proton, to be simultaneously attracted to a lone pair of electrons on a highly electronegative atom in a neighboring molecule.

① Hydrogen bonds are possible only with certain hydrogen-containing compounds.

② Only F, O and N easily meet the requirements for hydrogen-bond formation.

③ Weak hydrogen bonding is occasionally encountered between a H and Cl or S atom.

As we see for hydrogen fluoride in Fig.4-15. The main points established in the figure are outlined below.

① The alignment of HF dipoles places a H atom between two F atoms. Because of the very small size of the H atom, the dipoles come close together and produce strong dipole-dipole attractions.

② Although a H atom is covalently bonded to one F atom, it is also weakly bonded to the F atom of a nearby HF molecule. This occurs through a lone pair of electrons on the F atom. Each H atom acts as a bridge between two F atoms.

③ The bond angle between two F atoms bridged by a H atom (that is, the angle F—H···F—) is about 180°.

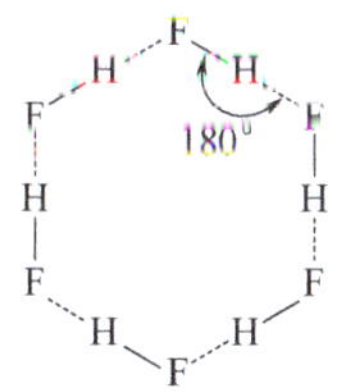

Fig.4-15 Hydrogen bonding in gaseous hydrogen fluoride

(In gaseous hydrogen fluoride, many of the HF molecules are associated into cyclic structures of the type pictured here)

In general, a hydrogen bond is depicted as X—H···Y—, where the three dots denote the hydrogen bond. The X—H fragment is typically referred to as the hydrogen bond donor since the fragment X—H provides the hydrogen as part of the hydrogen bond. The fragment Y—is known as the hydrogen bond acceptor since it accepts the hydrogen as part of the hydrogen bond. Fig.4-16 illustrates the definition of a hydrogen bond along with several examples.

Compared with other intermolecular forces, hydrogen bonds are relatively strong, having energies of the order of 15 to 40kJ/mol. By contrast, single covalent bonds (also known as intramolecular bonds) are much stronger still—greater than 150kJ/mol.

(2) Hydrogen bonding in water

Ordinary water is certainly the most common substance in which hydrogen bonding occurs. Fig.4-17 shows how one water molecule is held to four neighbors in a tetrahedral arrangement by hydrogen bonds. In ice, hydrogen bonds hold the water molecules in a rigid but rather open structure. As ice melts, only a fraction of the hydrogen bonds are broken. One indication of this is the relatively low heat of fusion of ice (6.01kJ/mol). It is much less than we would expect if all the hydrogen bonds were to break during melting.

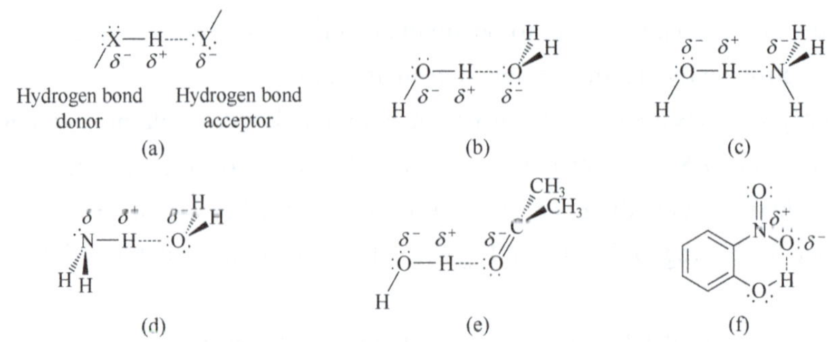

Hydrogen bond donor Hydrogen bond acceptor

(a) (b) (c)

(d) (e) (f)

Fig.4-16 The hydrogen bond illustrated

[(a) A hydrogen bond involves a hydrogen bond donor (shown on the left) and a hydrogen bond acceptor (shown on the right); (b) Hydrogen bonding between H_2O molecules represented by using Lewis structures; (c) In this hydrogen bond between water and ammonia molecules, the water molecule is the donor and the ammonia molecule is the acceptor; (d) For this hydrogen bond, the ammonia molecule is the donor and the water molecule is the acceptor; (e) Hydrogen bonding between water and acetone molecules; (f) An intramolecular hydrogen bond]

The open structure of ice shown in Fig.4-17(b) gives ice a low density. When ice melts, some of the hydrogen bonds are broken. This allows the water molecules to be more compactly arranged, accounting for the increase in density when ice melts. That is, the number of water molecules per unit volume is greater in the liquid than in the solid.

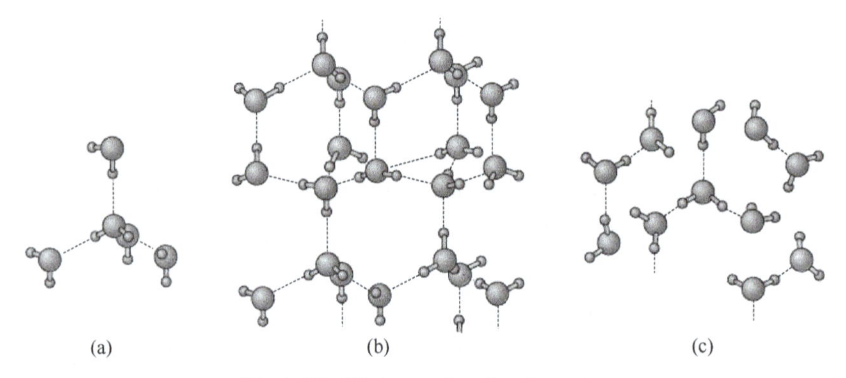

Fig.4-17 Hydrogen bonding in water

[(a) Each water molecule is linked to four others through hydrogen bonds. The arrangement is tetrahedral. Each H atom is situated along a line joining two O atoms, but closer to one O atom (100 pm) than to the other (180 pm); (b) For the crystal structure of ice. O atoms are arranged in bent hexagonal rings arranged in layers. This characteristic pattern is similar to the hexagonal shapes of snowflakes; (c) In the liquid, water molecules have hydrogen bonds to only some of their neighbors]

As liquid water is heated above the melting point, hydrogen bonds continue to break. The molecules become even more closely packed, and the density of the liquid water continues to increase. Liquid water attains its maximum density at 3.98℃. Above this temperature, the water behaves in a "normal" fashion: Its density decreases as temperature increases. The unusual freezing-point behavior of water explains why a freshwater lake freezes from the top down. When the water temperature falls below 4℃, the denser water sinks to the bottom of the lake and the colder surface water freezes. The ice over the top of the lake then tends to insulate the water below from further heat loss. This allows fish to survive the winter in a lake that has been frozen over. Without hydrogen bonding, all lakes would freeze from the bottom up; and fish, small bottom-feeding animals, and aquatic plants would not survive the winter.

Water molecules form cages

Water molecules form cages with each other through hydrogen bonds, encapsulating exogenous neutral molecules or ions (Cl_2, CH_4, Ar, Xe, *etc.*) in hydrates (molecular crystals) in cages.

The finding of large-scale nature gas hydrate (commonly called "flammable ice") reserves brings the study of hydrate into a new phase. Researchers think there could be immensely abundant gas hydrate reserves all around the world, possibly exceeding all other fossil fuels combined.

(3) Other properties affected by hydrogen bonding

Generally speaking, the greater the intermolecular forces, the higher the melting point (m.p.) and boiling point (b.p.) of the substance. Table 4-7 lists the boiling point and melting point data of hydrogen halide. It can be seen from the table that the melting point and boiling point of HCl, HBr, HI increase with the increase of the relative molecular mass. However, the boiling point of HF is particularly high, indicating that there are hydrogen bonding between HF molecules.

Table 4-7 The Boiling Point and Melting Point Data of Hydrogen Halide

Items	HF	HCl	HBr	HI
b. p./℃	19.9	−85.0	−66.7	−35.4
m. p./℃	−83.57	−114.18	−86.81	−50.79

Hydrogen bonding can also affect the solubility of substances. For example, ethanol can be miscible with water in any ratio because they can form hydrogen bonds with water molecules. Hydrogen bonding also has an important influence in biochemistry. The hydrogen bonds between protein molecules are conducive to the stable existence of their spatial structure; the base pairing and the formation of double helix structures in DNA also rely on the role of hydrogen bonds. Hydrogen bonding also plays an important role in molecular self-assembly. As shown in Fig.4-18, diacetylguanidine and phosphodiester molecules with intramolecular hydrogen bonding can self-assemble into supramolecular structures.

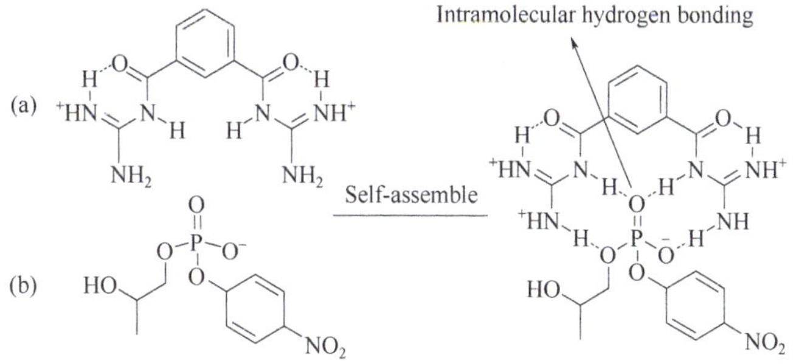

Fig.4-18 Self-assembly of diacetylguanidine (a) and phosphodiester (b)

Exercises

1. It is known that the bond energies of HCl, HBr and HI bonds are 431kJ/mol, 366kJ/mol, 299kJ/mol, respectively. Compare the thermal stability of HCl, HBr, and HI gas molecules.

2. Use the bond energy data to calculate the approximate value of $\Delta_r H_m^\ominus$ for the following gas reactions:

(1) $H_2 + Br_2 = 2HBr$;

(2) $C_2H_4 + Cl_2 = CH_2Cl\text{-}CH_2Cl$.

3. Explain the hybrid orbital types of the following substances and draw the geometric configuration of the molecule.

(1) CO_2 is linear, with a bond angle of 180°.

(2) BF_3 is a plane triangle with a bond angle of 120°.

(3) CH_4 is a regular tetrahedron with a bond angle of 109.5°.

4. Compare the polarity of the following pairs of molecules.

(1) HF___HCl (2) CO_2___CS_2 (3) CCl_4__$SiCl_4$ (4) BF_3__NF_3

(5) PH_3___NH_3 (6) H_2O___H_2S (7) OF_2 __H_2O (8) CH_4__CH_3Cl

5. There is little difference in electronegativity between Si and Sn. Please explain that SiF_4 is gaseous at room temperature but SnF_4 is solid.

6. Indicate which kinds of intermolecular forces exist between the following molecules.

(1) 1mole CCl_4 (2) 1mole H_2O (3) H_2O *vs* O_2

(4) HCl *vs* H_2O (5) 1mole CH_3Cl

7. Are there intermolecular hydrogen bonds between the following pairs of compounds? Why?

(1) CH_3CH_2OH and CH_3OCH_3

(2) CH_3NH_2 and CH_3OH

(3) CH_3CH_2SH and $(CH_3)_2NOH$

Chapter 5 Heat Effect of Chemical Reaction and Energy Utilization

Teaching contents	Learning requirements
Some terminology	Differentiate between open, closed and isolated systems, and between heat and work. Know the difference and connection between process function and state function. Determine the heat of reaction from data obtained by using a calorimeter. Use the correct equation for pressure-volume work to quantify an energy transfer occurring as a result of compression or expansion of gases
The first law of thermodynamics	Describe the first law of thermodynamics as a function of heat and work, and explain what is meant by the term state function. Distinguish between the constant volume heat of reaction (q_V) and the constant-pressure heat of reaction (q_P), and identify the relationships among q_V , q_P , the internal energy change (ΔU) and enthalpy change (ΔH)
Application of the first law to chemical and physical changes	Use Hess's law to determine an unknown heat of reaction. Use standard enthalpies of formation to determine an unknown heat of reaction. Use thermochemical data to assess fuels as energy sources, including fossil fuels and other alternatives

Fuel refers to a substance that can release energy through a combustion reaction, and is widely used in industrial and agricultural production, daily life, national defense and other fields. Firewood, coal, biogas, natural gas for cooking, petrol and diesel for cars and airplanes, and rocket propellants and nuclear fuel for nuclear-powered aircraft carriers are all inseparable from the application of theories related to reaction heat effects. Thermochemistry plays a very important role in rocket propulsion technology, especially the thrust of the rocket is closely related to the thermal efficiency of the propellant fuel. The theoretical calculation of the thermal effect of the propellant is a very important part of the propellant design process. After studying the content of this chapter, we can initially solve some thermal effect calculation problems, such as the reaction of rocket fuel hydrazine hydrate ($N_2H_4 \cdot H_2O$) with hydrogen peroxide (see Example 5-9 for the calculation process).

5.1 The First Law of Thermodynamics

5.1.1 Some Thermodynamic Terminology

(1) System and its surroundings

When studying thermodynamics problems, a part of an object is usually divided from other objects around it as the object of study. This part of the object is called a system. A system is the part of the universe chosen for study. It can be as large as all the oceans on Earth or as small as the contents of a beaker. The selection of the system is determined by the purpose of the study. Their sizes are suitable for the selected research methods. Most of the systems we will examine will be small and we will look at the transfer of energy (as heat and work) and matter between the system and its surroundings. The surroundings are that part of the universe outside the system with which the system interacts. There is not necessarily an obvious physical interface between the system and the surroundings. This interface can be actual or imaginary. For example, if a sucrose solution is placed in a beaker and the research object is a sucrose solution, then the sucrose solution is the system, and everything around the beaker is the surroundings. But if sucrose is the object of study, water and beakers are part of the environment, and the interface between water and sucrose can only be imagined.

In thermodynamic research, we can divide the system into three common systems according to the relationship between the system and the environment. They are open system, closed system and isolated system.

① Open system: An open system freely exchanges energy and matter with its surroundings. For example, a beaker contains water, and this cup of water is an open system. Because it is neither insulated to prevent the exchange of heat energy, nor closed to prevent the evaporation of water vapor. Chemical thermodynamics generally does not study open systems.

② Closed system: A closed system can exchange energy, but not matter, with its surroundings. For example, a tightly packed bottle contains water. Although this bottle of water does not evaporate, it exchanges heat with the outside world. The main research of chemical thermodynamics is the closed system. In chemical thermodynamics, not only to study the conversion and transfer of different energy forms between the system and the environment, but also to study the conversion of chemical energy into other forms of energy, such as heat and work.

③ Isolated system: An isolated system does not interact with its surroundings. For example, if there is hot water in an insulated container, its heat insulation and sealing are very good, and it can be regarded as an approximated isolated system. In fact, the isolated system is an ideal state. The thermal insulation of the container is not absolute. The temperature of the water in the bottle will still slowly decrease. After a period of time, the energy exchange between the system and the surroundings can be clearly displayed. On the contrary, if an explosion reaction is studied in a closed container with poor thermal insulation, because the reaction time is very short, the energy exchange between the system and the surroundings is extremely small, such a device with poor thermal insulation can still be regarded as an approximated isolated system under certain conditions.

Absolute isolated system does not exist. Because the relationship between the open system and its surroundings is too complicated, open systems are often not the main object of discussion in classical thermodynamics. And the object system what we usually discuss in chemical thermodynamics is the closed system. So in thermodynamic discussion, the system is referred to as closed system.

(2) State and state functions

The state of a thermodynamic system is a comprehensive manifestation of the physical and chemical properties of the system. To describe a system completely, we must indicate its temperature, its pressure, and the kinds and amounts of substances present. When we have done this, we have specified the state of the system. Every property that has a unique value for a specified state of a system is said to be a function of state, or a state function. In other words, it might be easier to understand. State is the sum of all the physical and chemical properties of the system, which reflects the existence of the system under certain conditions. The state function refers to one of those properties that make up a certain state. It describes a certain property of a system.

When all state functions are fixed, the state of the system is determined. As long as a state function in the system changes, the state of the system changes. This change in the system is called a process. The meaning of several important processes is listed below:

① Isothermal process: An isothermal process is a thermodynamic process, in which the temperature remains constant: $\Delta T = 0$. This typically occurs when a system is in contact with an outside thermal reservoir (heat bath), and the change will occur slowly enough to allow the system to continually adjust to the temperature of the reservoir through heat exchange.

② Isobaric process: An isobaric process, also called a constant-pressure process, is a thermodynamic process in which the pressure stays constant.

③ Isochoric process: An isochoric process, also called a constant-volume process, an isovolumetric process, or an isometric process, is a thermodynamic process in which the volume of the system undergoing such a process remains constant.

④ Adiabatic process: An adiabatic process is one that occurs without transfer of heat or matter between a thermodynamic system and its surroundings. In an adiabatic process, energy is transferred only as work. The adiabatic process provides a rigorous conceptual basis for the theory used to expound the first law of thermodynamics.

⑤ Cyclic process: The system starts from the initial state, goes through a series of changes, and then returns to the initial state.

In fact, the state functions of the system are not independent of each other, but interrelated. For example, for a single-component gas, there are four state functions that describe the state of the system: pressure, temperature, volume, and amount of substance. As long as the three state functions of pressure, temperature, and amount of substance are determined, the state of the system is determined.

The state function can be divided into two properties: extensive or intensive properties.

An extensive property is dependent on the quantity of matter observed. Mass and volume are both extensive properties. However, if we divide the mass of a substance by its volume, we obtain density, an intensive property.

$$\text{density}(d) = \frac{\text{mass}(m)}{\text{volume}(V)}$$

Intensive property is independent of the amount of matter observed. Thus, the density of pure water at 25℃ has a unique value, whether the sample fills a small beaker (small mass/small volume) or a swimming pool (large mass/large volume). Intensive properties are especially useful in chemical studies because they can often be used to identify substances.

$$V \propto \frac{nT}{P} \quad \text{and} \quad V = \frac{RnT}{P}$$

A gas whose behavior conforms to the ideal gas equation is called an ideal or perfect gas. Before we can apply equation, we need a value for the constant R, called the gas constant. One way to obtain this is to substitute into above equation, the molar volume of an ideal gas at 273.15K and 1atm. However, the value of R will depend on which units are used to express pressure and volume. With a molar volume of 22.4140L and pressure in atmospheres,

$$R = \frac{PV}{nT} = \frac{1\text{atm} \times 22.4140\text{L}}{1\text{mol} \times 273.15\text{K}} = 0.082057\text{atm} \cdot \text{L/(mol} \cdot \text{K)} \tag{5-1}$$

Using the SI units of m^3 for volume and Pa for pressure gives

$$R = \frac{PV}{nT} = \frac{101325\text{Pa} \times 2.24140 \times 10^{-2}\text{m}^3}{1\text{ mol} \times 273.15\text{ K}} = 8.314\text{ J/(mol} \cdot \text{K)} \tag{5-2}$$

Based on the above, we can summarize the three features of the state functions:

① A unique value. Each function has a unique value in each state. The value of a function of state depends on the state of the system, and not on how that state was established.

② The change in state function is only related to the initial state and the final state, and it has nothing to do with the specific process of state change.

③ The changes in state functions of the cyclic process are zero.

(3) Internal energy

When considering the energy of a system, we use the concept of internal energy, also called thermodynamic energy. The symbol of internal energy is U. It is the total energy (both kinetic and potential) in a system. It includes translational kinetic energy of molecules, the energy associated with molecular rotations and vibrations, the energy stored in chemical bonds and intermolecular attractions, and the energy associated with electrons in atoms. Internal energy also includes energy associated with the interactions of proton and neutron in atomic nucleus, although this component is unchanged in chemical reactions. Internal energy is usually used to represent the energy of the system.

The internal energy of a system is a function of state, although there is no simple measurement that we can use to establish its value. That means we cannot write down a value of U for a system. Fortunately, we don't have to know actual values of U.

Consider, for example, heating 10 gram of ice at 0℃ to a final temperature of 50℃. The internal energy of the ice at 0℃ has one unique value, U_1, while that of the liquid water at 50℃ has another, U_2. The difference in internal energy between these two states also has a unique value, $\Delta U = U_2 - U_1$, and this difference is something that we can measure. It is the quantity of energy that must be transferred from the surroundings to the system during the change from state 1 to state 2.

As a further illustration, imagine that a system changes from state 1 to state 2 and then back to state 1.

$$\text{state } 1(U_1) \xrightarrow{\Delta U} \text{state } 2(U_2) \xrightarrow{-\Delta U} \text{state } 1(U_1)$$

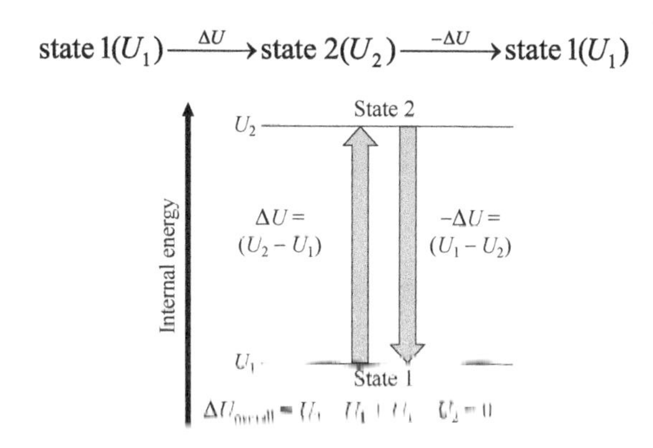

Because U has a unique value in each state, ΔU also has a unique value; it is U_2 minus U_1. The change in internal energy when the system is returned from state 2 to state 1 is $-\Delta U = U_1 - U_2$. Thus, the overall change in internal energy is

$$\Delta U + (-\Delta U) = (U_2 - U_1) + (U_1 - U_2) = 0$$

This means that the internal energy returns to its initial value of U_1. It must do, since it's a function of state. It is important to note here that when we reverse the direction of change, we change the sign of ΔU. This rule applies to all state functions.

(4) Heat and work

Heat and work are the means by which a system exchanges energy with its surroundings. Heat and work exist only during a change in the system. The transfer of energy between the system and it's surrounding can lead to changes in internal energy. The energy transfer caused by the temperature difference between the system and its surrounding is called heat. In addition to heat, other forms of energy transfer between the system and its surrounding are collectively referred to as work.

Path-Dependent Functions

There's a statement: "Unlike internal energy and changes in internal energy, heat and work are not functions of state." Is it true or false?

The statement is true. Heat and work are not functions of state. Their values depend on the path followed when a system undergoes a change. We call them path-dependent functions.

It can be verified by calculating and comparing the quantity of work done in the two different expansions of a gas. We know that there are many kinds of works. We just have to pick one of them to prove it.

In chemical thermodynamics, it is necessary to master the calculation of pressure-volume work. Let's review how to get it.

In the hypothetical apparatus pictured in Fig.5-1(a), a weightless piston is attached to a weightless wire support, to which a weightless pan is attached. On the pan are two same weights just sufficient to stop the gas from expanding. The gas is confined by the cylinder walls and piston, and the space above the piston is a vacuum. The cylinder is contained in a constant-temperature water bath, which keeps the temperature of the gas constant.

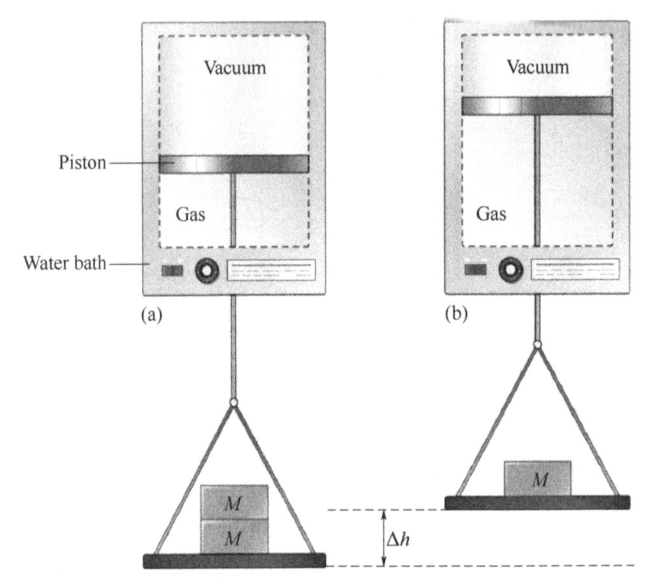

Fig.5-1 Pressure-volume work

Now imagine that one of the two weights is removed, leaving half the original mass on the pan. Let us call this remaining mass M. The gas will expand and the remaining weight will move against gravity, the situation represented by Fig.5-1(b). After the expansion, we find that the piston has risen through a vertical distance, that the volume of gas has doubled; and that the pressure of the gas has decreased.

Now let us see how pressure and volume enter into calculating, and how much pressure-volume work the expanding gas does.

First we can calculate the work done by the gas in moving the weight of mass M through a displacement Δh.

$$\text{work}(w) = \text{force}(M \cdot g) \times \text{distance}(\Delta h) = -Mg\Delta h$$

Where g is the acceleration due to gravity. The negative sign appears because the force is acting in a direction opposite to the piston's direction of motion.

We also know that

$$\text{pressure=force}(M \cdot g) / \text{area}(A)$$

So that if the expression for work is multiplied by A/A we get

$$w = -\frac{Mg}{A} \cdot \Delta hA = -P_{\text{ext}} \Delta V \tag{5-3}$$

The "pressure" part of the pressure-volume work is seen to be the external pressure (P_{ext}) on the gas, which is equal to the weight pulling down on the piston. Note that the product of the area (A) and height Δh is equal to a volume—the volume change, ΔV, produced by the expansion.

Two significant features to note in above equation (5-3) are the negative sign and the factor P_{ext}. When a gas expands, ΔV is positive and w is negative, signifying that energy leaves the system as work. When a gas is compressed, ΔV is negative and w is positive, signifying that energy (as work) enters the system. P_{ext} is the external pressure—the pressure against which a system expands or the applied pressure that compresses a system. If pressure is stated in bars (1bar = 10^5Pa) or atmospheres and volume in liters, the unit of work is bar·L or atm·L. However, the SI unit of work is the joule. To convert from bar·L to J, or from atm·L to J, we use one of the following relationships, both of which are exact.

$$1\text{bar}\cdot\text{L} = 100\text{J} \qquad 1\ \text{atm}\cdot\text{L} = 101.325\text{J}$$

Keep in mind: That a pressure of 100kPa is equal to 1bar.

Example 5-1 Suppose the gas in Fig.5-1 above is 0.100mol He at 298K, the two weights correspond to an external pressure of 2.40 atm in Fig.5-1(a), and the single weight in Fig.5-1(b) corresponds to an external pressure of 1.20 atm. How much work, in joules, is associated with the gas expansion at constant temperature?

Solution

First calculate the initial and final volumes.

$$V_{\text{initial}} = \frac{nRT}{P_{\text{i}}} = \frac{0.1\text{mol} \times 0.0821\text{atm} \cdot \text{L} / (\text{mol} \cdot \text{K}) \times 298\text{K}}{2.40\text{atm}} = 1.02\text{L}$$

$$V_{\text{final}} = \frac{nRT}{P_{\text{f}}} = \frac{0.1\text{mol} \times 0.0821\text{atm} \cdot \text{L} / (\text{mol} \cdot \text{K}) \times 298\text{K}}{1.20\text{atm}} = 2.04\text{L}$$

$$\Delta V = V_{\text{f}} - V_{\text{i}} = 2.04\text{L} - 1.02\text{L}=1.02\text{L}$$

$$w = -P_{\text{ext}} \times \Delta V = -1.20\text{atm} \times 1.02\text{L} \times \frac{101\text{J}}{1\text{atm} \cdot \text{L}} = -1.24 \times 10^2\,\text{J}$$

The negative value signifies that the expanding gas (*i.e.*, the system) does work on its surroundings. Please remember this result first.

Let's go back to the discussion of path-dependent functions.

Think of the 0.100mol of He at 298K and under a pressure of 2.40 atm as state 1, and under a pressure of 1.20 atm as state 2. The change from state 1 to state 2 occurred in a single step. Suppose that in another instance, we allowed the expansion to occur through an intermediate state pictured in the middle. That is, suppose the external pressure on the gas was first reduced from 2.40 atm to 1.80 atm (at which point, the gas volume would be 1.36L). Then, in a second stage, reduced from 1.80 atm to 1.20 atm, thereby arriving at state 2 (shown in Fig.5-2).

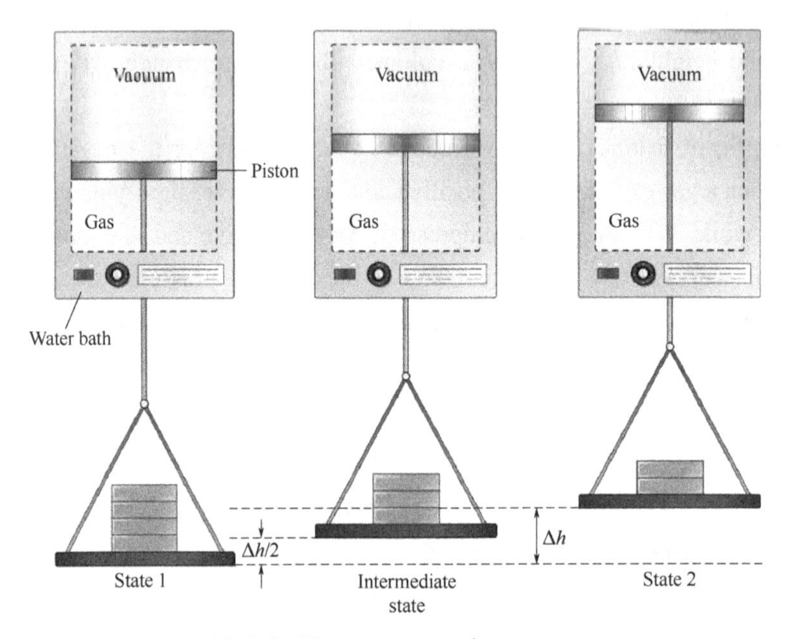

Fig.5-2 Two-stage expansion process

We calculated the amount of work done by the gas in a single-stage expansion in Example 5-1; it was $-1.24\times10^2\,\text{J}$. The amount of work done in the two-stage process is the sum of two terms: the pressure-volume work for each stage of expansion.

$$\begin{aligned} w &= -1.80\text{atm}\times(1.36\text{L}-1.02\text{L})-1.20\text{atm}\times(2.04\text{L}-1.36\text{L}) \\ &= -0.61\text{atm}\cdot\text{L}-0.82\text{atm}\cdot\text{L} \\ &= -1.43\text{atm}\cdot\text{L}\times\frac{101\text{J}}{1\text{atm}\cdot\text{L}} = -1.44\times10^2\,\text{J} \end{aligned}$$

The value of ΔU is the same for the single- and two-stage expansion processes because internal energy is a function of state. However, we see that slightly more work is done in the two-stage expansion. In fact, the work done by expansion is related to the number of expansion stages. The more stages, the more the work. If the system expands by indefinitely stages, it will do the maximum work. This process that can do the maximum work is an infinitely slow process, and each moment of the process is infinitely close to the equilibrium state. This process is called a reversible process thermodynamically. When this process is reversed, the system and it's surrounding can return to their initial state without leaving any traces, that is, the expansion process and the compression process have the same work value.

5.1.2 The First Law of Thermodynamics

Since the industrial revolution, a large number of steam engines have been used to provide industrial power to transform thermal energy into mechanical energy and electrical energy. During the transformation of these energies, will the energy decrease or increase? A lot of facts tell us: In an isolated system, various forms of energy can be transformed into each other, but the total energy inside the system is constant.

$$\Delta U_{\text{isolated system}} = U_{\text{final}} - U_{\text{initial}} = 0 \tag{5-4}$$

In most cases in thermodynamics, the system of interest is closed. So the internal energy changes when energy in the form of heat (q) is added or lost and work (w) is done by or on the system.

The relationship between heat (q), work (w) and changes in internal energy (ΔU) is dictated by the law of conservation of energy, expressed in the form known as the first law of thermodynamics.

ΔU_{system}, is calculated from the following equation:

$$\Delta U_{\text{closed system}} = q + w \tag{5-5}$$

where

q = energy in the form of heat exchanged between system and surroundings;

w = work done by or on the system.

When using the first law equation (5-5), we must keep these important points in mind:

① That heat is the disordered flow of energy and work is the ordered flow of energy.

② Any energy entering the system carries a positive sign. Thus, if heat is absorbed by the system, $q > 0$. If work is done on the system, $w > 0$.

③ Any energy leaving the system carries a negative sign. Thus, if heat is given off by the system, $q < 0$. If work is done by the system, $w < 0$.

④ In general, the internal energy of a system changes as a result of energy entering or leaving the system. If, on balance, more energy enters the system than leaves, ΔU is positive. If more energy leaves than enters, ΔU is negative.

Fig.5-3 shows three different paths leading to the same internal energy change in a system. In path (a), no work is done and $\Delta U = q$. In path (b), work is done by the system ($w < 0$). In path (c), the surroundings do work on the system ($w > 0$). This figure illustrates qualitatively that q and w are not state functions. If work differs in the two expansion processes, q must also differ, and in such a way that $q + w = \Delta U$ has a unique value, as required by the first law of thermodynamics.

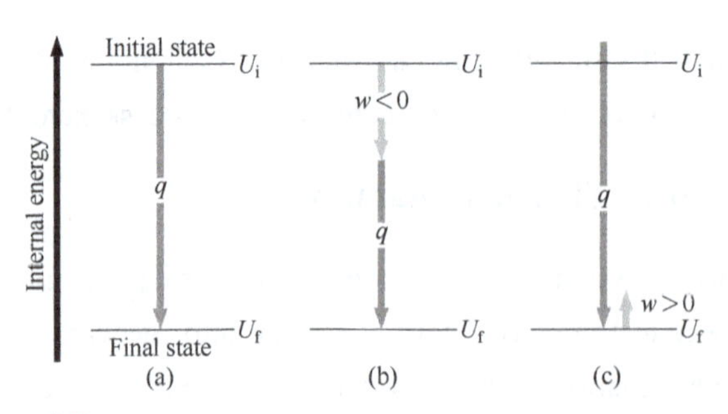

Fig.5-3 Three different paths leading to the same internal energy change in a system

Example 5-2 When the ideal gas in Example 5-1 expands to the final state in two different ways (one stage and two stages), how many joules of heat transfer are there between the system and its surrounding?

Solution Since the internal energy of the ideal gas is only related to temperature, the internal energy does not change when the ideal gas expands isothermally, that is, ΔU=0.

So for one stage gas expansion at constant temperature

$$W = -1.24\times10^2\text{J}$$

$$\Delta U = q + w = 0$$

$$q = 1.24\times10^2\text{J}$$

for two stages expansion $q = 1.44\times10^2\text{J}$

That is, the system does 1.44×10^2J work to its surrounding, the surrounding transfers 1.44×10^2J heat to the system.

Example 5-3 A gas, while expanding, absorbs 25J of heat and does 243J of work. What is ΔU for the gas?

Solution The key to problems of this type lies in assigning the correct signs to the quantities of heat and work. Because heat is absorbed by (enters) the system, q is positive. Because work done by the system represents energy leaving the system, w is negative. You may find it useful to represent the values of q and w with their correct signs. Then complete the algebra.

$$\Delta U = q + w = (+25\text{J}) + (-243\text{J}) = 25\text{J}-243\text{J} = -218\text{J}$$

The negative sign for the change in ΔU signifies that the system, in this case the gas, has lost energy.

5.2 Heat Interaction During Chemical Process

Another type of energy that contributes to internal energy is chemical energy. This is energy associated with chemical bonds and intermolecular attractions. If we think of a chemical reaction as a process in which some chemical bonds are broken and others are formed, then in general, we expect the chemical energy of a system to change as a result of a reaction. Furthermore, we might expect some of this energy change to appear as heat.

Applying the first law of thermodynamics to chemical reactions, the subject of discussing and calculating the heat of chemical reactions is called thermochemistry.

5.2.1 Heats of Reaction

The heat of reaction, q_{rxn} is the quantity of heat exchanged between a system and its surroundings when a chemical reaction occurs within the system at constant temperature. One of the most common reactions studied is the combustion reaction. This is such a common reaction that we often refer to the heat of combustion.

Chemical reactions are often carried out under constant volume or constant pressure conditions. Therefore, the heat of reactions are often divided into constant volume reaction heat and constant pressure reaction heat.

(1) Constant volume reaction heat

Fig.5-4 shows a bomb calorimeter. It is ideally suited for measuring the heat evolved in a combustion reaction. The system is everything within the double-walled outer jacket of the calorimeter. This includes the bomb and its contents, the water in which the bomb is immersed, the thermometer, the stirrer, and so on.

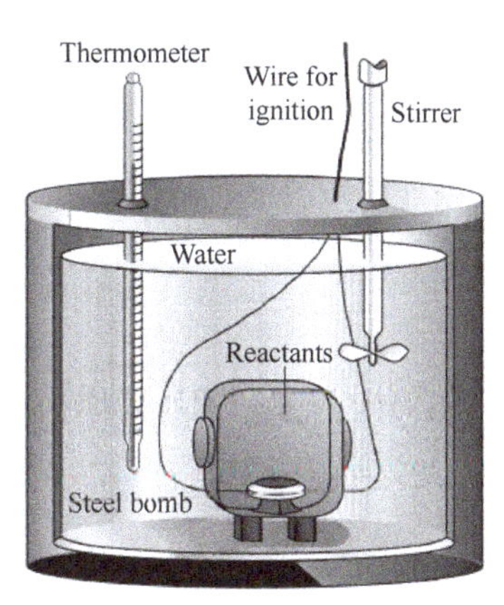

Fig.5-4 Internal structure of a bomb calorimeter

The system is isolated from its surroundings. When the combustion reaction occurs, chemical energy is converted to thermal energy, and the temperature of the system rises.

The heat of reaction is the quantity of heat that the system would have to lose to its surroundings to be restored to its initial temperature. This quantity of heat, in turn, is just the negative of the thermal energy gained by the calorimeter and its contents.

$$q_{rxn} = -q_{calorim} = q_{water} + q_{bomb}$$

When describing the heat of reaction, we can use the first law to illustrate the relationship between ΔU and q.

$$\Delta U = q + w$$

If the process is carried out in a container of constant volume (such as bomb calorimeter), the initial and final volumes of the system are the same. Because the volume is constant, $\Delta V=0$, no work is done, $w = -P_{ext}\Delta V=0$. If the heat transferred for this constant volume process is denoted as q_v , then

$$\Delta U = q_v \tag{5-6}$$

Then we get the following conclusion:

For a constant-volume process, the heat transferred is equal to the internal energy change, ΔU, of the system.

(2) Constant external pressure reaction heat

Chemical reactions are not ordinarily carried out at constant volume in bomb calorimeters. More often, they are carried out in beakers, flasks, and other containers open to the atmosphere and under the constant pressure of the atmosphere.

In many reactions carried out at constant pressure, a small amount of pressure-volume work is done as the system expands or contracts.

Let's assume the volume of the system changes from an initial volume V_i, to a final volume, V_f, under a constant external pressure, P_{ext}. If we represent the heat transferred in this constant-pressure process by q_P then, by the first law, we have

$$\Delta U = q_P - P_{ext}\Delta V = q_P - P_{ext}(V_f - V_i) = q_P - P_{ext}V_f + P_{ext}V_i$$

Now, we make some substitutions: U_f-U_i for ΔU; P_iV_i for $P_{ext}V_i$, and P_fV_f for $P_{ext}V_f$. $P_i=P_f=P_{ext}$

We obtain

$$U_f - U_i = q_P - P_fV_f + P_iV_i$$

which can be rearranged to give

$$q_P = (U_f + P_fV_f) - (U_i + P_iV_i)$$

The quantities U, P, and V are all state functions, so $U+PV$ must also be a state function. This state function is called the enthalpy, H, and is the sum of the internal energy and the pressure-volume product:

$$H = U + PV \tag{5-7}$$

The above expression (5-7) can be written in a very simple form by replacing $U_f + P_fV_f$ with H_f, the final enthalpy, and $U_i + P_iV_i$ with H_i, the initial enthalpy.

$$q_P = \Delta H \tag{5-8}$$

It is a simple way of expressing a very important idea:

For a constant-pressure process, such as a reaction occurring in a container open to the atmosphere, the heat transferred is equal to the enthalpy change of the system, ΔH . This equation (5-8) is a statement of the first law for a constant-pressure process. When $\Delta H < 0$, the system was exothermic; and when $\Delta H > 0$, the system was endothermic.

(3) Relationship between ΔU and ΔH

Let's explore further the relationship between ΔU and ΔH. For a constant pressure process, $q_P = \Delta H$ and $w = -P\Delta V$. By making these substitutions into the first law, we obtain

$$\Delta U = \Delta H - P\Delta V$$

The last term in this expression is the energy (work) associated with the change in volume of the system under a constant external pressure. Then we can use the ideal gas equation to write this alternative expression.

$$P\Delta V = RT(n_{f,gas} - n_{i,gas}) = \Delta n_{gas}RT$$

The $P\Delta V$ term is quite small compared to ΔU and that ΔU and ΔH are almost the same. An additional interesting fact here is that the volume of the system decreases as a consequence of the work done on the system by the surroundings.

The result obtained above can be expressed more generally as

$$\Delta H = \Delta U + \Delta n_{gas}RT \tag{5-9}$$

$$q_P = q_V + \Delta n_{gas}RT \tag{5-10}$$

In these expressions, Δn_{gas} is the change in the number of moles of gas. Using equations (5-9) and (5-10), it is relatively straightforward to establish the following points:

① For a given chemical change, the heat transferred at constant pressure (q_P) is equal to the heat transferred at constant volume (q_V) only if there is no net consumption or production of gas (*i.e.*, when $\Delta n_{gas} = 0$).

② Because RT is approximately equal to 2.5kJ/mol at 298K, the magnitude of the difference between q_P and q_V is typically only a few kJ.

5.2.2 Enthalpy Change of Reaction: $\Delta_r H_m$

We know that the energy released by the reaction is associated with the number of reactants that are involved and the number of products that are produced. So if we want to compare the energy release of different reactions we need to get a suitable state function. That is enthalpy of reaction, $\Delta_r H_m$. The subscript *r* means "reaction" and *m* refers to per mole of reaction.

For example, we saw that when 2 mole CO(g) and 1 mole O_2(g) react to give 2 mole CO_2(g) at constant pressure. The corresponding enthalpy change, denoted by $\Delta_r H_m$, is called the enthalpy of reaction. We can write:

$$2CO(g) + O_2(g) \xlongequal{\quad} 2CO_2(g) \qquad \Delta_r H_m = -566\text{kJ/mol}$$

One mole of reaction refers to the situation in which the extent of reaction is equal to one mole; that is, the conversion of 2 mole CO and 1 mole O_2 to 2 mole CO_2.

The extent of reaction, ξ, can be calculated by the following formula:

$$\xi = \frac{n_i(\xi) - n_i(0)}{\nu_i} = \frac{\Delta n_i}{\nu_i} \tag{5-11}$$

Where $n_i(\xi)$ is the amount of a certain substance when $\xi = 1$; $n_i(0)$ is the amount of a certain substance when $\xi = 0$; ν_i is the stoichiometric coefficient of the substance in the chemical reaction. Obviously, the unit of ξ is mol.

Example 5-4 Calculate ξ of the reaction below at time t.

$$N_2(g) + 3H_2(g) \longrightarrow 2NH_3(g)$$

		$N_2(g)$	$H_2(g)$	$NH_3(g)$
start	n_i /mol	3.0	10.0	0
t time	n_i /mol	2.0	7.0	2.0

Solution

$$\xi = \frac{\Delta n(N_2)}{\nu(N_2)} = \frac{\Delta n(H_2)}{\nu(H_2)} = \frac{\Delta n(NH_3)}{\nu(NH_3)}$$

$$= \frac{2.0 - 3.0}{-1} = \frac{7.0 - 10.0}{-3} = \frac{2.0 - 0}{2} = 1.0(\text{mol})$$

ξ =1.0mol indicates that one mole of the reaction was carried out according to the chemical reaction metric formula, which means that 1 mole of N_2 and 3 mole of H_2 reacted to produce 2 mole of NH_3.

If the reaction is written as follows

$$\frac{1}{2}N_2(g) + \frac{3}{2}H_2(g) = NH_3(g)$$

Then $\Delta n(N_2) = 2-3 = -1.0$mol, at this time

$$\xi = \frac{\Delta n(N_2)}{\nu(N_2)} = \frac{-1.0}{-1/2} = 2.0(\text{mol})$$

It can be seen from the above calculation that the ξ of all substances in the same chemical reaction are same. Therefore, the value of the extent of reaction has nothing to do with the change in the amount of the selected compound. However, it should be noted that, if a set of stoichiometric coefficients, ν_B, are different, it will inevitably lead to the different value of ξ.

Since the heat of a chemical reaction is related to the extent of reaction, ξ, $\Delta_r H$ is defined as the product of ξ and molar enthalpy change of the reaction under the isobaric condition

$$\Delta_r H = \xi \Delta_r H_m \tag{5-12}$$

The physical significance of enthalpy change

In the preceding discussion, we used the first law of thermodynamics, $\Delta U = q + w$, to show that $\Delta H = q_P$ (see equation 5-8). $\Delta H = q_P$ is just another form of the first law, one that is particularly convenient for constant pressure processes.

What is the physical significance or meaning of $\Delta_r H$? The answer to this question is remarkably simple: $\Delta_r H$ represents the heat transferred under constant pressure conditions. In other words, we use two different symbols, $\Delta_r H$ and q_P, to represent the same thing. One of these symbols refers to a property (H) of the system, and the other to something we can measure (q).

Does the property H have a simple physical or molecular interpretation? The answer to this question is, perhaps disappointingly, no. By definition, $H=U+PV$, each of U, P, and V are easily interpreted or explained. The internal energy, U, represents the total energy of a system, which is distributed among the various molecular motions and interactions. The pressure, P, of a system is the force per unit area exerted by the molecules of the system and V is just the volume occupied by the system. However, the combination $U+PV$ does not have any simple physical meaning or molecular interpretation. This combination of quantities is introduced for convenience only.

Example 5-5 In certain conditions, 25mL, $c(C_2O_4^{2-}) = 0.16\ mol/L$ of oxalic acid solution and 20mL, $c(MnO_4^-) = 0.08\ mol/L$ of potassium permanganate solution completely react. The calorimetric experiment shows that the reaction is exothermic 1.2kJ, what is the molar enthalpy change of the reaction?

Solution The chemical equation of this reaction is as follows:

$$C_2O_4^{2-} + \frac{2}{5}MnO_4^- + \frac{16}{5}H^+ = \!=\!= \frac{2}{5}Mn^{2+} + \frac{8}{5}H_2O(l) + 2CO_2(g)$$

Because

$$\Delta n\left(C_2O_4^{2-}\right) = -25 \times 0.001 \times 0.16 = -0.004(mol)$$

Then

$$\xi = \Delta n(C_2O_4^{2-}) / \nu(C_2O_4^{2-}) = -0.004/(-1) = 0.004(mol)$$

$$\Delta_r H_m = \Delta_r H / \xi = -1.2 / 0.004 = -300(kJ/mol)$$

5.2.3 Enthalpy Change Accompanying a Change in State of Matter

When a liquid is in contact with the atmosphere, energetic molecules at the surface of the liquid can overcome forces of attraction to their neighbors and pass into the gaseous, or vapor, state. We say that the liquid vaporizes. If the temperature of the liquid is to remain constant, the liquid must absorb heat from its surroundings to replace the energy carried off by the vaporizing molecules. The heat required to vaporize a fixed quantity of liquid is called the enthalpy (or heat) of vaporization. Usually the fixed quantity of liquid chosen is one mole, and we can call this quantity the molar enthalpy of vaporization. For example,

$$H_2O(l) = \!=\!= H_2O(g) \qquad \Delta_{vap}H = 44.0\ kJ/mol\ \text{at}\ 298.15K$$

We described the melting of a solid in a similar fashion. The energy requirement in this case is called the enthalpy (or heat) of fusion. For the melting of one mole of ice, we can write

$$H_2O(s) \longrightarrow H_2O(l) \qquad \Delta_{fus}H = 6.01\ \text{kJ/mol at } 273.15\text{K}$$

According to the International Union of Pure and Applied Chemistry (IUPAC), the subscript used to denote a chemical process should be used as a right subscript on the Δ symbol.

5.2.4 Standard States and Standard Enthalpies of Reaction

The measured enthalpy change for a reaction has a unique value only if the initial state (reactants) and final state (products) are precisely described. If we define a particular state as standard for the reactants and products, we can then say that the standard enthalpy change is the enthalpy change in a reaction in which the reactants and products are in their standard states. This so-called standard enthalpy of reaction is denoted with a degree symbol, $\Delta_r H_m^{\ominus}$.

The thermodynamic standard state is the state under the standard pressure 1bar (1bar = 10^5Pa). The standard state defined as follows:

① Standard state of a solid or liquid substance is the pure element or compound at a pressure of 1bar and the temperature of interest.

② For a solution, the standard state refers to the state when the concentration for each solute is 1mol/kg at a pressure of 1bar and the temperature of interest.

③ For a gas, the standard state is the pure gas behaving as an (hypothetical) ideal gas at a pressure of 1bar and the temperature of interest. The standard state of any component in a mixed ideal gas refers to the state where the partial pressure of the gas component is 1bar.

Although temperature is not part of the definition of a standard state, it still must be specified in tabulated values of $\Delta_r H_m^{\ominus}$, because $\Delta_r H_m^{\ominus}$ depends on temperature. The values given in this text are all for 298.15K (25℃) unless otherwise stated.

The IUPAC recommended that the standard state pressure be changed from 1atm to 1bar more than 30years ago, but some data tables are still based on the 1atm standard. Fortunately, the differences in values resulting from this change in standard state pressure are very small—almost always small enough to be ignored.

5.3 Hess's Law and Its Applications

5.3.1 Hess's Law and Applications

In 1840, the Russian scientist Herri Hess concluded an important law: If a process occurs in stages or steps (even if only hypothetically), the enthalpy change for the overall process is the sum of the enthalpy changes for the individual steps. This is so called Hess's law. Hess's law is simply a consequence of the state function property of enthalpy. Regardless of the path going from the initial state to the final state, $\Delta_r H_m$ (or $\Delta_r H_m^\ominus$ if the process is carried out under standard conditions) has the same value. One of the reasons that the enthalpy concept is so useful is that a large number of heats of reaction can be calculated from a small number of measurements. The Hess's law make this possible.

Let's look at the following features of enthalpy change.

① $\Delta_r H_m^\ominus$ depends on the way the reaction is written.

$$N_2(g) + O_2(g) \xlongequal{\quad} 2NO(g) \qquad \Delta_r H_m^\ominus = 180.50\text{kJ/mol}$$

To express the enthalpy change in terms of one mole of NO(g), we divide all coefficients and the $\Delta_r H_m^\ominus$ value by two.

$$\frac{1}{2}N_2(g) + \frac{1}{2}O_2(g) \xlongequal{\quad} NO(g) \qquad \Delta_r H_m^\ominus = \frac{1}{2} \times 180.50\ \text{kJ/mol} = 90.25\ \text{kJ/mol}$$

② $\Delta_r H_m^\ominus$ changes sign when a process is reversed.

$$NO(g) \xlongequal{\quad} \frac{1}{2}N_2(g) + \frac{1}{2}O_2(g) \qquad \Delta_r H_m^\ominus = -90.25\ \text{kJ/mol}$$

③ Hess's law of constant heat summation.

$$\frac{1}{2}N_2(g) + O_2(g) \xlongequal{\quad} NO_2(g) \qquad \Delta_r H_m^\ominus = ?$$

We can think of the reaction as proceeding in two steps: First we form NO(g) from $N_2(g)$ and $O_2(g)$ and then $NO_2(g)$ from NO(g) and $O_2(g)$. When the equations for these two steps are added together in the manner suggested in Fig.5-5, we get the overall result.

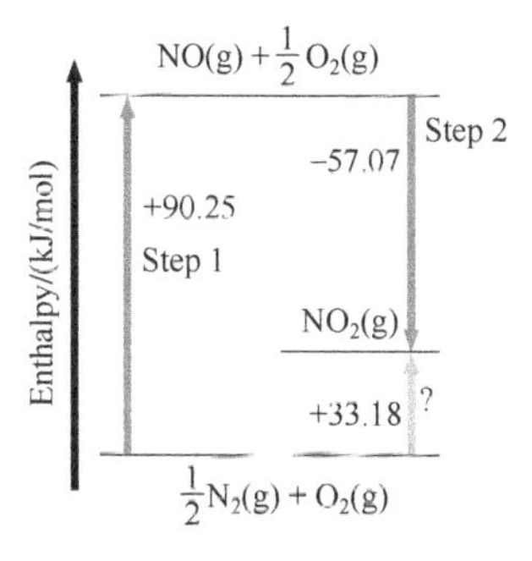

Fig.5-5 An enthalpy diagram illustrating Hess's law

$$\frac{1}{2}N_2(g) + O_2(g) \xlongequal{\quad} NO(g) + \frac{1}{2}O_2(g) \qquad \Delta_r H_m^\ominus = +90.25\ \text{kJ/mol}$$

$$NO(g) + \frac{1}{2}O_2(g) \xlongequal{\quad} NO_2(g) \qquad \Delta_r H_m^\ominus = -57.07\ \text{kJ/mol}$$

$$\frac{1}{2}N_2(g) + O_2(g) \xlongequal{\quad} NO_2(g) \qquad \Delta_r H_m^\ominus = +33.18\ \text{kJ/mol}$$

Note that in summing the two equations NO(g), a species that would have appeared on both sides of the overall equation was canceled out.

5.3.2 Basic Thermodynamics Data: Standard Molar Enthalpies of Formation

(1) Standard enthalpies of formation

Enthalpy is a function of state, so changes in enthalpy, ΔH, have unique values. We can deal just with these changes. Nevertheless, as with many other properties, it is still useful to have a starting point, a zero value.

Consider how do we mark the height of a mountain. Do we mean by this the vertical distance between the mountaintop and the center of Earth? Between the mountaintop and the deepest trench in the ocean? No. By agreement, we mean the vertical distance between the mountaintop and mean sea level. We arbitrarily assign to mean sea level an elevation of zero, and all other points on Earth are relative to this zero elevation.

The standard enthalpy of formation ($\Delta_f H_m^\ominus$) of a substance is the enthalpy change that occurs in the formation of one mole of the substance in the standard state from the reference forms of the elements in their standard states. Its symbol is $\Delta_f H_m^\ominus$ (formula, phase state, T) and unit is kJ/mol. The reference forms of the elements in all but a few cases are the most stable forms of the elements at one bar and the given temperature. For example, the simple substance of the reference forms of phosphorus is white phosphorus, P_4 (s, white), not red phosphorus or black phosphorus, which is more stable than it. The superscript symbol denotes that the enthalpy change is a standard enthalpy change, and the subscript f and m signify that the reaction is one in which one mole substance is formed from its elements. The formation of the reference form of an element from itself is no change at all. Therefore:

The standard enthalpy of formation is 0 for a pure element in its reference form at any temperature.

Appendix II of this book lists the standard enthalpies of formation data for common substances at 298.15K, 10^5Pa. We will use standard enthalpies of formation in a variety of calculations. Often, the first thing we must do is writing the chemical equation to which a $\Delta_f H_m^\ominus$ value applies, as in Example 5-6.

Example 5-6 Calculating the standard enthalpy of formation of $ZnSO_4$. The standard enthalpy of formation of $ZnSO_4$ cannot be measured directly, but can be obtained indirectly by the following four-step reaction:

$$Zn(s) + \frac{1}{2}O_2(g) \longrightarrow ZnO(s) \qquad \Delta_r H_m^\ominus(1) = -350.5\text{kJ/mol}$$

$$S(s) + O_2(g) \longrightarrow SO_2(g) \qquad \Delta_r H_m^\ominus(2) = -296.8\text{kJ/mol}$$

$$SO_2(g) + \frac{1}{2}O_2(g) \longrightarrow SO_3(g) \qquad \Delta_r H_m^\ominus(3) = -98.9\text{kJ/mol}$$

$$ZnO(s)+SO_3(g) \longrightarrow ZnSO_4(s) \qquad \Delta_r H_m^\ominus(4) = -236.6\text{kJ/mol}$$

Solution The four reactions add up to get the total reaction as

$$Zn(s)+S(s)+2O_2(g) \longrightarrow ZnSO_4(s)$$

Note that when writing the equation for the formation of $ZnSO_4$, the stoichiometric coefficient of $ZnSO_4$ is 1. According to Hess's law,

$$\Delta_f H_m^\ominus(ZnSO_4, s, 298K) = \Delta_r H_m^\ominus(\text{total reaction})$$

$$= \Delta_r H_m^\ominus(1) + \Delta_r H_m^\ominus(2) + \Delta_r H_m^\ominus(3) + \Delta_r H_m^\ominus(4) = -982.8 \text{ kJ/mol}$$

For ionic reactions in aqueous solution, the standard enthalpy of formation for hydrated ions is often used. The standard enthalpy of formation for hydrated ions refers to the change of standard molar enthalpy for hydrated ions B(aq) dissolved in a large amount of water (forms infinitely dilute solution) from a reference form of element under the temperature T and standard conditions. Its symbol is $\Delta_f H_m^\ominus$ (formula, ∞, aq, T) and unit is kJ/mol. The symbol ∞ is expressed as "in a large amount of water" or "infinitely dilute aqueous solution" and is often omitted.

The standard molar enthalpy of formation of hydronium ion is zero.

That is, at 298K, the standard molar reaction enthalpy of generating hydronium ions from elemental H_2 (g) is arbitrarily prescribed as zero in the standard state.

$$\frac{1}{2}H_2 + aq \longrightarrow H^+(aq) + e$$

$$\Delta_f H_m^\ominus(H^+, aq) = \Delta_f H_m^\ominus(H^+) = 0$$

(2) Calculating $\Delta_r H_m^\ominus$ from tabulated values of $\Delta_f H_m^\ominus$

For any chemical reaction, the type and number of atoms of the product and the reactant are the same, that is, starting from the same elements, reactants and products can be formed through different paths.

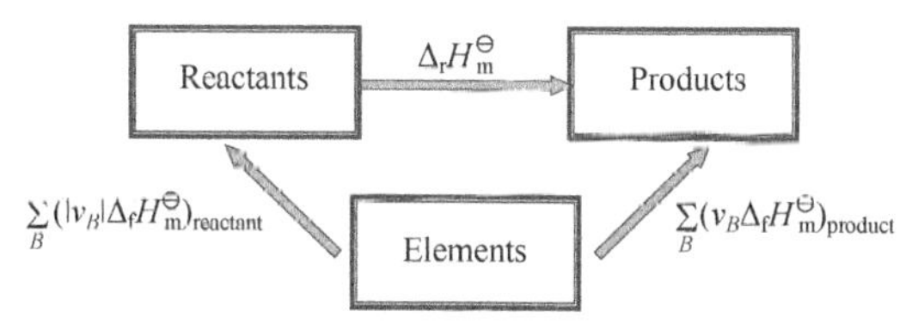

Fig.5-6 Calculating $\Delta_r H_m^\ominus$ from $\Delta_f H_m^\ominus$

As shown in Fig.5-6, according to Hess's law

$$\Delta_r H_m^\ominus = \sum_B(\nu_B \Delta_f H_m^\ominus)_{product} - \sum_B(|\nu_B| \Delta_f H_m^\ominus)_{reactant} \tag{5-13}$$

Equation (5-13) indicates that the standard molar enthalpy change of any constant pressure reaction is equal to the result that the sum of the standard molar formation enthalpies of all products minus the sum of the standard molar formation enthalpies of all reactants.

Example 5-7 The reaction of glucose supplying energy in the body is one of the most important biochemical reactions. Try to use the standard molar formation enthalpy data to calculate the thermal effect of the glucose oxidation reaction.

Solution First write out the thermochemical equation of the glucose oxidation reaction, and mark the standard molar formation enthalpy under each substance (Appendix II).

$$C_6H_{12}O_6(s)+6O_2(g) \Longrightarrow 6CO_2(g)+6H_2O(l)$$

$$\Delta_f H_m^\ominus(298.15\text{K}) / (\text{kJ/mol}) \quad -1273 \quad 0 \quad -393.5 \quad 285.8$$

$$\Delta_r H_m^\ominus = [6\Delta_f H_m^\ominus(CO_2,g)+6\Delta_f H_m^\ominus(H_2O,l)] \quad [\Delta_f H_m^\ominus(C_6H_{12}O_6,s)+6\Delta_f H_m^\ominus(O_2,g)$$

$$=[6\times(-393.5)+6\times(-285.8)]-[1\times(-1273)+6\times 0]=-2802.8(\text{kJ/mol})$$

The results show that the oxidation of glucose is a strong exothermic reaction. When each mole glucose is oxidized, about 2802.8kJ of heat released. The main food of human beings is starchy food, which is converted into glucose after being hydrolyzed in the body. Therefore, most of the calories needed by the human body are supplied by glucose.

Example 5-8 When the mixture of metallic aluminum powder and ferric oxide (called thermite) is ignited, the reaction emits a large amount of heat to melt the iron (the temperature can reach above 2000℃). This reaction is used in high temperature outdoor operations such as orbital welding. Calculate $\Delta_r H_m^\ominus$(298.15K) of the reaction between aluminum powder and ferric oxide.

Solution Write out the thermochemical equation between aluminum powder and ferric oxide, and mark the standard molar formation enthalpy under each substance (Appendix II).

$$2Al(s)+Fe_2O_3(s) \Longrightarrow Al_2O_3(s)+2Fe(s)$$

$$\Delta_f H_m^\ominus(298.15\text{K}) / (\text{kJ/mol}) \quad 0 \quad -822.1 \quad -1669.8 \quad 0$$

$$\Delta_r H_m^\ominus(298.15\text{K})=\Delta_f H_m^\ominus(Al_2O_3,s)-\Delta_f H_m^\ominus(Fe_2O_3,s)$$

$$=-1669.8-(-822.1)=-847.7(\text{kJ/mol})$$

Example 5-9 Thermochemistry plays a very important role in rocket propulsion technology. The magnitude of rocket thrust is closely related to the thermal efficiency of fuel. Try to calculate $\Delta_r H_m^\ominus$(298.15K) of the reaction between rocket fuel hydrazine hydrate ($N_2H_4 \cdot H_2O$) and hydrogen peroxide.

$$N_2H_4 \cdot H_2O(l)+2H_2O_2(l) \Longrightarrow N_2(g)+5\ H_2O(l)$$

Solution According to the chemical reaction of hydrazine hydrate and hydrogen peroxide, check the relevant standard molar formation enthalpy of each substance:

$$N_2H_4 \cdot H_2O(l)+2H_2O_2(l) \longrightarrow N_2(g)+5H_2O(l)$$

$$\Delta_f H_m^\ominus(298.15\,K)/(kJ/mol) \quad -242 \quad -188 \quad 0 \quad -242$$

$$\Delta_r H_m^\ominus(298.15K) = 5\Delta_f H_m^\ominus(H_2O,l) - 2\Delta_f H_m^\ominus(H_2O_2,l) - \Delta_f H_m^\ominus(N_2H_4 \cdot H_2O,l)$$
$$= 5\times(-242) - 2\times(-188) - (-242) = -592(kJ/mol)$$

5.3.3 Standard Molar Combustion Enthalpy

The standard molar combustion enthalpy of a substance is the standard molar reaction enthalpy change when a substance undergoes complete combustion with oxygen under standard conditions. The chemical reaction is typically a hydrocarbon or other organic molecule reacting with oxygen to form carbon dioxide and water and release heat. It is represented by a symbol $\Delta_c H_m^\ominus$ (formula, phase state, T) and the unit is kJ/mol. In designated products, C element in the reactant is oxidized to CO_2(g), H element is oxidized to H_2O(l), S element is oxidized to SO_2(g), and N element is oxidized to N_2(g). Since the reactants have been completely burned (or oxidized), the products after the reaction can no longer burn. Therefore, the above definition actually means that "The combustion enthalpy of all products is 0" in each combustion reaction.

$$\Delta_c H_m^\ominus(H_2O,l) = 0, \quad \Delta_c H_m^\ominus(CO_2,g) = 0$$

When writing the combustion reaction, make sure that the stoichiometric coefficient of the specified burning substance is −1. For example, the combustion reaction of CH_3OH(l) should be:

$$CH_3OH(l) + \frac{3}{2}O_2(g) \longrightarrow CO_2(g) + 2H_2O(g)$$

$$\Delta_c H_m^\ominus(CH_3OH,l) = -726.51kJ/mol$$

Example 5-10 Use the standard molar formation enthalpy data to calculate the standard molar combustion heat of acetylene, $\Delta_c H_m^\ominus$ (C_2H_2, g, 298.15K).

Solution Write out the chemical equation of acetylene combustion, and mark its standard enthalpy of formation under each substance (Appendix II).

$$C_2H_2(g) + \frac{5}{2}O_2(g) \longrightarrow 2CO_2(g)+H_2O(l)$$

$$\Delta_f H_m^\ominus(298.15\,K)/(kJ/mol) \quad 227.4 \quad 0 \quad -393.5 \quad -285.8$$

$$\Delta_c H_m^\ominus(C_2H_2,l) = \Delta_r H_m^\ominus = [2\Delta_f H_m^\ominus(CO_2,g) + \Delta_f H_m^\ominus(H_2O,l)] - \left[\Delta_f H_m^\ominus(C_2H_2,g) + \frac{5}{2}\Delta_f H_m^\ominus(O_2,g)\right]$$
$$= [2\times(-393.5) + (-285.8)] - (227.4+0) = 1300.2(kJ/mol)$$

The calculated standard molar enthalpy is the standard molar combustion heat of acetylene.

It should be noted here that if the above reaction equation is written as

$$2C_2H_2(g)+5O_2(g) = 4CO_2(g)+2H_2O(l)$$

What is the standard molar enthalpy change of above reaction? Is it the standard molar combustion heat of acetylene? Please think about it.

Example 5-11 The standard combustion enthalpies of C(graphite), $H_2(g)$ and $C_3H_8(g)$ are −393.5, −285.8, and −2219.9kJ/mol, respectively. Use these values to calculate $\Delta_r H_m^\ominus$ for reaction below

$$3C(graphite) + 4H_2(g) = C_3H_8(g) \qquad \Delta_r H_m^\ominus = ?$$

Solution To determine an enthalpy change with Hess's law, we need to combine the appropriate chemical equations. A good starting point is to write chemical equations for the given combustion reactions based on one mole of the indicated reactant.

Begin by writing the following equations

(a) $C_3H_8(g) + 5O_2(g) = 3CO_2(g) + 4H_2O(l) \quad \Delta_c H_m^\ominus(C_3H_8,g) = \Delta_r H_m^\ominus = -2219.9\text{kJ/mol}$

(b) $C(graphite) + O_2(g) = CO_2(g) \quad \Delta_c H_m^\ominus(C,graphite) = \Delta_r H_m^\ominus = -393.5\text{kJ/mol}$

(c) $H_2(g) + \frac{1}{2}O_2(g) = H_2O(l) \quad \Delta_c H_m^\ominus(H_2,g) = \Delta_r H_m^\ominus = -285.8\text{kJ/mol}$

Because our objective is to produce $C_3H_8(g)$, the next step is to find a reaction in which $C_3H_8(g)$ is formed—the reverse of reaction (a).

−(a): $3CO_2(g) + 4H_2O(l) = C_3H_8(g) + 5O_2(g) \quad \Delta_r H_m^\ominus = +2219.9\text{kJ/mol}$

To get the proper number of moles of each reactant, we must multiply equation (b) by three and equation (c) by four.

3×(b): $3C(graphite) + 3O_2(g) = 3CO_2(g) \quad \Delta_r H_m^\ominus = 3(-393.5) = -1181(\text{kJ/mol})$

4×(c): $4H_2(g) + 2O_2(g) = 4H_2O(l) \quad \Delta_r H_m^\ominus = 4(-285.8) = -1143(\text{kJ/mol})$

We can now combine the three modified equations.

−(a) $3CO_2(g) + 4H_2O(l) = C_3H_8(g) + 5O_2(g) \quad \Delta_r H_m^\ominus = +2219.9(\text{kJ/mol})$

3×(b) $3C(graphite) + 3O_2(g) = 3CO_2(g) \quad \Delta_r H_m^\ominus = 3(-393.5) = -1181(\text{kJ/mol})$

4×(c) $4H_2(g) + 2O_2(g) = 4H_2O(l) \quad \Delta_r H_m^\ominus = 4(-285.8) = -1143(\text{kJ/mol})$

$3C(graphite) + 4H_2(g) = C_3H_8(g) \quad \Delta_r H_m^\ominus = -104\text{kJ/mol}$

5.4 Fuels as Sources of Energy

One of the most important uses of thermochemical measurements and calculations is in assessing materials as energy sources. Energy sources, called fuels, are materials resource in nature that can provide humans with some form of energy. It is an important foundation for national economy, social development and people's living standards. It is the fundamental guarantee for the country's sustainable development. The energy industry relies heavily on chemical processes, and more than 90% of energy consumption relies on chemistry and chemical engineering technology.

According to the basic form of energy, it can be divided into primary energy and secondary energy. Primary energy refers to energy resources that exist in nature and have not been processed or converted, including renewable solar energy, wind energy, geothermal energy, ocean energy, biological energy, nuclear energy, and non-renewable coal, oil, and natural gas resources. Secondary energy refers to energy resources that are directly or indirectly converted from primary energy to other forms of energy, such as electricity, coal-gas, gasoline, diesel, coke, clean coal, lasers, and biogas.

Fossil fuels such as coal, oil, and natural gas has limited reserves and cannot be regenerated, so its exhaustion has become an irreversible trend. For this reason, new energy resources must be developed to meet the increasing demand for fuels from human development. The development of strategically important new energy sources, including solar energy, biomass energy, nuclear energy, natural gas hydrate and secondary energy sources such as hydrogen energy and fuel cells, requires chemists to propose new ideas, create new concepts, and develop new methods.

In addition, for a long time, due to the increasing consumption of fossil fuels and the continuous reduction of reserves, the world has serious environmental pollution, climate abnormalities and energy shortages. Controlling the chemical reaction of low grade fuels and improving energy conversion efficiency, protecting the environment and reducing energy costs are big challenges to the chemists all over the world.

China is in a rapid development and is under the dual pressures of continuous increasing on fuels demand and environmental protection. Therefore, it is an inevitable choice to develop clean energy technology and accelerate the development of clean energy. Renewable energy mainly refers to a type of energy resources that are abundant in resources such as solar energy, biomass energy, hydrogen energy, geothermal energy, and ocean energy, and can be recylced. The transformation and utilization of renewable energy has the characteristics of wide coverage, complex and changeable research objects, multiple interdisciplinary categories, and high degree of integration of disciplines. In this section, we focus on several common renewable energy sources for a brief introduction.

5.4.1 Solar Energy

Solar energy is the energy produced by the continuous nuclear fusion reaction process inside the sun. Although the energy radiated by the sun into the earth's atmosphere is only one out of 2.2 billion of its total radiant energy (approximately 3.75×10^{26}W), it is already as high as 173 000 TW. The solar energy arriving on the earth per second is equivalent to 5 million tons of standard coal. The total amount of solar energy resources is large, widely distributed, clean in use, and there is no problem of resource exhaustion. Since the beginning of the 21st century, there have been exciting new developments in the use of solar energy. The annual output of solar water heaters, solar cells and other products has maintained a growth rate of more than 30%, which is called "the world's fastest growing energy source". Solar energy conversion and utilization mainly refers to the use of solar radiation to achieve conversed energy processes such as heating, daylighting, hot water supply, power generation, water purification, and air conditioning and refrigeration, to meet the needs of people's lives, industrial applications and national defense technology. It includes the solar thermal conversion, photoelectric conversion and photochemical conversion, *etc*.

Solar thermal utilization refers to the conversion of solar energy into thermal energy for use, such as supplying hot water, thermal power generation, driving power plants, driving refrigeration cycles, seawater desalination, heating and enhanced natural ventilation, semiconductor thermoelectric power generation, *etc*. Photoelectric utilization is based on the photovoltaic effect, using photovoltaic materials to construct solar cells, and the solar cells convert the energy of sunlight directly into electrical energy.

Photovoltaic effect refers to an effect in which electromotive force and current are generated by the change of the charge distribution state in the object when the object is illuminated. When sunlight or other light shines on the semiconductor p-n junction, a voltage (called photovoltaic voltage) will appear on both sides of the p-n junction. If an external load loop is connected, a current will be generated in the loop.

Photochemical utilization includes conversion processes such as plant photosynthesis, solar photolysis of water to produce hydrogen, pyrolysis of water to produce hydrogen, and natural gas reforming.

Today, many countries are vigorously developing and utilizing solar energy resources. Europe, Australia, Israel, and Japan have increased their investment to actively explore effective ways to achieve large-scale utilization of solar energy. Germany and other European Union countries are vigorously supporting and developing renewable energy sources such as solar energy and wind energy as the main alternative energy sources to replace fossil fuels. The United States has mastered the technology of high-efficiency conversion of photovoltaic power generation and has implemented engineering in solar thermal power generation. The New Energy Policy in the United States has strongly promoted the in-depth research and development and large-scale application of renewable energy such as solar energy.

The development trends in solar energy utilization research are as follows:

① The multidisciplinary characteristics of solar energy utilization. The use of solar energy is closely related to disciplines such as physics, chemistry, optics, electricity, machinery, materials science, construction science, biological science, control theory, mathematical programming theory, meteorology, *etc.* It is a research field with strong comprehensiveness. In the interdisciplinary process, new disciplines and research directions may also be formed.

② The utilization of solar energy is developing towards high efficiency and low cost. Due to the low energy density of solar energy, the intermittent nature of cloudy, sunny, rainy and snowy, day and night, and seasonal changes, and the complex and diverse energy conversion equipment, the economic and large-scale utilization of solar energy can only be achieved by improving efficiency, so improving conversion efficiency has always been researched focus. The increase in conversion efficiency is related to the limit efficiency of the second law of thermodynamics. In addition, under the existing technical conditions, the use of cheap materials and simple process flow to achieve low-cost solar energy utilization without reducing efficiency is also an important research direction.

③ There are multiple technological paths in solar energy utilization research that compete and complement each other. From solar power generation to solar cooling, there are multiple technological paths to achieve these aims. Taking solar refrigeration as an example, there are multiple technological paths such as absorption, adsorption, solid dehumidification, and liquid dehumidification. They have certain competitive relationships, but they are not simple ones. They each have their own characteristics and applications, and they have certain complementary effects. Therefore, research on multiple technologies should be encouraged.

④ Multiple links of solar energy utilization are matched and optimized. From the collection, storage, and utilization of solar energy, there are differences in time, space, and capacity. According to different applications, such as usage, energy use grade, stability, economy, *etc.*, it is necessary to meet different needs through the selection of working parameters, technical paths, and equipment, and obtain the highest possible benefits.

⑤ The use of solar energy is complementary and optimized with other renewable energy and fossil energy. Because solar energy supply is affected by the climate, and there are day and night differences and seasonal differences, it needs to be used together with other renewable energy or fossil energy to achieve a reliable and stable energy supply. Therefore, with solar energy as the main energy source, multi-energy complementary high-efficiency energy system is also an important research field, and its goal is to maximize the solar energy utilization ratio.

Solar energy conversion and utilization have gone through multiple stages from demonstration utilization, special occasion utilization, partial utilization, to universal utilization and large-scale utilization. The main research directions can be divided into two categories: one is the key technology for the large-scale utilization of solar energy; the other is to explore new methods and materials for solar energy utilization, discover and solve new phenomena and new problems in the energy conversion process, especially to develop the thermodynamic optimization based on the phenomenon of solar energy conversion and utilization, the high efficiency of the energy conversion process, and the economicalization of energy utilization devices.

5.4.2　Biomass Energy

All non-fossil organic substances containing inherent chemical energy are called biomass, including various types of plants, municipal solid waste, urban sewage sludge, animal excrement, forestry and agricultural waste, and certain types of industrial organic waste.

Broadly speaking, biomass energy is energy that is indirectly derived from solar energy and stored in the form of organic matter. It is a renewable, naturally available, energy-rich, and carbon-containing resource that can replace fossil fuels. Since the production, conversion and utilization of biomass constitute a closed carbon cycle, its carbon-neutral characteristics will play an important role in mitigating global climate change. In addition, biomass has the advantages of less pollutants (less sulfur and nitrogen), relatively clean and cheap combustion, and the conversion of organic matter into fuel can reduce environmental pollution.

The total amount of biomass produced by the earth through the photosynthesis of green plants is about 140-180 billion tons (dry weight) each year, which contains energy equivalent to 10 times the current world's total energy consumption. As the world's largest agricultural country, China has abundant biomass energy resources, the main sources of which are agricultural and forestry waste, food processing waste, wood processing waste, and urban household waste. Agricultural and forestry waste is the main body of China's biomass resources. China produces approximately 650 million tons of agricultural straw each year. Together with firewood and forestry waste, the energy is equivalent to 460 million tons of standard coal. It is expected to increase to 904 million tons by 2050, which is equivalent to more than 600 million tons of standard coal. China annual forest consumables reach 210 million m^3, which is equivalent to 120 million tons of standard coal energy. In addition, the national annual output of municipal solid waste has exceeded 150 million tons, and the annual output will reach 210 million tons by 2020. If these wastes are incinerated for power generation or landfilled for gas power generation, the energy produced will be equivalent to 5 million tons of standard coal. It can also effectively reduce environmental pollution.

As a traditional energy source, biomass energy occupies an important position in the history of human development. From a global perspective, biomass energy still accounts for more than 35% of total renewable energy consumption and about 15% of primary energy consumption, but it is mainly used through traditional low-efficiency combustion mode. Full utilization of modern advanced high-efficiency biomass energy conversion and utilization technology will greatly increase the share and position of biomass energy in renewable energy and primary energy. Biomass resources have great potential and most of them are wastes that cannot be fully utilized. It can be predicted that biomass energy is most likely to become the first new energy in large-scale utilization.

There are many different ways or methods to convert biomass into useful energy. Currently, two main technologies are mainly used: thermochemical technology and biochemical technology. In addition, mechanical extraction (including esterification) is also a form of energy obtained from biomass. Common thermochemical techniques include three methods: combustion, gasification and liquefaction. Common biochemical technologies include ethanol fermentation, biogas fermentation, and microbial hydrogen production. Through the above methods, biomass can be converted into heat or power, fuel and chemical substances.

5.4.3 Hydrogen Energy

Hydrogen energy is the star of renewable energy. It refers to the energy released in a reaction dominated by hydrogen and its isotopes, or hydrogen in the process of state change. It can be produced from the thermonuclear reaction of hydrogen or from the chemical reaction between hydrogen and oxidant. The former is called thermonuclear energy or fusion energy, and its energy is very huge, usually in the category of nuclear energy; the latter is the chemical energy of fuel reaction, which is usually called hydrogen energy.

As an energy source, hydrogen has many advantages:

① Among all the elements, hydrogen has the lightest mass. It is the fuel with the largest calorific value except nuclear fuel. Its high calorific value is 142.35kJ/kg, which is 3 times the calorific value of gasoline;

② Hydrogen is the most abundant element in nature. It is estimated to be 75% of the mass of the universe. On the earth, the amount of natural hydrogen gas is rare, but hydrogen element is very abundant. Water is the most abundant hydrogen-containing substance. Followed by various fossil fuels (natural gas, coal, oil, *etc.*) and various biomass;

③ Hydrogen is non-toxic. Compared with other fuels, hydrogen and atmospheric oxygen will only produce water after combustion or reaction, so it is clean and pollution-free;

④ Hydrogen has good combustion performance, fast ignition, wide combustible range when mixed with air, high ignition points and fast burning speed;

⑤ Hydrogen energy can be used in various forms, which can produce heat energy through combustion, produce mechanical work in heat engines such as gas turbines and internal combustion engines, and can be used as fuel for fuel cells;

⑥ Hydrogen can exist in the form of gas, liquid or solid, metal hydride and adsorbed hydrogen, so it can adapt to the different needs of storage, transportation and various application environments.

Like electricity, hydrogen energy has no direct resource reserves and needs to be converted from other primary energy sources. Therefore, hydrogen energy is a secondary energy source. Compared with energy carriers such as electricity and heat, the biggest feature of hydrogen is that it can be stored in the form of chemical energy on a large scale. As a secondary energy source, hydrogen energy has many advantages and potential to support the sustainable development of energy. Hydrogen energy is not only of great significance to the future long-term energy system (mainly fusion nuclear energy and renewable energy), but also of important practical significance to the fossil energy system that mankind will still rely on for a long time. The connotation of hydrogen energy system can be understood as an energy system based on hydrogen energy production, storage, transportation, conversion and terminal utilization. In such a system, hydrogen, as an energy carrier, has become a currency or commodity for energy circulation. Hydrogen energy is not only a high-quality secondary energy (equivalent and complementary to electricity), but can also be directly used as a terminal fuel of various power or conversion devices.

In view of the important role of hydrogen energy in the future energy landscape, many countries are stepping up the deployment and implementation of hydrogen energy strategies, such as the US "FreedomCAR" plan for transportation machinery and the "FutureGen" plan for large-scale hydrogen production, and Japan's "NewSunshine" project and the "We-NET" system; the investment in hydrogen technology in the European "Framework" project has also shown an exponential increase.

Insufficient supply of conventional primary energy, shortage of liquid fuels, serious pollution caused by the fossil fuels use, pressure to reduce CO_2 emissions, and energy use problems in remote rural areas have put China energy system under multiple pressures. China Energy Development Strategy clearly states: "We must adjust the energy structure, increase alternative energy sources, and achieve sustainable development while improving energy efficiency, clean use of fossil energy, and reducing environmental pollution." Therefore, from completely relying on fossil energy to using renewable energy is the only way for China's energy structure adjustment.

The source of the hydrogen energy system can rely on fossil resources as well as renewable energy. In the process of the transition from fossil resources to renewable energy, in addition to changes in the source, other links including hydrogen separation, transportation, distribution, storage, conversion and application do not need to be changed. Therefore, with the help of hydrogen energy, a smooth transition from the fossil energy system to the renewable energy system can be realized without causing too much fluctuation in the energy system. It can be considered that hydrogen energy is an important bridge linking the transition from fossil energy to renewable energy and will fundamentally play an important role in solving the country's future energy supply and environmental problems.

In the 21st century, mankind will enter the era of hydrogen economy. However, to truly realize the widespread use of hydrogen as an energy source, a series of key scientific and technological issues, such as the large-scale production, storage, transportation, and efficient conversion and utilization of hydrogen, need to be solved. From the perspective of the rapid development of hydrogen energy in the world, the first 20-40years of the 21st century will be an important stage of commercialization and industrialization of various key hydrogen energy technologies, and major breakthroughs will be made in their technical practicability.

Water is one of the most abundant substances on earth, and the decomposition products are only hydrogen and oxygen, which is an ideal raw material for hydrogen production. From a thermodynamic point of view, water as a compound is very stable, and high energy is required to decompose water. Due to the limitation of thermodynamics, it is difficult to use thermal catalysis. However, water as an electrolyte is unstable. Theoretical calculations show that it takes only 1.23eV to electrolyze a molecule of water into hydrogen and oxygen in an electrolytic cell. Therefore, the production of hydrogen by hydrolysis is mainly accomplished by electrolysis. At present, the methods for producing hydrogen by hydrolysis mainly include two types of hydrogen production by electrolysis of water and hydrogen production by photolysis of water.

① After 200 years of development, the technology of hydrogen production by electrolysis has been quite mature. At present, 4% of the world's hydrogen production comes from electrolysis of water. However, because the current electricity is mainly derived from fossil energy, the power generation efficiency is low as 35%-40%, while the efficiency of industrial electrolyzed water is about 75%, so the total hydrogen production efficiency of electrolyzed water is 26%-30% and no more than 40%. To make electrolyzed water the main way to produce hydrogen in the future, reducing the energy consumption of electrolyzed water and lowering electricity prices are two important methods.

At present, research in this area includes the use of natural gas to assist water vapor electrolysis, the addition of ion activators to electrolyze water, and the use of renewable energy to generate electricity to electrolyze water.

② Photolysis of water to produce hydrogen. Solar energy is the cleanest and inexhaustible natural energy source. Hydrogen production by photolysis of water is the best way to convert and store solar energy photochemically, which is of great significance. However, the use of solar energy to produce hydrogen from water is a very difficult research field, and a large number of theoretical and engineering technical problems need to be solved. Solar energy splitting water to produce hydrogen can be carried out in two ways: photoelectrochemical cell method and semiconductor photocatalysis method. The US Department of Energy has set an efficiency target of 15% for the research on solar photolysis of water for hydrogen production, and a cost target of 10-15 US dollars per million British thermal units (MBtu, 1Btu=1055.06J). Although the current research has made great progress, the use of solar energy is still lower than the 10% conversion efficiency benchmark set by the US Department of Energy for commercialization. The development of efficient, stable, and inexpensive photocatalytic materials and reaction systems is the key for breakthrough.

In the long run, hydrogen production by hydrolysis is an ideal alternative technology for hydrogen production from fossil fuels. The key factors for using solar energy to produce hydrogen by photolysis or electrolysis of water are the efficiency and cost of light energy conversion. Future research mainly focuses on: design and development of highly efficient and stable catalytic materials and semiconductor materials; in-depth discussion of the mechanism of charge separation, the transmission and photoelectric conversion in photocatalytic process; research on renewable energy power generation and continuously reducing power generation cost.

In short, the current international development trend of hydrogen energy production technology is to provide more advanced, cheap and small-scale on-site hydrogen production and purification technology. It is an important requirement for the establishment of hydrogen refueling stations to provide dispersed hydrogen sources. Another trend is to provide advanced hydrogen energy production technology and achieve near-zero CO_2 emissions. From a longer-term perspective, the use of renewable energy to produce hydrogen is the ultimate alternative to fossil energy and a important method to solve energy and environmental problems. Making full use of various resources (including fossil energy, nuclear energy and renewable energy) and continuously developing low-cost and high-efficiency hydrogen production methods is the development trend of hydrogen production technology.

5.4.4 Natural Gas Hydrate

Natural gas hydrate (NGH) is a non-fixed stoichiometric clathrate crystal compound formed by the interaction of natural gas with water at a certain temperature and pressure. $1m^3$ of NGH can store 150-180m^3 of natural gas (under standard state). Because it can burn in fire, it is commonly called “flammable ice”. NGH in nature is stored in the land under permafrost and seabed sediments with a water depth greater than 300m. Natural gas hydrate reserves are large, widely distributed, and high energy density, therefore, NGH is regarded as the new resource for the 21st century based on the features of high quality and clean. The organic carbon accounts about 53.3% of the global NGH and its reserves are about twice as much as the total existing fossil fuels (oil, natural gas and coal). It’s the most powerful alternative energy for alleviating the energy crisis.

As a new energy source, the basic application categories of NGH include resource exploration, resource evaluation, accumulation mechanism, basic physical properties, mining, storage and transportation, environmental impact. NGH can not only solve the pressure of world energy demand, but also provide abundant resources for many industries, mainly involving storage and transportation, gas separation, power generation, manufacturing, public buildings, public transportation, *etc.* It is a comprehensive and intersecting energy resource.

In order to alleviate the contradiction between energy supply and demand, the exploration, development and utilization of NGH resources are important strategic choices in the new century. The exploitation and application of NGH resources will provide the world with sustainable energy resources and establish a low-cost and clean energy system. It has an important strategic position in ensuring world energy security, reducing greenhouse gas emissions, reducing pollution and protecting the environment. The United States, Japan, India, South Korea, Russia, Canada, Germany, Mexico and other countries have included NGH in their National Key Development Strategies from the perspective of national energy security, and have successively formulated NGH national research and development plans, and have invested lots of money in the basic and applied research of NGH. The United States and Japan have respectively established timetables for commercial trial mining in 2015 and 2016. The South China Sea and the Qinghai-Tibet Plateau are rich in NGH resources, and commercial exploitation of the resources is imperative.

Exercises

1. At 298K, the volume of a certain amount of H_2 is 15L, this gas

(1) at a constant temperature, resists the external pressure of 50kPa, and expands to a volume of 50L at one time;

(2) at a constant temperature, resists the external pressure of 100kPa, and expands to a volume of 50L at a time. Calculate the work of the two expansion processes.

2. Calculate the changes in the thermodynamic energy of the system under the following conditions:

(1) The system emits 2.5kJ of heat and does 500J of work to the environment;

(2) The system emits 650J of heat, and the environment does 350J to the system.

3. One mole ideal gas, after three steps expansion in constant temperature (constant volume heating and constant pressure cooling) returns to its initial state after completing one cycle. The whole process absorbs heat 100J, please find the w and ΔU of this process.

4. Which of the following reactions can release the most heat, Q?

(1) $CH_4(l)+2O_2(g) = CO_2(g)+2H_2O(g)$

(2) $CH_4(g)+2O_2(g) = CO_2(g)+2H_2O(g)$

(3) $CH_4(g)+2O_2(g) = CO_2(g)+2H_2O(l)$

(4) $CH_4(g)+\frac{3}{2}O_2(g) = CO(g)+2H_2O(l)$

5. Please prove the following expression: for the chemical reaction of ideal gas under constant temperature and constant pressure, $\Delta H = \Delta U + \Delta nRT$.

6. At 373K and 101.325kPa, the volume of 1mol $H_2O(l)$ is 0.018 8dm^3, water vapor is 30.2dm^3, and the heat of vaporization of water is 2.256kJ/g. Calculate the ΔH and ΔU, when 1mol of water becomes water vapor.

7. According to the standard enthalpy of formation of CH_4, CO, $H_2O(g)$ and CO_2 in the Appendix II. Calculate the reaction heat effects of 1m^3 CH_4 and 1m^3 CO respectively at 25℃

and 100kPa.

8. Triolein is a typical fat. When it is metabolized by the human body, the following reactions occur:

$$C_{57}H_{104}O_6(s)+80O_2(g) = 57CO_2(g)+52H_2O(l)$$

$$\Delta_r H_m^{\ominus} = -3.35\times10^4 \text{kJ/mol}$$

How many calories will be released when 1kg of this fat is consumed?

9. The following thermochemical equations are known:

$$Fe_2O_3(s)+3CO(g) = 2Fe(s)+3CO_2(g), \quad \Delta_r H_m^{\ominus}=-24.8\text{kJ/mol}$$

$$3Fe_2O_3(s)+CO(g) = 2Fe_3O_4(s)+CO_2(g), \quad \Delta_r H_m^{\ominus}=-47.2\text{kJ/mol}$$

$$Fe_3O_4(s)+CO(g) = 3FeO(s)+CO_2(g), \quad \Delta_r H_m^{\ominus}=19.4\text{kJ/mol}$$

Calculate the $\Delta_r H_m^{\ominus}$ of the reaction: $FeO(s)+CO(g) = Fe(s)+CO_2(g)$.

Chapter 6 Basic Principles of Chemical Reaction

Teaching contents	Learning requirements
Spontaneous reaction and entropy	Understand the characteristics and exothermic trend of spontaneous reaction, grasp the relevant definition of entropy and calculation of entropy change of chemical reaction, understand the thermodynamic significance of entropy and entropy change criterion of spontaneous reaction direction
Gibbs function change and reaction direction	Understand the origin of the change of Gibbs function, master the definition of $\Delta_f G_m^\ominus$ and relationship with $\Delta_r G_m^\ominus$, master the calculation of $\Delta_r G_m^\ominus$ under different temperatures and non-standard conditions, understand the thermodynamic meaning of Gibbs function and the judgment basis of spontaneous direction
Chemical equilibrium and movement	Grasp the significance of chemical equilibrium, the characteristics and calculation of standard equilibrium constant, the influence of concentration, pressure and temperature on chemical equilibrium, and understand the universality of equilibrium in natural law
Chemical reaction rate	Understand the concepts of collision theory and transition state theory, master the laws of concentration, temperature and catalyst effect on chemical reaction rate, and understand the difference and connection between chemical thermodynamics and chemical kinetics

6.1 Spontaneous Reaction and Entropy

In human life, the problem of food and clothing is the most basic one. So can food and textiles be produced in large quantities through chemical reactions using readily available raw materials?

Clothing: organic polymer is synthesized from organic small molecule.

$$n\,\overset{R_1}{\overset{|}{HC}}=\overset{R_2}{\overset{|}{CH}} = \!=\!= \left(\underset{H}{\overset{R_1}{\overset{|}{C}}}-\underset{H}{\overset{R_2}{\overset{|}{C}}}\right)_n \qquad \text{(A)}$$

Food: glucose and starch are produced by the reaction of CO_2 and H_2O.

$$6CO_2+6H_2O = 6O_2+C_6H_{12}O_6 \qquad \text{(B)}$$

A large number of experimental results show that the reaction (A) can be carried out spontaneously, and a large number of synthetic fibers such as acrylic fiber, chlorine fiber, polypropylene fiber and polyester fiber can be produced with petroleum as raw materials. Now the output of synthetic fiber has exceeded that of natural fiber. The problem of human wear can be solved by chemical production. But reaction (B) cannot be spontaneous, so how to use theory to judge whether a reaction can be spontaneous under certain conditions is very important.

6.1.1 Spontaneous Reaction and Its Thermal Effect

Under certain conditions, the system does not need any external force, and the process of automatically changing from one state to another is called spontaneous process. Spontaneous processes are abundant in nature, which can be physical or chemical processes. For example, water at a higher level can flow spontaneously to a lower level, and the difference in height or potential energy is the driving force for this process. Another example is to drop a drop of ink into a glass of clear water, after a period of time, the ink will spontaneously spread to the whole glass of water, and the color of water will change. For example, the following chemical reactions can be spontaneous

$$H^+(aq)+OH^-(aq) = H_2O(l)$$

$$C(s)+O_2(g) = CO_2(g)$$

$$Zn(s)+2H^+(aq) = Zn^{2+}(aq)+H_2(g)$$

The spontaneous process has the following characteristics:

① The spontaneous process is directional. Any spontaneous process is irreversible, that is to say, the reverse process of spontaneous process is not spontaneous. Water can flow spontaneously from high to low, but not from low to high. Ink can spread spontaneously, but not gather spontaneously. Acid and base can react spontaneously to form salt and water, but salt and water can't change into acid and base spontaneously.

② The spontaneous process can do work through a certain device. For example, electricity can be generated by water level difference, primary batteries can be formed by oxidation-reduction reaction, *etc.*

③ The spontaneous process can only be carried out to a certain extent. This limit is equilibrium. When the water level difference is equal to zero, the water will not flow, and when the solution concentration is uniform, the diffusion process will not be carried out.

Since the spontaneous process has directionality and certain limit, how to determine the direction and limit? In particular, how to determine the direction and limit of spontaneous chemical reactions?

As early as the 19th century, some chemists hoped to find a criterion for judging the direction of reaction. They found that many spontaneous reactions are exothermic, such as:

$$H^+(aq)+OH^-(aq) = H_2O(l) \qquad \Delta_r H_m^\ominus = -55.8\text{kJ/mol}$$

$$C(s)+O_2(g) = CO_2(g) \qquad \Delta_r H_m^\ominus = -393.5\text{kJ/mol}$$

$$Zn(s)+2H^+(aq) = Zn^{2+}(aq)+H_2(g) \qquad \Delta_r H_m^\ominus = -153.9\text{kJ/mol}$$

In 1878, M. Berthelot, a French chemist, and J. Thomsen, a Danish chemist, proposed that spontaneous chemical reactions tend to cause the system to emit the most heat. So, some people try to use the thermal effect or enthalpy change of the reaction as the judgment basis of spontaneous reaction.

However, subsequent studies found that some endothermic processes or reactions can also be spontaneous. For example:

① $H_2O(s) \Longrightarrow H_2O(l)$, $\Delta H>0kJ/mol$, when the temperature is higher than 273k, ice can spontaneously turn into water;

② Dissolution of NH_4Cl: $NH_4Cl \Longrightarrow NH_4^+(aq) + Cl^-(aq)$, $\Delta H = 9.76kJ/mol$, it can be carried out spontaneously under certain conditions;

③ Reaction of calcining limestone in industry: $CaCO_3(s) \Longrightarrow CaO(s)+CO_2(g)$, $\Delta H>0kJ/mol$. At 101.325kPa and 1183K (910℃), $CaCO_3$ can decompose spontaneously and violently to form CaO and CO_2.

Obviously, these situations cannot be explained only by the enthalpy change of reaction or process. This shows that there are other factors besides enthalpy change to judge whether a reaction or process can be spontaneous under given conditions. Heat release is only one of the factors, not the only one.

6.1.2 Entropy and the Direction of Spontaneous Reaction

(1) Entropy and number of microstates

From the perspective of transformation research, it is found that there are many spontaneous processes that are closely related to the increase of system chaos. For example, the diffusion of gas, the diffusion of red ink in water, *etc*., but it is impossible for the diffused gas or liquid to return to the state before diffusion spontaneously. In daily life or work, similar examples can be seen everywhere, such as the melting of ice, the evaporation of water, the dissolution of solid substances in water, the dissolution of insoluble hydroxides in acid, *etc*. This shows that the process can spontaneously go in the direction of increasing the degree of disorder, or that the system tends to the maximum degree of disorder.

Chemical reaction system is a kind of thermodynamic system, which is composed of a large number of particles. These particles are micro particles, but the properties of micro particles must be reflected in the macro properties. In other words, the macroscopic properties of the thermodynamic system are related to the microscopic properties of the microscopic particles in the system. For example, Fig.6-1 (a) shows O_2 in the flask on the left and N_2 in the flask on the right. After opening the piston, the gas on both sides will be completely mixed after a period of time [see Fig. 6-1 (b)]. At this time, the degree of confusion increases and the system reaches a stable state. Conversely, the mixture (O_2 and N_2) in Fig.6-1(b) will not spontaneously separate in Fig.6-1(a), because the disorder degree of the system will decrease and the stability of the system will be reduced.

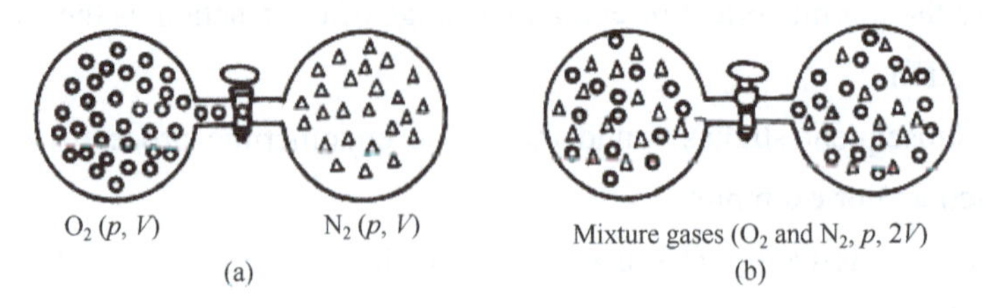

Fig.6-1　Mixing process of ideal gas at constant temperature and pressure

The disorder degree of the system can be expressed by entropy. Entropy, S, is a thermodynamic property that is related to the way in which the energy of a system is distributed among the available energy levels. In 1878, L. E. Boltzman proposed a quantitative relationship between the number of microstates, W and S (also known as Boltzman formula):

$$S = k \ln W \tag{6-1}$$

Formula (6-1) shows that entropy is a measure of the disorder degree of the system. In the relationship, where S is entropy, k_B is the Boltzmann constant, and W is the number of microstates. the Boltzmann constant is related to the gas constant R and Avogadro's number, N_A, by the expression $k_B = R/N_A$. We can think of k_B as the gas constant per molecule.

(2) Standard molar entropy and its change

The disorder degree of the micro particles in the system is related to the aggregation state and temperature of the matter. It can be assumed that at absolute zero, the particles in the complete crystal of pure matter are arranged orderly on the lattice, the number of microscopic states is 1, and the entropy corresponding to this state should be zero. That is to say, "under absolute zero, the entropy value of the complete crystal of all pure substances and compounds is zero", which is the third law of thermodynamics. The third law of thermodynamics is only a theoretical inference, because up to now, we can't reach absolute zero. On this basis, the entropy of matter at other temperatures can be determined.

If a pure substance is heated from 0K to T K, the entropy change ΔS of the process is:

$$\Delta S = S_T - S_0 = S_T \tag{6-2}$$

Where S_T is called the absolute entropy of the substance. At a certain temperature (usually 298K), the absolute entropy of 1mol of a substance B ($\nu_B = 1$) under the standard state ($p^{\ominus} = 100$kPa) is the standard molar entropy, which is represented by the symbol $S_m^{\ominus}$ (B, phase state, T), and the unit is J/(mol·K). Obviously, the standard molar entropy $S_m^{\ominus}$ (B, phase state, T) of all substances (including simple substance) at 298K is greater than zero. According to the first law of thermodynamics, the standard molar entropy of a substance is equal to 0. However, similar to the standard molar enthalpy, the standard molar entropy of H^+ in the standard state of water at 298K is zero for hydrated ions due to the existence of both positive and negative ions, that is $S_m^\ominus$ (H^+, aq, 298K)=0J/(mol·K), thus, the standard entropy (relative value) of other hydrated ions at 298.15K can be obtained. See Appendix Ⅱ.

Through the analysis of the standard molar entropy of some substances, some rules can be obtained:

① Entropy is related to the aggregation state of matter. When the same material is at the same temperature, the gas entropy is the largest, the liquid is the second, and the solid is the smallest, that is, $S_m^\ominus$ (*B*, s, 298K) < $S_m^\ominus$ (*B*, l, 298K) < $S_m^\ominus$ (*B*, g, 298K).

② The entropy of the same matter increases with the increase of temperature in the same aggregation state, that is, *S* (high temperature) > *S* (low temperature).

③ When the temperature and aggregation state are the same, the standard molar entropy $S_m^\ominus$ of substances with similar molecular structure is similar. For example, $S_m^\ominus$ (CO, g, 298K) = 197.7J/(mol·K), $S_m^\ominus$ (N_2, g, 298K) = 191.6J/(mol·K).

④ The standard molar entropy $S_m^\ominus$ of substances with similar molecular structure but different relative molecular weight increases with the increase of relative molecular weight. Such as gaseous hydrogen halide, $S_m^\ominus$ (HF, g, 298K) < $S_m^\ominus$ (HCl, g, 298K) < $S_m^\ominus$ (HBr, g, 298K).

⑤ In the case of solid, the entropy of hard solid (such as diamond) is lower than that of soft solid (such as graphite), $S_m^\ominus$ (C, graphite, 298K) > $S_m^\ominus$ (C, diamond, 298K).

⑥ In the same aggregation state, the entropy of the mixture or solution is often higher than that of the corresponding pure substance, *viz*. *S* (mixture) > *S* (pure substance).

It can be seen that the standard molar entropy of matter is closely related to the aggregation state, temperature and microstructure. According to the above laws, we can get a useful rule for qualitative judgment of process entropy change: for physical or chemical changes, if a process or reaction causes the increase of gas molecular number, the entropy value will increase, that is, $\Delta S > 0$; otherwise, if thc gas molecular number decreases, $\Delta S < 0$.

Because entropy is a state function, the entropy change of a chemical reaction is equal to the sum of the product of the absolute entropy and coefficient of the product minus the sum of the reactant of the absolute entropy and coefficient of the reactant, that is:

$$\Delta_r S_m^\ominus = \sum_B (\nu_B S_m^\ominus)_{\text{product}} - \sum_B (|\nu_B| S_m^\ominus)_{\text{reactant}} \tag{6-3}$$

Example 6-1 Calculate the entropy change $\Delta_r S_m^\ominus$ of reaction

$$2H_2(g) + O_2(g) = 2H_2O(l).$$

Solution Refer to Appendix Ⅱ to get the standard molar entropy $S_m^\ominus$ of each substance as follows

$$2H_2(g) + O_2(g) \xlongequal{\quad} 2H_2O(l)$$

	$2H_2(g)$	$O_2(g)$	$2H_2O(l)$
$S_m^\ominus$ /[J/(mol·K)]	130.7	205.5	70.0

$$\begin{aligned}\Delta_r S_m^\ominus &= 2S_m^\ominus(H_2O,l) - [2S_m^\ominus(H_2,g) + S_m^\ominus(O_2,g)] \\ &= 2\times70.0-(2\times130.7+205.5) \\ &= -326.9[J/(mol\cdot K)]\end{aligned}$$

(3) Entropy criterion of spontaneous process direction

The results of thermodynamics study show that in isolated system, the spontaneous process always goes in the direction of entropy increase until the system reaches equilibrium, which is called entropy increase principle. According to this principle, we can deduce the entropy criterion for isolated system:

$\Delta S_{univ} > 0$, positive spontaneity of process or reaction

$\Delta S_{univ} = 0$, process or reaction system in equilibrium

$\Delta S_{univ} < 0$, reverse spontaneity of process or reaction

According to the entropy increase principle, to judge the spontaneity of a reaction or process, we must first calculate ΔS_{univ}. The combination of the system and the surroundings is the universe (univ = sys + surr). Hence, the total entropy change in the system ΔS_{sys} and surroundings ΔS_{surr}, equals the change in entropy in the universe, ΔS_{univ}. That is:

$$\Delta S_{univ} = \Delta S_{sys} + \Delta S_{surr} \tag{6-4}$$

Although the spontaneity of a reaction can be judged according to the $\Delta S_{univ} > 0$, the entropy change of the system and the environment should be considered in the calculation of ΔS_{univ}, and the entropy change of the environment should be calculated separately, which undoubtedly increases the calculation workload. Moreover, most of the chemical reactions are carried out under constant temperature and pressure, and they are not isolated systems. It is necessary to introduce a new convenient state function to judge the direction of spontaneous reaction.

Example 6-2 Calculate the entropy change $\Delta_r S_m^\ominus$ and the enthalpy change $\Delta_r H_m^\ominus$ of the thermal decomposition reaction of $CaCO_3$ (aragonite) at 298.15K, and analyze the spontaneity of the reaction.

Solution Refer to Appendix Ⅱ to get the standard molar entropy $S_m^\ominus$ and the standard molar enthalpy $\Delta_f H_m^\ominus$ of each substance as follows

$$CaCO_3(s) \xlongequal{\quad} CaO(s) + CO_2(g)$$

	$CaCO_3(s)$	$CaO(s)$	$CO_2(g)$
$\Delta_f H_m^\ominus$ /(kJ/mol)	−1207.8	−634.9	−393.5
$S_m^\ominus$ /[J/(mol·K)]	88	38.1	213.8

$$\Delta_r H_m^\ominus = [\Delta_f H_m^\ominus(CO_2,g) + \Delta_f H_m^\ominus(CaO,s)] - \Delta_f H_m^\ominus(CaCO_3,s)$$
$$= (-393.5) + (-634.9) - (-1207.8)$$
$$= 179.4(kJ/mol)$$
$$\Delta_r S_m^\ominus = [S_m^\ominus(CO_2,g) + S_m^\ominus(CaO,s)] - S_m^\ominus(CaCO_3,s)$$
$$=213.8 + 38.1 - 88$$
$$=163.9[J/(mol \cdot K)]$$

At 298.15K, $\Delta_r H_m^\ominus$ is a positive value, indicating that the reaction is endothermic. From the point of view that the system tends to obtain energy, endothermic is not conducive to spontaneous reaction. However, $\Delta_r S_m^\ominus$ becomes positive at this temperature, indicating that the entropy of the system increases in the reaction process. It can be seen that only $\Delta_r H_m^\ominus$ or $\Delta_r S_m^\ominus$ can not accurately judge the spontaneity of this reaction, and they should be considered comprehensively.

6.2 Gibbs Function Changes and Judgment of Spontaneous Direction of Reaction

Copper products are exposed to flowing atmosphere for a long time at room temperature, and their surface is gradually covered with a layer of black copper oxide CuO. When the product is heated above a certain temperature, the black copper oxide changes into red cuprous oxide Cu_2O. At higher temperatures, the red oxide disappears. If we want to obtain Cu_2O red coating by artificial archaize, what temperature should we choose?

6.2.1 Gibbs Function Changes and Gibbs Function Criterion

In 1876, J. W. Gibbs, a famous American physicochemical, put forward a new state function which integrates the relationship among enthalpy, entropy and temperature of the system. It is called Gibbs function (also called Gibbs free energy). The symbol is G, which is defined as follows:

$$G \equiv H - TS \tag{6-5}$$

The change value of Gibbs function of a chemical reaction or process under constant temperature and pressure is:

$$\Delta G = \Delta H - T\Delta S \tag{6-6}$$

Equation (6-6) is called Gibbs isothermal equation. According to the principle of energy reduction and entropy increase, the spontaneity of the reaction or process can be determined by the Gibbs function change of the system under the condition of constant temperature and pressure, and when the system does not do non-volumework:

$$\begin{cases} \Delta G < 0\text{, spontaneous process} \\ \Delta G = 0\text{, equilibrium state} \\ \Delta G > 0\text{, reverse spontaneous process} \end{cases}$$

It is shown that the Gibbs function of the system is always reduced ($\Delta G < 0$) without any non-volumework and at constant temperature and pressure. This criterion can be used to judge the direction of reaction in closed system.

Because the general chemical reactions are carried out under constant pressure, and only ΔH and ΔS of the system are used to calculate ΔG, it is much more convenient to distinguish the direction of spontaneous reaction by ΔG.

G has the following properties.

① G is the state function. Because of $G \equiv H-TS$, and H, T, S are all state function, G is a state function too. The value of ΔG is only related to the initial and final states of the system, but not to the path.

② ΔG is a measure of the system's useful work. The enthalpy change of a reaction or process can be divided into two parts: one is used to maintain the temperature of the system and change the disorder degree of the system. This part of energy cannot be used to change into another form of energy, so this part of energy is called binding energy; the other part of enthalpy change is energy that can be used to do useful work, that is, ΔG.

③ ΔG is the driving force of spontaneous process. According to the Gibbs function change ΔG, under the condition of constant temperature, constant pressure and volumework only, the direction of spontaneous process is the direction of decreasing Gibbs function. That is to say, the reason why the system spontaneously changes from one state to another is that there is the difference of Gibbs function between the two states. Just as there is temperature difference, there will be heat transfer, water level difference , and water flow; ΔG is also a driving force of spontaneous process. The spontaneous process always goes from the state of big G to the state of small G until ΔG = 0, which means reaching the equilibrium state. In other words, the larger the Gibbs function is, the more unstable the system is; there is a tendency of spontaneous transition to a state with small Gibbs function, so the state with small Gibbs function is relatively stable. Therefore, Gibbs function is also a measure of the stability of the system.

6.2.2 Standard Molar Gibbs Function Variation of Reaction and Its Calculation

Like the enthalpy change of reaction, the absolute value of Gibbs function of the system can not be measured, and the change value of Gibbs function is the focus of thermodynamics. The change of Gibbs function is related to the amount of substance consumed in the reaction and the conditions under which the reaction proceeds.

$\Delta_r G_m^\ominus$ (in kJ/mol) is the standard Gibbs function change of the reaction, when a chemical reaction completes the conversion from reactant to product (*i.e.* the extent of reaction is 1mol) according to the reaction formula in the standard state at a certain temperature.

(1) Calculation of $\Delta_r G_m^\ominus$ by Gibbs formula

In the standard state, the Gibbs isothermal equation of equation (6-6) can be expressed as

$$\Delta_r G_m^\ominus = \Delta_r H_m^\ominus - T\Delta_r S_m^\ominus \tag{6-7}$$

According to the knowledge of the first law of thermodynamics, $\Delta_r H_m^\ominus$ and $\Delta_r S_m^\ominus$ at 298.15K can be calculated, and then $\Delta_r G_m^\ominus$ at 298.15K can be easily calculated by substituting them into formula (6-7).

Example 6-3 Acrylonitrile is the raw material for making acrylic fiber, which can be synthesized in one step with propylene and ammonia. Now the following conditions are known, please calculate the $\Delta_r G_m^\ominus$ at 298.15K.

	$C_3H_6(g)$ +	$NH_3(g)$ +	$\frac{3}{2}O_2(g)$ ═══	$CH_2{=}CH{-}CN(g)$ +	$3H_2O(g)$
$\Delta_f H_m^\ominus$/(kJ/mol)	20.0	−45.9	0	184.9	−241.8
$S_m^\ominus$/[J/(mol·K)]	267	192.8	205.2	273.9	188.8

Solution

$$\Delta_r H_m^\ominus = [\Delta_f H_m^\ominus(C_3H_3N,g) + 3\Delta_f H_m^\ominus(H_2O,g)] - [\Delta_f H_m^\ominus(C_3H_6,g) + \Delta_f H_m^\ominus(NH_3,g)]$$
$$= [184.9 + 3 \times (-241.8)] - [20.0 + (-45.9) + 0]$$
$$= -514.6(\text{kJ/mol})$$

$$\Delta_r S_m^\ominus = [S_m^\ominus(C_3H_3N,g) + 3S_m^\ominus(H_2O,g)] - \left[S_m^\ominus(C_3H_6,g) + S_m^\ominus(NH_3,g) + \frac{3}{2}S_m^\ominus(O_2,g)\right]$$
$$= [273.9 + 3 \times 188.8] - [267 + 192.8 + 1.5 \times 205.2]$$
$$= 72.7[\text{J/(mol·K)}]$$

$$\Delta_r G_m^\ominus = \Delta_r H_m^\ominus - T\Delta_r S_m^\ominus$$
$$= -514.6 - 298 \times 0.001 \times 72.7$$
$$= -536.3(\text{kJ/mol}) < 0$$

So, at 298.15K, the reaction can be spontaneous.

It should be noted that only the unit of $\Delta_r S_m^\ominus$ is J/(mol·K). When the known data is substituted into Gibbs formula for calculation, the data unit should be unified.

(2) Calculate the $\Delta_r G_m^\ominus$ of the reaction according to $\Delta_f G_m^\ominus$

In thermodynamics, it is specified that the standard molar Gibbs function change of reaction when the most stable single substance generates 1mol of compound *B* under the condition of temperature *T* and pressure $p^\ominus$, which is called the standard molar Gibbs function change of substance B, recorded as $\Delta_f G_m^\ominus$ (B, phase state, T), and the unit is kJ/mol. $\Delta_f G_m^\ominus$ of the most stable simple substance is zero. At present $\Delta_f G_m^\ominus$ of many substances has been determined, see Appendix Ⅱ.

The data listed in Appendix Ⅱ are the values of $\Delta_f H_m^\ominus$, $S_m^\ominus$ and $\Delta_f G_m^\ominus$ of each substance at 298.15K. The reaction $\Delta_r G_m^\ominus$ at 298.15K can be calculated directly by the formula:

$$\Delta_r G_m^\ominus = \sum_B (\nu_B \Delta_f G_m^\ominus)_{\text{products}} - \sum_B (|\nu_B| \Delta_f G_m^\ominus)_{\text{reactants}} \tag{6-8}$$

Example 6-4 There are toxic gases NO and CO in automobile exhaust gases. One of the schemes to remove these two toxic gases is to use the following reactions:

$$NO(g) + CO(g) \xlongequal{\quad} CO_2(g) + \frac{1}{2}N_2(g)$$

Using $\Delta_f G_m^\ominus$ to calculate the $\Delta_r G_m^\ominus$ of the reaction at 298.15K.

Solution $$NO(g) + CO(g) \xlongequal{\quad} CO_2(g) + \frac{1}{2}N_2(g)$$

$\Delta_f G_m^\ominus$/(kJ/mol) 87.6 −137.2 −394.4 0

$$\Delta_r G_m^\ominus = \left[\Delta_f G_m^\ominus(CO_2,g) + \frac{1}{2}\Delta_f G_m^\ominus(N_2,g)\right] - [\Delta_f G_m^\ominus(NO,g) + \Delta_f G_m^\ominus(CO,g)]$$
$$= [(-394.4)+0]-[(-137.2)+87.6]$$
$$= -344.8(\text{kJ/mol}) < 0$$

So at 298.15K, the reaction can be spontaneous under the standard state.

It should be pointed out that $\Delta_r G_m^\ominus$, as a criterion, can only judge whether a reaction can be carried out spontaneously in a standard state, and can not explain how the reaction will be carried out. $\Delta_r G_m^\ominus < 0$ indicates that the reaction can be spontaneous under constant temperature and pressure, but the actual reaction rate is unknown. In fact, it is very difficult to use the reaction in the above example to purify automobile exhaust. The main reason is the reaction rate. The way to solve this problem is to find efficient and cheap catalyst.

(3) Direct calculation of $\Delta_r G_m^\ominus$ according to Hess's law

Because G is a state function, Hess's law is also applicable to the calculation of Gibbs function variation $\Delta_r G_m^\ominus$ of chemical reaction.

Example 6-5 Calculate the $\Delta_r G_m^\ominus$ of the following reactions at 298.15K.

$$CH_2{=}CH_2(g) + O_2(g) \xlongequal{\quad} CH_3COOH(g) \tag{1}$$

Known: $CH_2{=}CH_2(g) + \frac{1}{2}O_2(g) \xlongequal{\quad} CH_3CHO(g)$, $\Delta_r G_m^\ominus(2) = -201.4$kJ/mol (2)

$$CH_3CHO(g) + \frac{1}{2}O_2(g) = CH_3COOH(g), \quad \Delta_r G_m^\ominus(3) = -241.4\text{kJ/mol} \tag{3}$$

Solution

$$CH_2{=}CH_2(g) + \frac{1}{2}O_2(g) = CH_3CHO(g)$$

$$+)\quad CH_3CHO(g) + \frac{1}{2}O_2(g) = CH_3COOH(g)$$

$$CH_2{=}CH_2(g) + O_2(g) = CH_3COOH(g)$$

That is : (1) = (2) + (3)

According to Hess's law

$$\Delta_r G_m^\ominus(1) = \Delta_r G_m^\ominus(2) + \Delta_r G_m^\ominus(3) = (-201.4) + (-241.2) = -442.6(\text{kJ/mol})$$

6.2.3 Application of Gibbs Formula

It must be noted that only $\Delta_f H_m^\ominus$, $S_m^\ominus$ and $\Delta_f G_m^\ominus$ at 298.15K can be found in the thermodynamic data sheet, but not all chemical reactions happen to be carried out at 298.15K. So at other temperatures, how is $\Delta_r G_m^\ominus$ calculated? How to judge the direction of spontaneous reaction?

For the standard state at other temperatures T, when there is no phase change, $\Delta_r H_m^\ominus(T)$ and $\Delta_r S_m^\ominus(T)$ have little change with temperature, so it can be approximately considered that:

$$\Delta_r H_m^\ominus(T) \approx \Delta_r H_m^\ominus(298.15\text{K})$$

$$\Delta_r S_m^\ominus(T) \approx \Delta_r H_m^\ominus(298.15\text{K})$$

So $\Delta_r G_m^\ominus(T)$ at other temperatures T can still be calculated by Gibbs formula, that is:

$$\Delta_r G_m^\ominus(T) = \Delta_r H_m^\ominus(298.15\text{K}) - T\Delta_r S_m^\ominus(298.15\text{K}) \tag{6-9}$$

Gibbs formula has the following applications:

(1) Judge the influence of temperature on the direction of chemical reaction

Gibbs formula reflects the relationship among reaction temperature T, enthalpy change ΔH, entropy change ΔS, and Gibbs function change ΔG. Because the positive and negative of ΔG determines the direction of spontaneous reaction, and ΔG is closely related to the temperature, the Gibbs formula can be used to calculate the change of Gibbs function of the reaction, as well as to discuss the relationship between the temperature and the direction of spontaneous reaction. Gibbs formula is also applicable to non-standard state, as long as the enthalpy change ΔH and entropy change ΔS are in the same state. For the convenience of discussion, we take the standard state as an example.

According to the different numerical symbols of $\Delta_r H_m^\ominus$ and $\Delta_r S_m^\ominus$, there are four possible situations when discussing the influence of temperature on the spontaneous direction of chemical reaction.

① $\Delta_r H_m^\ominus < 0$, $\Delta_r S_m^\ominus > 0$. This is a process of exothermic and entropy increasing. Whether from the principle of minimum energy or from the principle of entropy increase, it is conducive to the forward reaction. The Gibbs formula also shows that the $\Delta_r G_m^\ominus$ of the reaction is negative at any temperature, so the reaction can be spontaneous at any temperature. For example:

$$C_6H_{12}O_6(s) + 6O_2 \xlongequal{} 6CO_2\ (g) + 6H_2O(l)$$

$$H_2(g) + Cl_2(g) \xlongequal{} 2HCl(g)$$

$$C(s) + O_2(g) \xlongequal{} CO_2(g)$$

② $\Delta_r H_m^\ominus < 0$, $\Delta_r S_m^\ominus < 0$. This is a process of exothermic and entropy reduction. At this time, temperature will play a major role, because only when $|\Delta_r H_m^\ominus| > |T\Delta_r S_m^\ominus|$, $\Delta_r G_m^\ominus < 0$. Therefore, the lower the temperature, the more favorable for this process. The formation of water into ice is an example of this process. Water freezes to give off heat, $\Delta_r H_m^\ominus < 0$; However, the water molecules become more orderly and less chaotic during the freezing process, $\Delta_r S_m^\ominus < 0$. To ensure $\Delta_r G_m^\ominus < 0$, the temperature should not be too high. At 100kPa, water can only freeze when the temperature is lower than 273.15K, and when the temperature is higher than 273.15K, $\Delta_r G_m^\ominus > 0$ water can not freeze spontaneously. There are a lot of such reactions in industrial production and real life. For example:

$$N_2(g) + 3H_2(g) \xlongequal{} 2NH_3(g)$$

$$2H(g) \xlongequal{} H_2(g)$$

$$CaO(s) + CO_2(g) \xlongequal{} CaCO_3(s)$$

③ $\Delta_r H_m^\ominus > 0$, $\Delta_r S_m^\ominus > 0$. This is a process of endothermic and entropy increasing. The same as above, temperature is an important factor. To ensure $\Delta_r G_m^\ominus < 0$, temperature must be high enough. Because only when $|\Delta_r H_m^\ominus| < |T\Delta_r S_m^\ominus|$, $\Delta_r G_m^\ominus < 0$. That is to say, the reaction can be spontaneous at high temperature: the higher the temperature, the more favorable the reaction. Ice melting and water evaporation belong to this kind of process. For example:

$$CaCO_3(s) \xlongequal{} CaO(s) + CO_2(g)$$

$$2NaHCO_3(s) \xlongequal{} Na_2CO_3(s) + CO_2(g) + H_2O(g)$$

$$2H_2O(g) \xlongequal{} 2\ H_2(g) + O_2(g)$$

④ $\Delta_r H_m^\ominus > 0$, $\Delta_r S_m^\ominus < 0$ Both of these factors are unfavorable to spontaneous process. No matter what temperature, it is always $\Delta_r G_m^\ominus > 0$, so the reaction can not be positive spontaneous. Because the reaction can not be spontaneous, we must add energy, such as light, to promote the reaction forward. In nature, photosynthesis and ozonation all belong to this situation. For example:

$$N_2(g) + 3Cl_2(g) \longrightarrow 2NCl_3(g)$$

$$3O_2(g) \longrightarrow 2O_3(g)$$

$$6CO_2\,(g) + 6H_2O(l) \longrightarrow C_6H_{12}O_6(s) + 6O_2(g)$$

See Table 6-1 for the summary of the above situations.

Table 6-1 The Effect of Temperature on the Direction of Chemical Reaction

Form	$\Delta_r H_m^\ominus$	$\Delta_r S_m^\ominus$	$\Delta_r G_m^\ominus(T) = \Delta_r H_m^\ominus - T\Delta_r S_m^\ominus$	Spontaneity of Reaction
①	−	+	Always $\Delta_r G_m^\ominus < 0$	At any temperature, the reaction is spontaneous
②	−	−	At low temperature, $\Delta_r G_m^\ominus < 0$	At low temperature, the reaction proceeds spontaneously
③	+	+	At high temperature, $\Delta_r G_m^\ominus < 0$	At high temperature, the reaction proceeds spontaneously
④	+	−	Always $\Delta_r G_m^\ominus > 0$	At any temperature, the reaction is nonspontaneous

(2) Estimate the temperature of spontaneous reaction in standard state, *i.e.* judge the turning temperature of chemical reaction

$\Delta_r H_m^\ominus < 0$, $\Delta_r S_m^\ominus < 0$, which means the reaction can be spontaneous at low temperature but not at high temperature; conversely, $\Delta_r H_m^\ominus > 0$, $\Delta_r S_m^\ominus > 0$, which means the reaction can be spontaneous at high temperature but not at low temperature. In both cases, the temperature of spontaneous reaction can be estimated, which is usually called turning temperature, *i.e.* T_{tur}.

$$\Delta_r G_m^\ominus(T) = \Delta_r H_m^\ominus - T\Delta_r S_m^\ominus$$

$$T_{tur} = \frac{\Delta_r H_m^\ominus}{\Delta_r S_m^\ominus} \tag{6-10}$$

It should be noted that the common dimension of $\Delta_r H_m^\ominus$ is kJ/mol, and the common dimension of $\Delta_r S_m^\ominus$ is J/(mol·K), which requires a unified unit in calculation.

Example 6-6 It is known that $\Delta_r H_m^\ominus$=242.6kJ/mol, $\Delta_r S_m^\ominus$=107.3J/(mol·K), calculation of the minimum temperature of chlorine decomposition reaction.

Solution $$Cl_2(g) \longrightarrow 2Cl(g)$$

According to formula (6-10), the lowest temperature of spontaneous reaction is calculated, *i.e*:

$$T > T_{tur} = \frac{\Delta_r H_m^\ominus}{\Delta_r S_m^\ominus} = \frac{242.6 \times 1000}{107.3}\text{K} = 2261\text{K}$$

The lowest spontaneous temperature of the reaction is 2261K.

Generally, decomposition reactions are endothermic and entropy increasing reactions, so they have the lowest decomposition temperature. When the temperature is higher than decomposition temperature, the $\Delta_r G_m^\ominus$ of decomposition reaction changes from positive to negative, and the reaction is spontaneous. For example, the decomposition reaction of $CaCO_3$

is spontaneous when the temperature is higher than 1183K.

(3) Estimation of phase transition temperature and entropy change in standard state

By using Gibbs formula, we can also calculate the temperature of normal phase transformation (such as the freezing point and boiling point of substance in standard state) and the entropy change of phase transformation. Because the two phases are in equilibrium during normal phase transition, when $\Delta_r G_m^\ominus (T)=0$, then

$$\Delta_r H_m^\ominus - T\Delta_r S_m^\ominus = 0$$

$$T(\text{phase transformation})=\frac{\Delta_r H_m^\ominus}{\Delta_r S_m^\ominus} \tag{6-11}$$

$$\Delta_r S_m^\ominus(\text{phase transformation})=\frac{\Delta_r H_m^\ominus}{T} \tag{6-12}$$

For the normal phase transformation process, $\Delta_r H_m^\ominus$ and T are easy to be determined, so formula (6-12) is often used to calculate $\Delta_r S_m^\ominus$.

Example 6-7 It is known that at 100kPa and 373K, 1mol of water evaporates into water vapor. Calculate the entropy change of this phase transition process

$$H_2O(l) = H_2O(g), \quad \Delta_r H_{m,vap}^\ominus = 40.67\text{kJ/mol}$$

Solution According to formula (6-12),

$$\Delta_r S_{m,vap}^\ominus = \frac{\Delta_r H_{m,vap}^\ominus}{T_{vap}} = \frac{40.67 \times 1000}{373} = 109.03[\text{J/(mol} \cdot \text{K)}]$$

6.2.4 Gibbs Function Changes under Nonstandard Condition

At any temperature T and pressure p (*i.e.* nonstandard condition), whether the reaction or process can be spontaneous or not depends on the Gibbs function changes in nonstandard condition. In the actual reaction, $\Delta_r G_m(T)$ will change with the change of partial pressure (for gas) or concentration (for solution) of reactants and products. There is a certain mathematical relation between $\Delta_r G_m(T)$ and $\Delta_r G_m^\ominus(T)$, and the relation formula can be deduced from the correlation formula of chemical thermodynamics.

A chemical reaction at any temperature T:

$$a\text{A} + b\text{B} = g\text{G} + d\text{D}$$

The relation between Gibbs function changes at nonstandard conditions, $\Delta_r G_m(T)$,and the standard Gibbs function changes, $\Delta_r G_m^\ominus(T)$, must be considered. This relationship is expressed as follows:

$$\Delta_r G_m(T) = \Delta_r G_m^\ominus(T) + RT\ln Q \tag{6-13}$$

Equation (6-13) is called thermodynamic isothermal equation. Where Q is the reaction quotient; R is the ideal gas constant, and the value is equal to 8.314J/(mol·K).

Gas reaction:

$$aA(g) + bB(g) = gG(g) + dD(g)$$

$$Q = \frac{(p_G / p^\ominus)^g (p_D / p^\ominus)^d}{(p_A / p^\ominus)^a (p_B / p^\ominus)^b} \tag{6-14}$$

In the formula, p_A, p_B, p_G and p_D represent the partial pressure of gaseous substances A, B, G and D under any conditions, respectively; $p^\ominus$ is the standard pressure; $p / p^\ominus$ is the relative partial pressure.

Reaction in solution:

$$aA(aq) + bB(aq) = gG(aq) + dD(aq)$$

$$Q = \frac{(c_G / c^\ominus)^g (c_D / c^\ominus)^d}{(c_A / c^\ominus)^a (c_B / c^\ominus)^b} \tag{6-15}$$

In the formula, c_A, c_B, c_G and c_D are the volume molar concentrations of substances A, B, G and D under the given conditions respectively; $c^\ominus$ is the standard volume molar concentration ($c^\ominus$=1mol/L); $c / c^\ominus$ is the relative concentration.

It can be seen from equations (6-14) and (6-15) that the unit of reaction quotient is 1. If there is solid substance in the reaction formula, the relative concentration (or partial pressure) of solid substance is not included in the formulas.

According to formula (6-13), the Gibbs function changes $\Delta_r G_m(T)$ of chemical reaction, under any specified p and T can be calculated, and the direction of reaction can be determined, namely:

$$\begin{cases} \Delta_r G_m(T) < 0, & \text{spontaneous process} \\ \Delta_r G_m(T) = 0, & \text{equilibrium state} \\ \Delta_r G_m(T) > 0, & \text{reverse spontaneous process} \end{cases}$$

Example 6-8 It is known that the partial pressure of CO_2 in the atmosphere is 0.03kPa, judge whether the decomposition reaction of $CaCO_3$ can be spontaneous in the atmosphere at 298K. If the reaction is spontaneous in air, how much should the temperature be?

Solution In example 6-2, $\Delta_r H_m^\ominus$ and $\Delta_r S_m^\ominus$ of the decomposition reaction of $CaCO_3$ is calculated,

$$CaCO_3(s) \xlongequal{\quad} CaO(s) + CO_2(g)$$

$$\begin{aligned} \Delta_r G_m^\ominus(T) &= \Delta_r H_m^\ominus - 298.15 \times \Delta_r S_m^\ominus \\ &= 179.4 - 298.15 \times 163.9 \times 0.001 \\ &= 130.6 (\text{kJ/mol}) \end{aligned}$$

$CaCO_3$ and CaO are solid, so $Q = p_{CO_2} / p^{\ominus} = 0.0003$

$$\Delta_r G_m(298.15\text{K}) = \Delta_r G_m^{\ominus}(298.15\text{K}) + RT\ln Q$$
$$= 130.6 + 8.314\times0.001\times298.15\times\ln0.0003$$
$$= 121.8(\text{kJ/mol}) > 0$$

Therefore, $CaCO_3$ cannot decompose spontaneously in the atmosphere at 298K. For the spontaneous decomposition of $CaCO_3$ in the atmosphere, $\Delta_r G_m(T) < 0$ is necessary, that is:

$$\Delta_r G_m(T) = \Delta_r G_m^{\ominus}(T) + RT\ln Q < 0$$

$$\Delta_r G_m^{\ominus}(T) \approx \Delta_r H_m^{\ominus}(298.15\text{K}) - T\times\Delta_r S_m^{\ominus}(298.15\text{K})$$
$$\approx 179.4 - T\times163.9\times0.001$$

$$\Delta_r G_m(T) = \Delta_r G_m^{\ominus}(T) + RT\ln Q$$
$$= (179.4 - T\times163.9\times0.001) + 8.314\times0.001\times T\ln0.0003$$

Make $\Delta_r G_m(T) < 0$, get $T > 775.6$K.

It should be pointed out that when $\Delta_r G_m(T) > 0$, it only means that the reaction can not be spontaneous under specific conditions, which does not mean that the reaction can not be spontaneous under other conditions. If we change the reaction temperature, pressure or composition to make $\Delta_r G_m(T) < 0$, then the reaction can still be spontaneous. For example, in the process of reaction, the product is constantly removed to reduce its partial pressure or concentration, or the reactant is increased to increase its partial pressure or concentration, so as to change the value of $\Delta_r G_m(T)$ to be less than zero; or by changing the temperature, then changing the balance constant, so that the value of $\Delta_r G_m(T)$ is reduced to less than zero, then the positive reaction can be spontaneous.

In addition, it should be pointed out that the equilibrium state is not that the chemical reaction stops, but that the concentration of reactants and products will not change due to the equal rate of forward reaction and reverse reaction. Once the external conditions (temperature, pressure, *etc.*) change, the equilibrium will be destroyed immediately, and the system will move to another equilibrium state. We are studying and trying to explore the deeper content of chemical equilibrium, that is, we should find the "joystick" that controls the direction of transformation of the reaction freely, so that the balance can be transformed to the favorable direction we are looking for.

6.3 Chemical Equilibrium

Under the same condition, the reaction that can be carried out both in the positive direction and in the reverse direction is called reversible reaction. Most chemical reactions are reversible. Under a given conditions, when the forward direction and reverse direction have equal rates, the reaction will reach equilibrium. Chemical equilibrium has two important characteristics: first, as long as the external conditions remain unchanged, the concentration or partial pressure of each substance in the reaction will stay constant after equilibrium. No matter how long it takes, this state will not change, and the products will not increase, that is to say, the reaction has reached the limit. Second, chemical equilibrium is dynamic. From a macro perspective, when the chemical reaction reaches the equilibrium state, the reaction seems to "stop"; but from a micro perspective, the forward and reverse reactions will never end, only their reaction rates are the same, so the concentration or partial pressure of each substance no longer changes with time.

6.3.1 The Equilibrium Constant

(1) The equilibrium constant $K_T^{\ominus}$

① The equilibrium constant $K_T^{\ominus}$ and $\Delta_r G_m^{\ominus}(T)$. At a certain temperature, when a chemical reaction reaches equilibrium state, $\Delta_r G_m(T) = 0$, according to formula (6-13), there are

$$\Delta_r G_m(T) = \Delta_r G_m^{\ominus}(T) + RT \ln Q_{eq} = 0$$

$$\Delta_r G_m^{\ominus}(T) = -RT \ln Q_{eq}$$

Where Q_{eq} is the reaction quotient in equilibrium.

Since the standard Gibbs function change $\Delta_r G_m^{\ominus}(T)$ of the reaction is a constant at a certain temperature, the reaction quotient Q_{eq} in the above equation is also a constant. Let this constant be $K_T^{\ominus}$, *i.e* $K_T^{\ominus} = Q_{eq}$

So

$$\Delta_r G_m^{\ominus}(T) = -RT \ln K_T^{\ominus} \tag{6-16}$$

Where $K_T^{\ominus}$ is called the standard equilibrium constant (also called the thermodynamic equilibrium constant), and its unit is 1.

The value of the equilibrium constant indicates the extent to which reactants are converted into products in a chemical reaction. If the equilibrium constant is large, it indicates that the amount of products present at equilibrium is much greater than the amount of reactants. A very large equilibrium constant, greater than 10^{10}, for example, means that for all intents and purposes the reaction goes to completion. Conversely, when K is very small, less than 10^{-10}, very little product is formed and virtually no visible reaction occurs. If $K = 1$, the equilibrium mixture contains approximately equal amounts of reactants and products. These general ideas allow us to quickly estimate the composition of an equilibrium mixture.

By using the equilibrium constant of a reaction and knowing the amount of reactants at the beginning, the amount of reactants and products and the conversion rate of reactants can be calculated. The conversion rate of a reactant refers to the percentage of the converted amount of the reactant in its initial amount.

② Expression of equilibrium constant $K_T^\ominus$. Q_{eq} is the reaction quotient in equilibrium. According to the formula (6-14) and (6-15) of the reaction quotient, as long as the relative partial pressure (or concentration) of each substance in equilibrium is substituted into the above formula, the expression of the standard equilibrium constant can be obtained.

For any reversible gas reaction:

$$aA(g) + bB(g) = gG(g) + dD(g)$$

At a given temperature, when the reaction reaches equilibrium, there are

$$K_T^\ominus = \frac{(p_G / p^\ominus)^g (p_D / p^\ominus)^d}{(p_A / p^\ominus)^a (p_B / p^\ominus)^b} \tag{6-17}$$

In the formula, p_A, p_B, p_G and p_D respectively represent the partial pressure of gaseous substances A, B, G and D under equilibrium conditions.

Reaction in solution:

$$aA(aq) + bB(aq) = gG(aq) + dD(aq)$$

$$K_T^\ominus = \frac{(c_G / c^\ominus)^g (c_D / c^\ominus)^d}{(c_A / c^\ominus)^a (c_B / c^\ominus)^b} \tag{6-18}$$

In the formula, c_A, c_B, c_G and c_D are the volume molar concentrations of substances A, B, G and D under equilibrium conditions.

For the expression of standard equilibrium constant, several points should be noted:

i. In the expression of standard equilibrium constant, the concentration (or partial pressure) of each substance is the relative concentration (or partial pressure) at equilibrium.

ii. $K_T^\ominus$ is related to temperature, but not to the partial pressure or concentration of the substance. Therefore, when writing the expression of the standard equilibrium constant, the temperature should be indicated generally. If the temperature is not indicated, it usually means that the temperature is equal to room temperature 298.15K.

iii. The solid or pure liquid substances in the reaction are not included in the standard equilibrium constant expression, or can be expressed as 1. For example, reaction

$$CO_2(g) + C(s) \Longrightarrow 2CO(g)$$

Its standard equilibrium constant expression is:

$$K_T^\ominus = \frac{(p_{CO} / p^\ominus)^2}{p_{CO_2} / p^\ominus}$$

iv. $K_T^\ominus$ is related to the formulation of reaction equation. Because of the different writing methods of reaction equation, the change value of standard Gibbs function $\Delta_r G_m^\ominus(T)$

of reaction is different, and by $\Delta_r G_m^\ominus(T) = -RT\ln K_T^\ominus$, the value of $K_T^\ominus$ is different.

For example, the reaction of SO_2 oxidation to SO_3, the reaction equation can be written as

$$2SO_2(g) + O_2(g) = 2SO_3(g) \qquad (1)$$

or as

$$SO_2(g) + \frac{1}{2}O_2(g) = SO_3(g) \qquad (2)$$

obviously $$K_{T(1)}^\ominus = [K_{T(2)}^\ominus]^2$$

v. If there are both gas and solution in the reaction, for gaseous substances, the relative pressure at equilibrium is included in the expression of standard equilibrium constant, and for solutions, the relative concentration at equilibrium is included in the expression.

For example, reaction aA(s) + bB(aq) $=$ gG(g) + dD(l), the expression of standard equilibrium constant is:

$$K_T^\ominus = \frac{(p_G / p^\ominus)^g}{(c_B / c^\ominus)^b}$$

The unit of $\Delta_r G_m^\ominus(T)$ is kJ/mol, while the unit of R is J/(mol·K). It should be paid attention to the unity of units when calculating.

(2) Calculation of standard equilibrium constant

① Calculation according to the composition of reaction system in equilibrium.

Example 6-9 It is known that at 400℃ and 10×101.325kPa, a synthetic ammonia reaction takes place. The volume ratio of hydrogen and nitrogen is 3 : 1. After the reaction reaches equilibrium, the volume percentage of ammonia is 3.9%. Calculate the standard equilibrium constant of ammonia synthesis reaction under this condition.

Solution The total pressure is

$$p_{\text{total}} = p_{NH_3} + p_{H_2} + p_{N_2} = 10 \times 101.325\text{kPa}$$

Because the volume percentage of gas is equal to its mole fraction, according to the Dalton's law of partial pressure , the partial pressure of NH_3 in the mixed gas at equilibrium is

$$p_{NH_3} = V_{NH_3} \times p_{\text{total}} = 3.9\% \times 10 \times 101.325 - 39.52\text{kPa}$$

The total pressure of H_2 and N_2 in the mixture is

$$p_{H_2} + p_{N_2} = p_{\text{total}} - p_{NH_3} = 10 \times 101.325 - 39.52 = 973.73\text{kPa}$$

Suppose the amount of N_2 participating in the reaction is x mol, according to the equation

$$N_2 + 3H_2 = 2NH_3$$

Ratio of moles at the beginning 1 : 3

Set value in the reaction x : $3x$

Ratio of moles at the equilibrium (1−x): (3−3x)

So

$$p_{H_2}=\frac{3}{4}(p_{H_2}+p_{N_2})=\frac{3}{4}\times 973.73=730.3(\text{kPa})$$

$$p_{N_2}=\frac{1}{4}(p_{H_2}+p_{N_2})=\frac{1}{4}\times 973.73-243.4(\text{kPa})$$

According to the expression of the standard equilibrium constant, it can be calculated

$$K_T^{\ominus}=\frac{(p_{NH_3}/p^{\ominus})^2}{(p_{N_2}/p^{\ominus})(p_{H_2}/p^{\ominus})^3}=\frac{(39.52/100)^2}{(243.4/100)(730.3/100)^3}=1.6\times 10^{-4}$$

② Calculate according to $\Delta_r G_m^{\ominus}(T)=-RT\ln K_T^{\ominus}$

Example 6-10 There is an important chemical reaction when processing steel parts at high temperature, as follows:

$$C(s)+CO_2(g) \Longrightarrow 2CO(g)$$

Calculation or estimation: the standard equilibrium constants $K_T^{\ominus}$ of this reaction at 298.15K and 1173K, and the significance of decarbonization at high temperature.

Solution Solid carbon is set as graphite

$$C(s)+CO_2(g) \Longrightarrow 2CO(g)$$

$$\Delta_r H_m^{\ominus}=2\Delta_f H_{m,CO}^{\ominus}-\Delta_f H_{m,CO_2}^{\ominus}=2\times(-110.5)-(-393.5)=172.5(\text{kJ/mol})$$

$$\Delta_r S_m^{\ominus}=2S_{m,CO}^{\ominus}-S_{m,CO_2}^{\ominus}-S_{m,C}^{\ominus}=2\times 197.6-5.7-213.6=175.9[\text{J}/(\text{mol}\cdot\text{K})]$$

(a) Calculate $\Delta_r G_m^{\ominus}$ of reaction at 298.15K and 1173K

$$\begin{aligned}\Delta_r G_m^{\ominus}(298.15\text{K})&=\Delta_r H_m^{\ominus}(298.15\text{K})-T\Delta_r S_m^{\ominus}(298.15\text{K})\\&=172.5-298.15\times 0.1759\\&=120.1(\text{kJ/mol})\end{aligned}$$

$$\begin{aligned}\Delta_r G_m^{\ominus}(1173\text{K})&=\Delta_r H_m^{\ominus}(298.15\text{K})-T\Delta_r S_m^{\ominus}(298.15\text{K})\\&=172.5-1173\times 0.1759\\&=-33.8\ (\text{kJ/mol})\end{aligned}$$

(b) Calculate $K^{\ominus}$ of reaction at 298.15K and 1173K

T=298.15K, $\ln K^{\ominus}=-\Delta_r G_m^{\ominus}(298.15\text{K})/RT$

$$=-120.1\times 1000/(8.314\times 298)=-48.45$$

$$K^{\ominus}=9.1\times 10^{-22}$$

T=1173K, $\ln K^{\ominus}=-\Delta_r G_m^{\ominus}(1173\text{K})/RT$

$$=33.8\times 1000/(8.314\times 1173)=3.466$$

$$K^{\ominus} = 32$$

The results show that when the reaction temperature increases from room temperature to higher temperature, the algebraic value of $\Delta_r G_m^{\ominus}$ decreases rapidly, the reaction changes from non spontaneous to spontaneous, and the $K^{\ominus}$ value of the reaction increases significantly. If there is CO_2 in the processing atmosphere at high temperature, CO_2 will react with carbon (in the form of graphite carbon or Fe_3C) on the surface of steel parts, resulting in decarburization and oxidation.

③ Using the relationship between the standard equilibrium constants at different temperatures

The relationship between the change of temperature and the standard equilibrium constant of the reaction can be derived from equations (6-7) and (6-16). For any given constant temperature and pressure chemical reaction, there are

$$\Delta_r G_m^{\ominus}(T) = -RT \ln K_T^{\ominus}$$

$$\Delta_r G_m^{\ominus}(T) = \Delta_r H_m^{\ominus} - T\Delta_r S_m^{\ominus}$$

The above two formulas are subtracted from each other and sorted out,

$$\ln K_T^{\ominus} = -\frac{\Delta_r H_m^{\ominus}}{RT} + \frac{\Delta_r S_m^{\ominus}}{R}$$

If the standard equilibrium constant of a reversible reaction is $K_{T_1}^{\ominus}$ at temperature T_1 and $K_{T_2}^{\ominus}$ at temperature T_2, then

$$\ln K_{T_1}^{\ominus} = -\frac{\Delta_r H_m^{\ominus}}{RT_1} + \frac{\Delta_r S_m^{\ominus}}{R}$$

$$\ln K_{T_2}^{\ominus} = -\frac{\Delta_r H_m^{\ominus}}{RT_2} + \frac{\Delta_r S_m^{\ominus}}{R}$$

Since $\Delta_r H_m^{\ominus}$ and $\Delta_r S_m^{\ominus}$ have little change with temperature, they can be approximately regarded as constants, and the two formulas can be reduced to:

$$\ln K_{T_2}^{\ominus} - \ln K_{T_1}^{\ominus} = \ln\frac{K_{T_2}^{\ominus}}{K_{T_1}^{\ominus}} = \frac{\Delta_r H_m^{\ominus}}{R}\left(\frac{1}{T_1} - \frac{1}{T_2}\right)$$

or

$$\ln\frac{K_{T_2}^{\ominus}}{K_{T_1}^{\ominus}} = \frac{\Delta_r H_m^{\ominus}}{R} \times \left(\frac{T_2 - T_1}{T_2 T_1}\right) \qquad (6\text{-}19)$$

Example 6-11 It is assumed that $\Delta_r H_m^{\ominus}$ of reaction

$$2SO_3\,(g) = 2SO_2\,(g) + O_2(g)$$

does not change with temperature. Calculate the standard equilibrium constant of the reaction at 100kPa, 600℃ according to the following data.

	$2SO_3(g) \rightleftharpoons$	$2SO_2(g)$ +	$O_2(g)$
$\Delta_f H_m^\ominus$/(kJ/mol)	−395.7	−296.8	0.0
$\Delta_f G_m^\ominus$/(kJ/mol)	−371.1	−300.1	0.0

Solution

$$\Delta_r H_m^\ominus(298\text{K}) = 2\Delta_f H_m^\ominus(SO_2) + \Delta_f H_m^\ominus(O_2) - 2\Delta_f H_m^\ominus(SO_3)$$
$$= 2\times(-296.8) + 0 - 2\times(-395.7) = 197.8(\text{kJ/mol})$$

$$\Delta_r G_m^\ominus(298\text{K}) = 2\Delta_f G_m^\ominus(SO_2) + \Delta_f G_m^\ominus(O_2) - 2\Delta_f G_m^\ominus(SO_3)$$
$$= 2\times(-300.1) + 0 - 2\times(-371.1) = 142(\text{kJ/mol})$$

According to $\Delta_r G_m^\ominus(T) = -RT\ln K_T^\ominus$

$$\ln K_T^\ominus = \frac{-\Delta_r G_m^\ominus}{RT} = \frac{-142\times10^3}{8.314\times298} = -57.31$$

$$K_{298}^\ominus = 1.29\times10^{-25}$$

According to Eq. (6-19), at 100kPa, 600℃,

$$\ln\frac{K_{298}^\ominus}{K_{873}^\ominus} = \frac{\Delta_r H_m^\ominus}{R}\times\left(\frac{T_1-T_2}{T_1T_2}\right) = \frac{197.8\times10^3}{8.314}\times\frac{298-873}{298\times873} = -52.58$$

$$\frac{K_{298}^\ominus}{K_{873}^\ominus} = 1.46\times10^{-23}$$

$$K_{873}^\ominus = \frac{1.29\times10^{-25}}{1.46\times10^{-23}} = 8.84\times10^{-3}$$

6.3.2 The Main Factors Affecting the Chemical Equilibrium

All equilibrium is conditional, and chemical equilibrium can only be maintained under certain conditions. If the conditions affecting the equilibrium change, the equilibrium state will change accordingly, that is, the original equilibrium will be destroyed, and a new equilibrium will be established. This process of transition from one equilibrium state to another is called the shift of chemical equilibrium. Three factors will affect the equilibrium, they are concentration, pressure and temperature.

(1) Concentration and total pressure

When the reversible reaction reaches equilibrium and the concentration of substance or the total pressure of gas changes, the shift of chemical equilibrium can be determined according to $\Delta_r G_m^\ominus(T)$ in the reaction process. Because from formular (6-13) and (6-16), we can get

$$\Delta_r G_m = -RT\ln K_T^\ominus + RT\ln Q = RT\ln\left(\frac{Q}{K_T^\ominus}\right) \tag{6-20}$$

Q is the reaction quotient. It has similar formation with $K_T^\ominus$, with equilibrium concentrations replaced by concentrations at any moment.

$Q = K_T^\ominus$, Q won't change, $\Delta_r G_m(T) = 0$. The reaction is already at equilibrium.

$Q < K_T^\ominus$, Q will increase, $\Delta_r G_m(T) < 0$. The reaction goes forward.

$Q > K_T^\ominus$, Q will decrease, $\Delta_r G_m(T) > 0$. The reaction goes backward.

① Concentration. At constant temperature, a solution reacts as

$$a\text{A(aq)}+b\text{B(aq)} \xlongequal{\quad} g\text{G(aq)}+d\text{D(aq)}$$

when the reaction reaches equilibrium, $\Delta_r G_m(T) = 0$, there are

$$Q_{eq} = K_T^\ominus = \frac{(c_G/c^\ominus)^g (c_D/c^\ominus)^d}{(c_A/c^\ominus)^a (c_B/c^\ominus)^b}$$

When the reactant concentration increases from c_A and c_B to c'_A and c'_B, since $c'_A > c_A$, $c'_B > c_B$, it can be known that

$$Q = \frac{(c_G/c^\ominus)^g (c_D/c^\ominus)^d}{(c_A/c^\ominus)^a (c_B/c^\ominus)^b} < K_T^\ominus$$

From formula (6-20), we can get

$$\Delta_r G_m = RT\ln\left(\frac{Q}{K_T^\ominus}\right) < 0$$

It means that the reaction goes forward.

It can be seen that when the reactant concentration increases at constant temperature, the reaction will shift to the right to produce more products. On the contrary, when the product concentration increases, $Q > K_T^\ominus$, then $\Delta_r G_m(T) > 0$, so the reaction will shift to the the left, *i.e.* towards the reactant.

Example 6-12 Water gas conversion reaction

$$CO(g)+H_2O(g) \xlongequal{\quad} CO_2(g)+H_2(g)$$

It is known that when the temperature is 1073K, $K^\ominus = 1.0$. If 100kPa CO gas and 300kPa water vapor are introduced into a constant volume closed vessel to make the reaction, try to determine the partial pressure of each gas and the conversion rate of CO when the reaction reaches the equilibrium state. In the above-mentioned equilibrium reaction system, keep the temperature and volume unchanged, add water vapor to increase its pressure by 400kPa, and determine the direction of equilibrium shift through calculation.

Solution Because it is a constant volume reaction, according to $pV = nRT$, ($\Delta pV = \Delta nRT$), the ratio of the change value of the partial pressure of each substance is equal to the ratio of the corresponding stoichiometric number, *i.e.*, assuming that $p_{CO_2} = p_{H_2} = x$ in equilibrium, we can get

	$CO(g) +$	$H_2O(g)$	$\rightleftharpoons CO_2(g)$	$+ H_2(g)$
Initial pressure value /kPa	100	300	0	0
Change pressure value /kPa	$-x$	$-x$	$+x$	$+x$
Equilibrium pressure value/kPa	$100-x$	$300-x$	x	x

$$K^{\ominus} = \frac{\left(\frac{p_{H_2}}{p^{\ominus}}\right)\left(\frac{p_{CO_2}}{p^{\ominus}}\right)}{\left(\frac{p_{CO}}{p^{\ominus}}\right)\left(\frac{p_{H_2O}}{p^{\ominus}}\right)}$$

$$= \frac{(x/100)^2}{[(100-x)/100][(300-x)/100]} = \frac{x^2}{(100-x)(300-x)} = 1.0$$

get $x = 75\text{kPa}$

Therefore, the partial pressure of H_2 and CO_2 at equilibrium is $p_{H_2} = p_{CO_2} = 75\text{kPa}$

The partial pressure of CO: $p_{CO} = 25\text{kPa}$

The partial pressure of H_2O: $p_{H_2O} = 225\text{kPa}$

The conversion rate of CO $= \frac{75}{100} \times 100\% = 75\%$

After adding water vapor, $p_{H_2O} = 225 + 400 = 625\text{kPa}$

$$Q \approx \frac{(75/100)^2}{(25/100)(625/100)} = 0.36$$

② Total pressure. For the reaction between liquid and solid, the change of total pressure has no effect on the equilibrium approximately, because the pressure has little effect on the volume of liquid or solid matter. Therefore, in the following discussion of the effect of total pressure on chemical equilibrium, only reactions involving gaseous substances are considered.

At a constant temperature, a gas reaction

$$aA(g) + bB(g) \rightleftharpoons gG(g) + dD(g)$$

when the reaction reaches equilibrium, $\Delta_r G_m(T) = 0$, there are

$$Q_{eq} = K_T^{\ominus} = \frac{(p_G/p^{\ominus})^g (p_D/p^{\ominus})^d}{(p_A/p^{\ominus})^a (p_B/p^{\ominus})^b}$$

When the total pressure changes m times ($m > 1$ when the total pressure increases and $0 < m < 1$ when the total pressure decreases), the partial pressure of each gas will change m times according to the Dalton's law of partial pressure. Now we can get

$$Q=\frac{(mp_{\mathrm{G}}/p^{\ominus})^{g}(mp_{\mathrm{D}}/p^{\ominus})^{d}}{(mp_{\mathrm{A}}/p^{\ominus})^{a}(mp_{\mathrm{B}}/p^{\ominus})^{b}}=\frac{(p_{\mathrm{G}}/p^{\ominus})^{g}(p_{\mathrm{D}}/p^{\ominus})^{d}}{(p_{\mathrm{A}}/p^{\ominus})^{a}(p_{\mathrm{B}}/p^{\ominus})^{b}}m^{(g+d)-(a+b)}=K_{T}^{\ominus}m^{\Delta n}$$

Namely:
$$Q=K_{T}^{\ominus}m^{\Delta n} \tag{6-21}$$

Where $(a+b)$ is the total number of reactant gas molecules and $(g+d)$ is the total number of product molecules. It can be seen from the above formula that when the total pressure increases by m times, if:

(i) $(a+b)>(g+d)$, *i.e.* $\Delta n<0$, we can get $Q<K_{T}^{\ominus}$, then $\Delta_{\mathrm{r}}G_{\mathrm{m}}(T)<0$. At this time, the equilibrium shifts to the right, *i.e.* to the direction where the total number of gas molecules is less;

(ii) $(a+b)<(g+d)$, *i.e.* $\Delta n>0$, we can get $Q>K_{T}^{\ominus}$, then $\Delta_{\mathrm{r}}G_{\mathrm{m}}(T)>0$. At this time, the equilibrium shifts to the left, *i.e.* to the direction where the total number of gas molecules is less;

(iii) $(a+b)=(g+d)$, *i.e.* $\Delta n=0$, we can get $Q=K_{T}^{\ominus}$, then $\Delta_{\mathrm{r}}G_{\mathrm{m}}(T)=0$. At this time, the reaction is in the original equilibrium state, that is, the chemical equilibrium does not move.

It can be seen that when the total pressure is increased at constant temperature, the equilibrium moves towards the direction of decreasing the total number of gas molecules. Similarly, when the total pressure is reduced, the equilibrium will move in the direction of increasing the total number of molecules. For the same total number of gas molecules ($\Delta n=0$) before and after the reaction, the equilibrium does not move no matter the pressure is increased or decreased.

Example 6-13 Synthetic ammonia reaction

$$N_2+3H_2 = 2NH_3$$

When the reaction reaches equilibrium at a certain temperature, if the total pressure of the equilibrium system is reduced to half of the original value, analyze and judge how the chemical equilibrium will move according to equation (6-21).

Solution In the chemical equilibrium state, set the partial pressure of each component to be $p_{\mathrm{H_2}}$, $p_{\mathrm{N_2}}$, $p_{\mathrm{NH_3}}$. When the total pressure is reduced to half of the original, *i.e.* $m=1/2$, according to formula (6-21), we can get

$$Q=\frac{(p_{\mathrm{NH_3}}/p^{\ominus})^{2}}{(p_{\mathrm{N_2}}/p^{\ominus})\,(p_{\mathrm{H_2}}/p^{\ominus})^{3}}\left(\frac{1}{2}\right)^{2-(1+3)}=4K_{T}^{\ominus}$$

That is
$$Q>K_{T}^{\ominus},\quad \Delta_{\mathrm{r}}G_{\mathrm{m}}>0$$

So the equilibrium moves to the left (*i.e.* to the direction where the total number of gas molecules increases).

Example 6-14 Conversion reaction of CO and H_2O in water gas

$$CO(g)+H_2O(g) \longrightarrow CO_2(g) + H_2(g)$$

At a certain temperature, the reaction has reached equilibrium. Judge the moving direction of chemical equilibrium when the total pressure of the system is doubled.

Solution

$$\Delta n = (1+1)-(1+1) = 0$$

$$Q = K_T^{\ominus} m^{\Delta n} = K_T^{\ominus}$$

$$\Delta_r G_m = 0$$

So the system is still in equilibrium.

Changing the concentration of a gas is actually equivalent to changing the pressure of the gas, so for a reaction involving a gas, changing the pressure of a substance and changing the concentration of a substance have the same effect on the direction of equilibrium movement.

(2) Temperature

The effect of temperature on the equilibrium system is essentially different from that of concentration and pressure. After the chemical reaction reaches equilibrium, changing the concentration or pressure does not change the equilibrium constant $K_T^{\ominus}$, but changes the reaction quotient Q to make $Q \neq K_T^{\ominus}$, leading to the equilibrium movement; changing the temperature is to change $K_T^{\ominus}$ to make $Q \neq K_T^{\ominus}$, leading to the equilibrium movement.

By formula (6-19)

$$\ln\frac{K_{T_2}^{\ominus}}{K_{T_1}^{\ominus}} = \frac{\Delta_r H_m^{\ominus}}{R} \times \left(\frac{T_2 - T_1}{T_2 T_1}\right)$$

It can be concluded that:

(i) If the reaction is endothermic, *i.e.* $\Delta_r H_m^{\ominus} > 0$, when the temperature rises ($T_2 > T_1$), the equilibrium constant becomes larger ($K_{T_2}^{\ominus} > K_{T_1}^{\ominus}$), while the original equilibrium is $Q = K_T^{\ominus}$, so $Q < K_T^{\ominus}$, the equilibrium moves to the right (*i.e.* the endothermic direction).

(ii) If the reaction is exothermic, *i.e* $\Delta_r H_m^{\ominus} < 0$, when the temperature rises, the equilibrium constant decreases ($K_{T_2}^{\ominus} < K_{T_1}^{\ominus}$), then the equilibrium moves to the left (*i.e.* the direction of heat absorption).

It can be seen that when the temperature is increased, the equilibrium moves to the direction of heat absorption; otherwise, when the temperature is reduced, the equilibrium moves to the direction of heat release.

Example 6-15 In the chemical heat treatment, the following reactions exist in the high temperature gas carburizing process: $2CO(g) \longrightarrow C(s) + CO_2(g)$. Calculation or estimation:

the standard equilibrium constants $K_T^\ominus$ of this reaction at 298.15K and 1173K, and the significance of decarbonization at high temperature.

Solution Look up the relevant data of each substance at 298K in the appendix, and calculate:

$$\Delta_r H_m^\ominus(298\text{K}) = -172.5\text{kJ/mol}$$

$$\Delta_r S_m^\ominus(298\text{K}) = -175.9\text{J/(mol}\cdot\text{K)}$$

At 298K,

$$\Delta_r G_m^\ominus(298\text{K}) = \Delta_r H_m^\ominus(298\text{K}) - T\Delta_r S_m^\ominus(298\text{K})$$
$$= -172.5 - 298\times0.001\times(-175.9)$$
$$= -120.1(\text{kJ/mol})$$

From $\Delta_r G_m^\ominus(T) = -RT\ln K_T^\ominus$

$$\ln K_T^\ominus = \frac{-\Delta_r G_m^\ominus}{RT} = \frac{120.1\times10^3}{8.314\times298} = 48.5$$

$$K_{298}^\ominus = 1.1\times10^{21}$$

At 1173K, from Eq. (6-19)

$$\ln\frac{K_{1173}^\ominus}{K_{298}^\ominus} = \frac{\Delta_r H_m^\ominus}{R}\times\left(\frac{T_2 - T_1}{T_2 T_1}\right) = \frac{-172.5\times10^3}{8.314}\times\frac{1173-298}{1173\times298}$$

$$K_{1173}^\ominus / K_{298}^\ominus = 2.8\times10^{-23}$$

$$K_{1173}^\ominus = 3.1\times10^{-2}$$

The calculation of thermodynamics provides a theoretical basis for carburizing process. According to the above calculation results, the standard equilibrium constant of this reaction decreases sharply at high temperature, and the equilibrium moves strongly to the left, which is not conducive to carburizing. Therefore, a lower carburizing temperature should be selected. When the reaction temperature is lower than 273K, the amount of carburization is large, but the reaction rate is slow; when the reaction temperature is higher than 1173K, the amount of carburization is too small, which is not suitable.

In addition, please compare the difference and connection between Example 6-15 and Example 6-10, understand the change of thermodynamic state function of forward and reverse reaction, as well as the selection basis of carburizing and decarbonizing process conditions.

The effects of concentration, total pressure and temperature on the equilibrium were discussed. If the concentration of reactants is increased in the equilibrium system, the equilibrium will move to the direction of reducing the concentration of reactants; if the total pressure of the equilibrium system is increased, the equilibrium will move to the direction of reducing the total number of gas molecules, that is to say, under the condition of constant volume, the equilibrium will move to the direction of reducing the total pressure. If the temperature is increased (heated), the equilibrium moves in the direction of the temperature reduction (heat absorption). In a word, the law of equilibrium movement can be summarized as follows: if one of the conditions of changing the equilibrium system is added, such as concentration, total pressure or temperature, the equilibrium will move in the direction of weakening the change. This law is called Le Châtelier's principle.

It should be noted that the principle of equilibrium movement applies only to systems that are in equilibrium but not to non-equilibrium systems.

In the actual production, it is often necessary to consider the two factors of speed and balance, and choose the most appropriate conditions. For example, in the reaction of SO_2 to SO_3, because of exothermic reaction, in terms of balance, if the temperature is reduced, the percentage of SO_2 to SO_3 in the system can be increased. However, when the temperature is low, the reaction rate is small and the time needed to reach the equilibrium is long. Therefore, the temperature should not be too low or too high. The temperature should be controlled in an appropriate range according to the specific conditions. At present, in the process of sulfuric acid production by contact method, the SO_2 conversion temperature is controlled at 400-500℃. In terms of pressure, increasing the total pressure can improve the conversion rate of SO_2, and increasing the total pressure is also beneficial to increase the reaction rate. However, the conversion rate of SO_2 under atmospheric pressure has been very high, and a lot of power needs to be consumed under pressure, and the equipment, materials and operation requirements are much more complex, so atmospheric pressure conversion is currently used in production. In addition, excessive oxygen (air) was added and V_2O_5 was used as catalyst.

Example 6-16 The pressure of water vapor is 611Pa at 0℃. Calculation: heat of vaporization of water, steam pressure of water at 50℃.

Solution $$H_2O(l) = H_2O(g)$$

(i) It is known that T_1 = 273k, p_1 = 611Pa; T_2 = 373k (when water is boiling), p_2 = 101325Pa.

By $$K_T^\ominus = p_{H_2O} / p^\ominus$$

$$K_{T_1}^\ominus = p_1 / p^\ominus = 611/(100\times10^3) = 6.11\times10^{-3}$$

$$K_{T_2}^\ominus = p_2 / p^\ominus = 101325/(100\times10^3) = 1.01325$$

According to the formula (6-19) $$\ln\frac{K_{T_2}^\ominus}{K_{T_1}^\ominus} = \frac{\Delta_r H_m^\ominus}{R}\times\left(\frac{T_2-T_1}{T_2T_1}\right)$$

$$\ln\frac{6.11\times10^{-3}}{1.01325} = \frac{\Delta_r H_m^\ominus}{8.314}\frac{(273-373)}{273\times373} = -5.11$$

$$\Delta_r H_m^\ominus(T) = 43261\text{J/mol} = 43.26\text{kJ/mol}$$

(ii) $T_1 = 273+50 = 323\text{K}$, $p_1 = p_{\text{H}_2\text{O},323\text{K}}$, $T_2=373\text{K}$, $p_2=101325\text{Pa}$

$$\ln\frac{p_1/p^\ominus}{101325/p^\ominus} = \frac{43.26\times10^3}{8.314}\times\frac{323-373}{323\times373} = -2.16$$

$$p_1=11679\text{Pa}$$

That is, the steam pressure of water is 11679Pa at 50℃.

6.4 Chemical Reaction Rate

First, let's look at the following two examples.

Example A Formation reaction of water:

$$H_2(g) + \frac{1}{2}O_2(g) = H_2O(l) \qquad \Delta_r G_m^\ominus = -285\text{kJ/mol}(<<0)$$

$$H^+(aq) + OH^-(aq) = H_2O(l) \qquad \Delta_r G_m^\ominus = -79.9\text{kJ/mol}$$

The spontaneous trend of the former reaction is much greater than that of the latter. If we put hydrogen and oxygen in a closed container at room temperature, we can't detect the formation of water in the past many years, that is, the reaction rate is too small; while the spontaneous trend of neutralization of hydrochloric acid and sodium hydroxide is relatively small, but it can be completed in an instant. How can the difference of reaction rate be explained by the basic principle of chemical reaction?

Example B: A set of equipment for esterification of acetic acid and butanol to produce butyl acetate has accumulated a large number of basic tests and production data after one year's trial operation. It is planned to increase the reaction temperature to speed up the production progress. Can we predict the reaction time needed to reach the specified conversion rate through the existing test data? Can the cooling water system of the unit ensure the heat balance?

To solve these problems, we need to discuss the mechanism of chemical reaction, the rate of chemical reaction and its influencing factors from the point of view of chemical kinetics. The mechanism of chemical reaction is to discuss how the reaction happened and how it went on. Generally speaking, the reactant molecules do not react directly to produce products as soon as they contact each other. Instead, they need to go through a number of reaction steps and generate a number of intermediate substances before they can be gradually transformed into products. The research of chemical reaction rate mainly includes: the mathematical concept and measurement method of reaction rate, the influence of various factors (including reactant concentration, temperature, catalyst, medium, light, sound, *etc.*) on chemical reaction rate.

The value of studying chemical kinetics is obvious. For example, in the production practice, people always hope that the chemical reaction will be carried out according to the expected path and rate. The so-called pathway refers to obtaining as many main products as possible to inhibit the formation of by-products; the so-called rate refers to the completion of the reaction in the desired time. Therefore, in order to control the reaction and obtain satisfactory products, it is necessary to study chemical kinetics.

6.4.1 Reaction Mechanisms

We know that the essence of chemical reaction is the fracture of old bond and the formation of new bond. However, how do old bonds break and how do new bonds form? At present, there are many theories describing the process of chemical reactions, one of which is the collision theory, the other is the transition state theory or activated complex theory.

(1) Collision theory

Matter molecules are always in constant thermal motion. According to the molecular energy distribution of a certain gas, as shown in Fig.6-2, raising T is like stretching a rubber graph. The same number of molecules are spread out over a wider range of energy. The area under each curve is the same. Each curve is for the same total number of molecules, The relationship between the relative number of molecules with higher energy n (the right half of the curve) and the temperature T is approximately in accordance with the Boltzmann distribution, *i.e*

$$n = Z\exp(-E / kT) \qquad (6\text{-}22)$$

Where, n is the percentage of molecule with certain energy E in the system, Z is a proportional constant, E is a certain energy, T is the thermodynamic temperature (K), k is the Boltzmann constant (k=1.3806503×10^{-23}J/K).

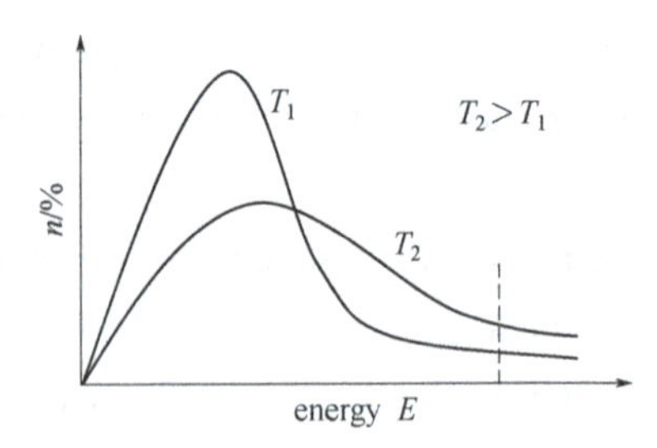

Fig.6-2 Schematic diagram of molecular energy distribution of gas

According to collision theory, molecules can only react through collision. But it's not just collisions that can react. We know that the occurrence of chemical reaction is a process of old bond breaking and new bond forming. The old bond breaking is in the front and the new bond forming is in the back. Therefore, only the collision between high-energy molecules can destroy the old chemical bond, and then form a new chemical bond and take place chemical reaction. We call the collision that can take place chemical reaction as effective collision. The high energy molecules that have effective collisions are called activated molecules. In fact, in order to meet the energy factor of breaking chemical bond, it is necessary to have an effective collision in a certain geometric orientation. When the chemical reaction occurs, the old bond does not break completely in the collision process, but first forms an intermediate substance called an activator.

In short, according to the collision theory, reactant molecules must have enough minimum energy and collide with each other in a proper direction to lead to effective collision. In general, a higher proportion of active molecules leads to a faster chemical reaction rate by increasing the probability of effective collision.

It has been shown that there are few chances of effective collision (*i.e.* the collision that leads to chemical bond breaking and chemical reaction) when several specific molecules collide at the same time. Such as the following reactions (1):

$$2NO + 2H_2 \xlongequal{} N_2 + 2H_2O \tag{1}$$

The probability of two NO molecules and two H_2 molecules (four molecules in total) colliding at the same time is much smaller than the probability of two NO molecules colliding with one H_2 molecule. Therefore, the reaction first generates N_2 and intermediate material H_2O_2, and then H_2O_2 collides with another H_2 molecule to generate two H_2O molecules, *i.e.* reaction (2):

$$2NO + H_2 \xlongequal{} N_2 + H_2O_2 \tag{2}$$

and reaction (3):

$$H_2 + H_2O_2 \xlongequal{} 2H_2O \tag{3}$$

The above reactions (2) and (3) are all completed by one-step collision of reactant molecules. In chemical kinetics, this "one-step reaction" is called elementary reaction, which is also called simple reaction. If a reaction is completed by more than two elementary reactions, it is called complex reaction, such as the above-mentioned reaction (1). The number of reactants in the elementary reaction is called the molecularity. The reaction with only one reactant molecule is called unimolecular reaction, and the one with two reactant molecules is called bimolecular reaction, and so on. Therefore, reaction (2) is a trimolecular reaction and reaction (3) is a bimolecular reaction. Studies have shown that there are few trimolecular reactions and no more than three reactions have been found.

The thermodynamic definition of activation energy is the difference between the average energy of reactant molecules in which effective collision occurs and the average energy of all molecules in the system. Let the activation energy of a chemical reaction be E_a. According to formula (6-22), the percentage of the activated molecules at a certain temperature is directly proportional to exp $(-E_a/RT)$. Here, the activation energy E_a is in kJ/mol, R is the molar gas constant, $R = k \times N_0 = 1.3806503 \times 10^{-23}(J/K) \times 6.02214199 \times 10^{23} mol^{-1} \approx 8.3145 J/(K \cdot mol)$, where k and N_0 are Boltzmann constant and Avogadro constant respectively. Obviously, for a given chemical reaction, the activation energy is a certain value, but because T is in the index, when the temperature increases, the number of activated molecules will increase sharply, resulting in a greatly accelerated reaction rate.

The change of molecular energy during the reaction is shown in Fig.6-3. In the figure, E_1 represents the average energy of reactant molecule, E_2 represents the average energy of product molecule, and E_x represents the average energy of intermediate substance (activator). Obviously, $E_{a_1} = E_x - E_1$ is the activation energy of the positive reaction; $E_{a_2} = E_x - E_2$ is the activation energy of the inverse reaction. It can also be seen in the figure that the enthalpy change of chemical reaction is:

$$\Delta_r H_m = E_2 - E_1 = E_{a_1} - E_{a_2}$$

That is, the change of reaction enthalpy is equal to the difference of activation energy between the positive and negative reactions.

The activation energy of general chemical reaction is between 60-240kJ/mol. Of course, the smaller the activation energy of the reaction, the smaller the energy needed for effective collision; the reaction rate is greater.

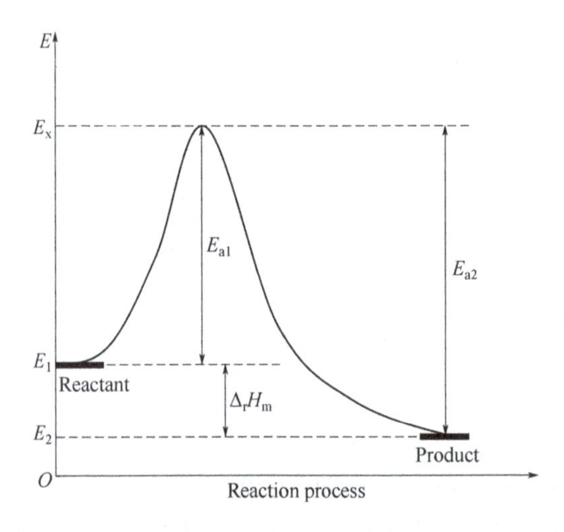

Fig.6-3 Reaction process potential energy diagram

(2) Transition state theory

In the 1930s, with the development of quantum mechanics and statistical mechanics, Eyring *et al.* put forward the transition state theory of reaction rate, also known as activation complex theory. This theory focuses on the study of the relationship between the energy of various interactions and molecular structure in the process of reactant molecules approaching each other and molecular valence bond rearrangement. It is considered that in the process of reactant molecules approaching each other and valence bond rearrangement, an intermediate transition state is formed first, which is called activated complex, and then decomposed into product molecules. This transition state is similar to the “activator” in collision theory. The rate at which reactant molecules pass through the transition state is the reaction rate.

The transition state theory mainly studies the structure, formation and decomposition of activated complexes, which is based on the potential energy surface of chemical reactions. The shape of the simple potential energy surface is like a saddle, as shown in Fig.6-4. The potential energy surface describes the change of the energy of the system with its position when the close molecules or atoms are in different positions in space, indicating the relationship between the potential energy of the whole molecule and the relative positions of the atoms in the molecule. Figuratively speaking, the simple potential energy surface is like the ground surface near a ridge between two mountains. The reactant energy on one side of the ridge needs to be increased, and the reactant can become a product only after crossing the ridge. In the process of the reactant molecules approaching each other, the lowest energy path (*i.e.* a certain spatial geometric orientation, which is equivalent to moving from the depression rather than from the hillside to the ridge) leads to the transition state. The transition state has the lowest energy on the ridge, that is, the state with the lowest height on the ridge, which is commonly called the saddle point on the potential energy surface. However, as the saddle point is still on the mountain ridge, compared with the reactants and products at the foot of the mountain, it is always in a high-energy state, very unstable and easy to slide down the mountain ridge. If it slides to one side of the reactant, the transition state returns to the original reactant; if it slides to the other side, it becomes a product and initiates a chemical reaction.

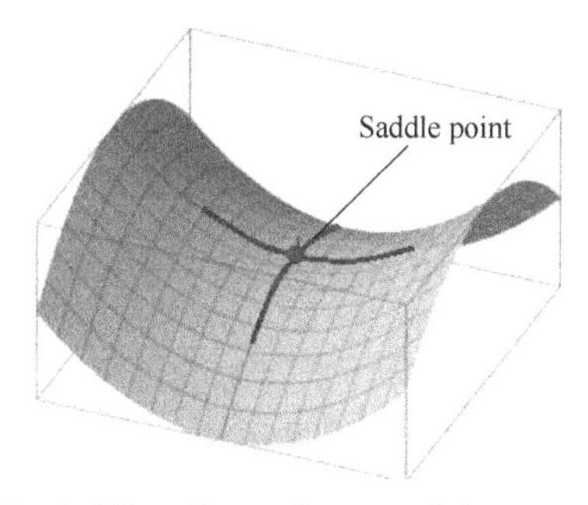

Fig.6-4 Saddle point on the potential energy surface

The geometry of transition state can be determined by quantum chemistry. This is because at the saddle point, the potential energy surface only bulges in the only direction along the depression, while it is concave in other directions. As a criterion, the position of the saddle point can be determined, and the geometry of the molecule corresponding to the transition state can be obtained. Quantum chemistry method can also calculate the lowest energy path along the mountain depression (called intrinsic reaction coordinates) from the transition state, and then confirm which reactants and products (or intermediate products) are connected along the transition state, so as to confirm the elementary reaction and reaction process.

In recent years, quantum chemistry combined with statistical mechanics and other methods, coupled with modern experimental techniques, is widely used by scientists, and has revealed many chemical reaction mechanisms.

It should be understood that a large number of experimental facts show that most chemical reactions are not simple ones, but complex ones. The so-called "reaction mechanism" is to decompose a specific complex reaction into a series of combinations of elementary reactions according to its reaction process. But up to now, there are only a few chemical reactions mechanisms that can be fully understood. This is mainly because the experimental technology for studying the reaction mechanism can not meet the requirements and can not directly track the chemical reaction, that is, it can not be observed how the reaction happens step by step. It is not difficult to understand this point, because the linear size of the molecule or atom we studied is about 10^{-8}cm, and the actual time required for the intermolecular reaction is about 10^{-13}s. To directly observe the reaction process between one molecule and another, we need the experimental means of linear size resolution of 10^{-9}cm and time resolution of 10^{-14}s. Before the emergence of laser, the time resolution can only reach the order of millisecond. After the advent of laser, it can now reach the order of femtosecond (10^{-15}s), which makes it possible to directly observe the most basic dynamic process in chemical reaction.

The application of molecular beam and laser technology not only makes it possible to observe the dynamic behavior of chemical process at the molecular level, but also to study the rate and micro process of the product from one quantum state to another. Therefore, chemical kinetics has entered the level of state-state chemistry, which plays an important role in understanding the micro mechanism of chemical reaction and controlling the chemical reaction and reaction pathway better.

6.4.2 Effect of Concentration on Reaction Rate

(1) Reaction rate

Chemical reaction rate refers to the change of concentration of reactant or product in unit time, that is, the average rate of reaction. However, the rate of most chemical reactions will change with time even when the external conditions are not changed. Therefore, the instantaneous rate of reaction is usually expressed in the form of derivative $\mathrm{d}c/\mathrm{d}t$. For example, there is the simplest single molecule reaction: A → B. The curve of concentration c of substances A and B with time t is shown in Fig.6-5. The slope of the tangent of a point in the graph is equal to the instantaneous rate of the reaction at time t.

At present, the rate of change of reaction progress ξ with time is generally used to define the reaction rate, which is called conversion rate. Namely

$$r = \frac{\mathrm{d}\xi}{\mathrm{d}t} \quad (\mathrm{mol/s})$$

By introducing the definition of reaction progress, $\mathrm{d}\xi = \mathrm{d}n_\mathrm{B}/\nu_\mathrm{B}$, we get:

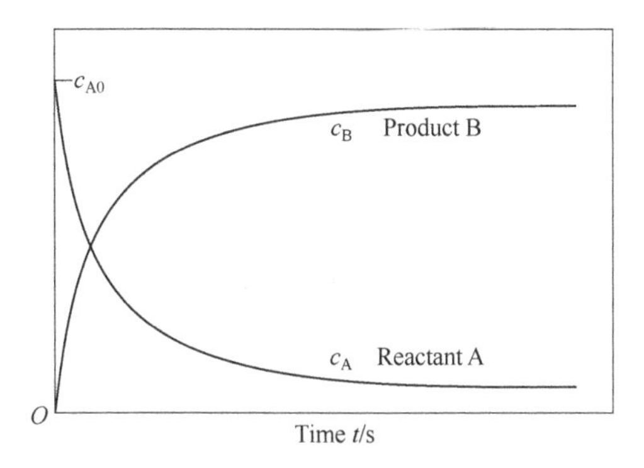

Fig.6-5 Change of concentration with reaction time

$$r = \frac{1}{\nu_B}\frac{dn_B}{dt} \text{(mol/s)} \tag{6-23}$$

Because it is not convenient to measure the change of substance content in the reaction system than to measure the change of concentration, for the chemical reaction system with constant volume V, the conversion rate per unit volume is usually used to express the chemical reaction rate, *i.e.*

$$v = r/V$$

For a chemical reaction

$$aA + bB \xlongequal{\quad} gG + dD$$

The reaction rate is defined as:

$$v = -\frac{1}{a}\frac{dc_A}{dt} = -\frac{1}{b}\frac{dc_B}{dt} = \frac{1}{g}\frac{dc_G}{dt} = \frac{1}{d}\frac{dc_D}{dt} \tag{6-24}$$

It is impossible to measure the minimum change value dc_B of substance B in a very short time dt by experimental means. The common method to determine the chemical reaction rate is to measure the concentration change of a substance B in a short time interval Δt. Therefore, the reaction rate measured experimentally is usually the average rate within Δt. If the time interval Δt is much less than the time when the reaction continues until the equilibrium is reached, the average rate can be approximately considered as the instantaneous rate of the reaction at this time.

(2) Rate equation

For a chemical reaction

$$aA + bB \xlongequal{\quad} gG + dD$$

The rate equation is as follows:

$$v = kc_A{}^m c_B{}^n \tag{6-25}$$

Where c_A and c_B refer to the concentration of solute in gas or solution. For a pure liquid or solid, their concentration is 1. k is called the rate constant, m and n are the power of their concentration terms, and the specific value should be determined by experiments.

The ratio constant k in the rate equation is the reaction rate when the concentration of each reactant is unit concentration. Therefore, for the same chemical reaction, k is a fixed value under the same temperature and catalyst conditions, which does not change with the change of reactant concentration. It should be pointed out that the unit of reaction rate constant k is related to m and n of rate equation, and the unit is different when m and n of reaction are different. For example, if $m = 1$, $n = 1$, the unit of reaction k is L/(mol/s).

Generally speaking, if the k value of a reaction is large under certain conditions, the reaction rate will be large. This is because in the general reaction system, the concentration of the substance will not change too much.

(3) Reaction order

The sum of the exponents of each reactant concentration in the rate equation is called the reaction order. In formula (6-25), m is called the reaction order of reaction to reactant A, n is called the reaction order of reaction to reactant B, and $m + n$ is called the total reaction order, or simply called the reaction order. Generally, m and n can not be determined by the stoichiometric number of chemical reactions, but only by experiments and reaction mechanism studies. Once the reaction order is determined, the reaction rate equation is also determined.

The reaction order shows the effect of reactant concentration on the reaction rate. The greater the values of m and n, the greater the effect of reactant concentration on the reaction rate. Common chemical reactions include zero order reaction, first order reaction, second order reaction and third order reaction, as well as fractional order reaction.

Zero order reaction refers to the chemical reaction that the reaction rate has nothing to do with the concentration of reactants. The decomposition reaction of pure solid or pure liquid substances, such as the decomposition of hydrogen iodide, is the zero order reaction; the heterogeneous reactions on the surface, such as enzyme catalytic reaction and photosensitive reaction, are also zero order reactions.

The first-order reaction refers to the chemical reaction whose reaction rate is directly proportional to the first power of reactant concentration, such as gas decomposition, radioactive element decay, *etc*. The decomposition of hydrogen peroxide is a first-order reaction; the decomposition of acetaldehyde is a typical fractional order reaction, with a reaction order of 2/3.

Some metals are oxidized (rusted) in the air to form oxide film. The thicker the film, the slower the growth rate of the film. This is a negative first-order reaction. Some reactions can not even determine the reaction order, which shows the complexity of chemical reaction mechanism.

There is a characteristic concentration time relationship in all reactions. For example, for a first-order reaction:

$$v = -\frac{dc_A}{dt} = kc_A$$

$$\frac{dc_A}{c_A} = -kdt$$

Set the concentration of A at the starting time as $c_{A,0}$, and the concentration of A at the time of t as c_A, and integrate the above formula:

$$\int_{c_{A0}}^{c_A} \frac{dc_A}{c_A} = -\int_0^t kdt$$

$$\lg\frac{c_A}{c_{A0}} = -\frac{k}{2.303}t \qquad (6\text{-}26)$$

According to the same mathematical derivation, the concentration time relations of zero order reaction, second order reaction and third order reaction can be obtained.

zero order reaction: $c_A = c_{A0} - kt$

second order reaction: $\frac{1}{c_A} = \frac{1}{c_{A0}} + kt$

third order reaction: $\frac{1}{c_A{}^2} = \frac{1}{c_{A0}{}^2} + 2kt$

Example 6-17 Radioactive nuclear decay reactions are all first-order reactions. It is customary to use half-life to express the rate of nuclear decay. Gamma rays produced by radioactive ^{60}Co are widely used in cancer treatment, with a half-life of 5.26years. The intensity of radioactive substances is expressed in Curie (1 Curie = 37GBq). When a hospital purchases a cobalt source containing 20 Curies, how much will it remain after 10years?

Solution According to $\lg\frac{c_A}{c_{A0}} = -\frac{k}{2.303}t$

$$\lg\frac{1}{2} = -\frac{k}{2.303}\times 5.26$$

$$k = 0.132a^{-1}$$

$$\lg\frac{[^{60}Co]}{[^{60}Co]_0} = -\frac{0.132}{2.303}t$$

$$\lg[^{60}Co] - \lg 20 = -\frac{0.132}{2.303}\times 10$$

$$[^{60}Co] = 5.3 \quad \text{Curie}$$

That is to say, the remaining amount of cobalt source after 10years is 5.3 Curie.

Reaction order is a concept defined by experimental study of chemical reaction rate, which should be determined according to the experiment. In fact, as many steps are involved in a process, some are very fast and some are slow. The speed of the whole reaction cannot be faster than the slowest step. So the slowest step is the rate-determining step. The speed of this step is considered as the speed of the whole process.

In a multistep reaction:

Overall $2NO + 2H_2 = N_2 + 2H_2O$

Steps:

1st step $2NO + H_2 = N_2 + H_2O_2$

2nd step $H_2 + H_2O_2 = 2H_2O$

The first step is slow reaction, and the second step is fast reaction. In order to carry out the second step, it is necessary to wait for the H_2O_2 generated in the first step. Because the rate of formation of H_2O_2 is slow, the first step is the rate-determining step, which determines the rate of the whole reaction. That is to say, the total reaction rate is approximately equal to that of the first step. Therefore, the rate equation of the total reaction follows the concentration relationship of the elementary reaction in the decisive step, which is expressed as follows:

$$v=k \cdot c^2(NO) \cdot c(H_2)$$

The reaction order can be an integer, a fraction or a decimal. This is because, when more than one step of a chemical reaction is a slow reaction, or when the difference in the rate of several slow elementary reactions is not particularly obvious, the reaction rate equation measured in the experiment is the comprehensive result of these slow reactions, and the index of some substances in the formula may appear a fraction.

The order of chemical reaction can be determined by experiment. By changing the concentration of reactant and measuring the reaction rate under two different concentration conditions, the reaction order of reactant can be obtained. After the index of reactant concentration term is determined, the reaction rate constant can be calculated and the reaction rate equation can be determined.

(4) Rate equation of elementary reaction—law of mass action

A large number of experiments show that under certain conditions, the reaction rate of elementary reaction (or simple reaction) is directly proportional to the power of the concentration of reactant (that is, the continuous product of the concentration with the absolute value of the stoichiometric number of reactant in the equation as the index). This quantitative relationship is called the law of mass action. That is, for elementary reaction:

$$aA + bB = gG + dD$$

The rate equation is as follows:

$$v = kc_A{}^a c_B{}^b \tag{6-27}$$

Equation (6-27) is called the law of mass action. It is only applicable to elementary reactions, but not to non elementary reactions. That is to say, the power of each substance concentration term in the elementary reaction rate equation is equal to the respective coefficient in the reaction equation. Of course, there are also some non elementary reactions.

In their rate equation, the index of concentration term is exactly equal to the coefficient of each substance in the reaction equation, but it can not be concluded that the reaction is elementary reaction according to this conclusion.

For the reaction involving gaseous substances, the effect of total pressure on the reaction rate is essentially the effect of concentration on the reaction rate. This is because, according to the ideal gas state equation $p_B = n_B RT/V$, at a certain temperature, increasing the total pressure of a certain amount of gaseous reactants can reduce the volume, increase the concentration n_B/V, and accelerate the reaction rate. On the contrary, to reduce the total pressure is to reduce the gas concentration, resulting in a decrease in the reaction rate.

Chemical reactions are usually reversible. The above discussion is only about positive reactions. Strictly speaking, the net rate of reaction should be equal to the rate of positive reaction minus the rate of reverse reaction. Of course, the rate equation of elementary reaction in the reverse direction still satisfies the law of mass action and the above discussion on positive reaction.

(5) Factors affecting reaction rate

The above discussion about the effect of concentration on the reaction rate is aimed at the reaction in gas mixture or solution, which is called homogeneous chemical reaction. For the reaction with solid, there are other factors besides the effect of concentration on the reaction rate.

The reaction with solid participation belongs to heterogeneous system. In the heterogeneous system, the reaction takes place at the interface between phases. Because the reactants can only contact each other at the interface. Therefore, the reaction rate of the heterogeneous system is not only related to the concentration, but also to the area of the phase interface. For example, the reaction of coke combustion is:

$$C(s) + O_2(g) \longequal CO_2(g)$$

If coal is replaced by pulverized coal, the contact surface of reactants can be increased to speed up the reaction. In addition, the reaction rate of heterogeneous system is also related to the rate of reactant diffusion to the surface and the rate of product diffusion from the surface. That is to say, diffusion makes the reactants which have not yet reacted enter into the interface continuously, and makes the products which have been produced leave the interface continuously. Stirring or blowing can accelerate the diffusion process, thus speeding up the reaction rate.

6.4.3 Effect of Temperature on Reaction Rate

Temperature is an important factor affecting the chemical reaction rate. The effect of temperature on the reaction rate varies with the difference of specific reaction. But for most of the reactions, the reaction rate generally increases with the increase of temperature. There have been many empirical laws about the effect of temperature on the reaction rate. The first one is an approximate law summed up by van't Hoff according to the experiment: in a certain temperature range, the reaction rate increases 2-4 times for every 10℃ rise of temperature. Although this empirical rule is not accurate, it can be used for rough estimation when the data is lacking. Then, the most influential one is Arrhenius formula.

(1) Arrhenius empirical formula

In the previous discussion of the effect of concentration on the reaction rate, it is assumed that the temperature is a constant value. According to the reaction rate equation, it is assumed that the reactant concentration is a constant value, when the temperature changes, the reaction rate will change, essentially the rate constant k has changed. The effect of temperature on the reaction rate is more significant than that of concentration. Generally speaking, the reaction rate constant k increases rapidly with the increase of temperature. At the end of 19th century, Arrhenius summarized a lot of experimental data and put forward an empirical formula to describe the reaction rate constant k and temperature T, which is called Arrhenius formula or Arrhenius equation, that is

$$k=Z\exp(-E_a/RT) \tag{6-28}$$

In the formula, E_a is called experimental activation energy or apparent activation energy, which can be generally regarded as a constant independent of temperature, with the unit of kJ/mol; Z is a constant, which is called pre exponential Arrhenius factor or frequency factor. It can be seen from equation (6-28) that k is exponentially related to T, so it is also called the exponential law of reaction rate.

In fact, the increase of temperature accelerates the reaction rate, which is consistent with the increase of temperature increases the number of active molecules in the reaction system [compare Fig. 6-3 and formula (6-22)]. Because in formula (6-22), exp $(-E/RT)$ (also known as Boltzmann factor) is the proportion factor of molecules with energy E (now active molecules with energy E_a) to the total molecular number. The Arrhenius formula is compared with the collision theory of the reaction mechanism. The frequency factor Z in formula (6-28) is equivalent to the collision frequency of the active molecule in a certain direction. Generally speaking, the size of Z is related to the temperature, but compared with the temperature T on the index, the effect of T on Z can often be ignored.

For most chemical reactions, the relationship between reaction rate constant and temperature can satisfy Arrhenius formula. However, some reactions do not, such as the ordinary explosion reaction, when the temperature rises to a certain limit, the reaction rate can tend to infinity; some chemical reaction rate decreases with the increase of temperature. These reactions which do not conform to Arrhenius formula are called anti Arrhenius reactions. At present, four kinds of such reactions have been found, which will not be discussed here.

(2) Relationship between rate constant and temperature

Take logarithm on both sides of Arrhenius formula (6-28), and get:

$$\ln\frac{k}{[k]} = -\frac{E_a}{RT} + \beta \tag{6-29}$$

Where $[k]$ is the unit of reaction rate constant k, $k/[k]$ becomes a pure number, and the value β is $\ln Z$. If the effect of temperature on Z is ignored, β is a constant.

It can be seen from equation (6-29) that $\ln k$ and $1/T$ are in a straight line relationship. If $\ln k$ is used as the ordinate and $1/T$ as the abscissa, a straight line can be obtained. The slope of the straight line is $\alpha = -E_a/R$. According to this relationship, the activation energy of chemical reaction can be measured accurately by experimental method.

For example, the rate constants of N_2O_5 decomposition reaction in CCl_4 liquid at different temperatures are shown in Table 6-2.

$$N_2O_5 = N_2O_4 + 0.5O_2$$

According to the data in Table 6-2, a straight line is obtained with slope α is -1.24×10^4, as shown in Fig.6-6. So the activation energy of this reaction:

$$E_a = -\alpha R = -(-1.24\times10^4)\times8.3145 = 1.03\times10^5 \text{(J/mol)}$$

Table 6-2 The Rate Constants of N_2O_5 Decomposition Reaction at Different Temperatures

t/℃	T/K	$T^{-1}\times10^3$/K^{-1}	$k\times10^5$/s^{-1}	$\ln(k/[k])$
65	338	2.96	487	−5.32
55	328	3.05	150	−6.50
45	318	3.14	49.8	−7.60
35	308	3.25	13.5	−8.91
25	298	3.36	3.46	−10.3
0	273	3.66	0.0787	−14.1

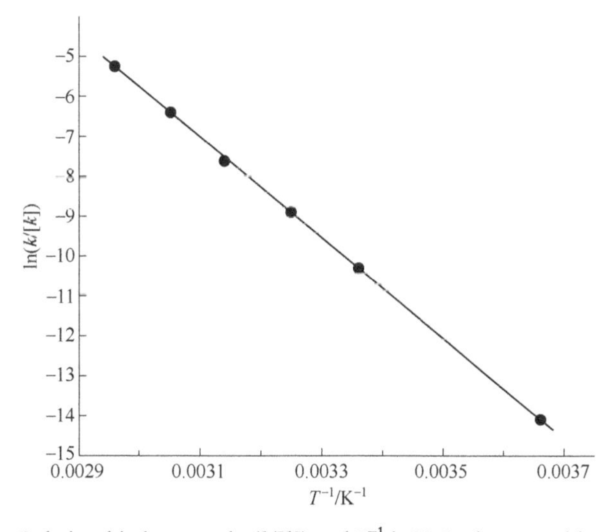

Fig.6-6 Relationship between ln ($k/[k]$) and T^{-1} in N_2O_5 decomposition reaction

When the extended line intersects the ordinate axis, the intercept β ($\ln Z$) of the line on the ordinate (corresponding to $1/T$ is zero) is 31.4, so $Z = 4.33\times10^{13}$.

Two Arrhenius formulas (6-29) at different temperatures T_1 and T_2 are subtracted from each other. Assuming that β is independent of temperature, we can get:

$$\ln\frac{k_2}{k_1} = \frac{E_a}{R}\times\frac{T_2 - T_1}{T_1 T_2} \tag{6-30}$$

If the activation energy of one reaction and the reaction rate constant at a certain temperature are known, the rate constant at another temperature can be calculated by equation (6-30).

Example 6-10 For the decomposition reaction of N_2O_5, calculate the change of reaction rate when the temperature rises 10℃.

Solution According to the law of mass action, when the concentration is constant

$$v = kc$$

Where c is the ratio constant related to concentration. If v_1, v_2, k_1 and k_2 are used to represent the rate and rate constants at T_1 and T_2 respectively, then

$$v_1 = k_1 c,\quad v_2 = k_2 c$$

According to Eq. (6-30),

$$\ln\frac{v_2}{v_1} = \ln\frac{k_2}{k_1} = \frac{E_a}{R}\times\frac{T_2 - T_1}{T_2 T_1}$$

Substituting $E_a/R = -\alpha = 1.24\times10^4$, $T_1 = 283\text{K}$, $T_2 = 293\text{K}$ into the above formula, we can get

$$\ln\frac{v_2}{v_1} = \ln\frac{k_2}{k_1} = 1.24\times10^4\times\frac{293-283}{293\times283} = 1.4954$$

So $$v_2/v_1 = 4.46$$

The reaction rate or rate constant is not only related to the temperature, but also to the activation energy of the reaction. If the reaction temperature increases by 10℃, according to formula (6-30), assuming that Z does not change with T, the $k_{(T+10)}/k_T$ value of the reaction with different activation energy at different temperatures can also be calculated. An example is shown in Table 6-3.

It can be seen from Table 6-3 that in each column of data, the larger the activation energy is, the greater the multiple of the increase of the rate constant will be when the temperature increases by 10℃; for the same activation energy reaction (each row), the influence of temperature change on the reaction rate will be greater at low temperature than at high temperature.

In addition, because the activation energy of the positive and reverse reactions in the same chemical reaction equation is generally different, the change of temperature will affect the equilibrium constant and make the equilibrium move, and further understanding can be made from the change of the positive and reverse reaction rates (*i.e.* from the perspective of chemical kinetics).

Table 6-3 $k_{(T+10)}/k_T$ of Reaction with Different Activation Energy E_a at Different Temperatures

E_a/(kJ/mol)	$k_{(T+10)}/k_T$					
	273K	373K	473K	573K	673K	773K
80.0	3.47	1.96	1.52	1.33	1.23	1.17
100.0	4.74	2.32	1.69	1.43	1.30	1.22
150.0	10.30	3.53	2.20	1.72	1.48	1.35
200.0	22.42	5.38	2.86	2.05	1.69	1.49
300.0	106.18	12.47	4.85	2.94	2.19	1.81
400.0	502.82	28.93	8.20	4.22	2.85	2.21

6.4.4 Effect of Catalyst on Reaction Rate

A large number of facts show that adding catalyst is one of the most effective methods to accelerate the reaction rate. This is why catalysts are used in 80% - 85% of chemical reactions. Many familiar reactions, such as ammonia synthesis, sulfuric acid synthesis, nitric acid synthesis, vinyl chloride synthesis and polymer polymerization, use catalysts. Its purpose is to speed up the reaction rate and improve the production efficiency.

Traditionally, a substance that can speed up the reaction, but its chemical composition, quantity and chemical properties do not change before and after the reaction is called catalyst. Generally speaking, catalyst refers to positive catalyst. The role of catalyst in changing the rate of chemical reaction is called catalysis. The reaction with catalyst is called catalytic reaction.

In contrast to catalysts, substances that slow down the reaction rate are called inhibitors. The "negative catalyst" used in the past is no longer recognized by the International Union of Pure and Applied Chemistry (IUPAC), so it is necessary to use "inhibitor" instead of "negative catalyst", and "catalyst" only refers to the substance that can accelerate the reaction rate. For example, adding 0.01%-0.02% *n*-propyl gallate to edible oil can effectively prevent rancidity. Here, *n*-propyl gallate is an inhibitor.

(1) Basic principles of catalysis

The results show that the effect of catalyst on the reaction rate is different from that of concentration and temperature. The effect of concentration or temperature on the reaction rate generally does not change the reaction mechanism, while the effect of catalyst on the reaction rate is achieved by changing the reaction mechanism. Generally, with the participation of catalyst, the reaction is usually divided into several steps. The activation energy of each step is not large, and the total activation energy is smaller than that without catalyst, so the reaction rate is accelerated. For example, the synthesis of ammonia (see Fig.6-7). When there is catalyst, the reaction may be carried out in two steps. The apparent activation energy of catalytic reaction is smaller than that of non catalytic reaction, so the catalytic reaction rate is significantly faster than that of non catalytic reaction. It should be noted that Fig.6-7 only shows that in the presence of catalyst, the reaction has changed the way, that is to say, a shortcut requiring low activation energy is taken, and the activation energy of catalytic reaction and non catalytic reaction can only be determined by experiment.

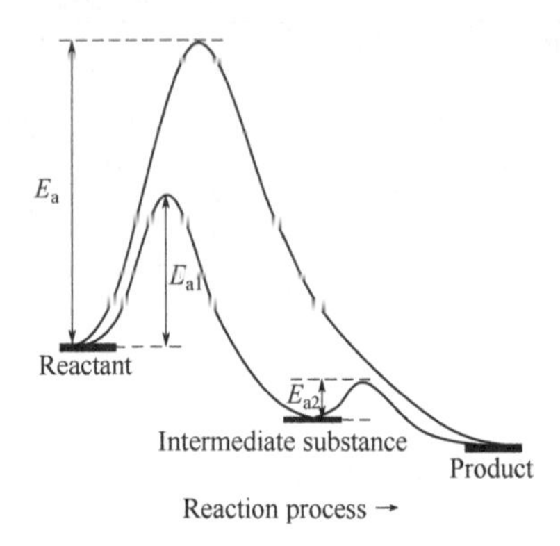

Fig.6-7 Schematic diagram of activation energy of synthetic ammonia reaction

From the point of view of material structure, most catalysts are transition metal compounds. Their active d electrons often relax the chemical bond of reactant molecules when they interact with reactant molecules (such as adsorption), thus changing the original reaction path, reducing the activation energy, resulting in the relative increase of the percentage of activated molecules.

Table 6-4 shows the experimental values of activation energy of three reactions in the presence and absence of catalyst.

Table 6-4 Activation Energy of Catalytic Reaction and Non Catalytic Reaction

Reaction	Activation energy/(kJ/mol)		Catalysis
	Non catalytic reaction	Catalytic reaction	
$2HI = H_2 + I_2$	184.1	104.6	Au
$2H_2O = 2H_2 + O_2$	244.8	136	Pt
$3H_2 + N_2 = 2NH_3$	334.7	167.4	$Fe\text{-}Al_2O_3\text{-}K_2O$

The data in Table 6-4 shows that the presence of catalyst significantly reduces the activation energy of the reaction. Because the activation energy is in the position of negative index, the reaction rate can be greatly accelerated.

Example 6-19 Calculate the ratio of reaction rate constant under catalytic and non catalytic conditions. In Table 6-4, the decomposition of hydrogen iodide is carried out at 503K, and assuming Z is constant.

Solution According to formula (6-28), when there is no catalyst, there are:

$$k = Ze^{-\frac{E_a}{RT}} = Ze^{-\frac{184100}{8.314\times 503}}$$

With catalyst:

$$k_{\text{cat}} = Ze^{-\frac{E_{\text{a,cat}}}{RT}} = Ze^{-\frac{104600}{8.314\times503}}$$

$$\frac{k_{\text{cat}}}{k} = \text{e}^{-\frac{104600}{8.314\times503}} / \text{e}^{-\frac{184100}{8.314\times503}} = 1.8\times10^8$$

The results show that the difference of reaction rate is 180 million times. The rate change caused by the addition of catalyst is difficult to achieve by changing the concentration and temperature.

The amount of catalyst has an effect on the reaction rate. When the amount of reactant is enough and the catalyst is in full contact with reactant, the higher the concentration of catalyst is, the higher the chemical reaction rate is. Therefore, in a certain range of dosage, with the increase of catalyst dosage, the reaction rate increases; when the dosage increases to a certain value, the reaction rate does not change.

(2) Effect of catalyst on chemical equilibrium

The catalyst can reduce the activation energy of both positive and reverse reactions. This shows that the catalyst can not only accelerate the positive reaction rate, but also accelerate the reverse reaction rate. Under certain conditions, the excellent catalyst for positive reaction is also excellent for reverse reaction. For example, iron, platinum, nickel and other metals are not only good dehydrogenation catalysts, but also good hydrogenation catalysts. Therefore, the catalyst can shorten the time to reach equilibrium.

However, it should be emphasized that the catalyst does not change the equilibrium state ($K^{\ominus}$ is constant). For a reversible reaction, the initial and final states are independent of the existence of catalyst. Therefore, the standard Gibbs function of the reaction is a fixed value, and the standard equilibrium constant is a fixed value naturally.

It can also be seen from Fig.6-7 that the presence of catalyst does not change the relative energy of reactants and products. That is to say, whether a reaction is carried out with or without catalyst, the initial and final states of the system will not change. Therefore, catalyst can not change $\Delta_r H_m$ and $\Delta_r G_m$ of a reaction. This also shows that the catalyst can only accelerate the reaction that can be spontaneous in thermodynamics, that is, the reaction of $\Delta_r G_m < 0$; for the reaction that can not be spontaneous by thermodynamic calculation, that is, the reaction of $\Delta_r G_m > 0$, is still not spontaneous after using the catalyst.

(3) Basic characteristics of catalyst

① Before and after the reaction, although the composition of catalyst, the amount of substance and the chemical properties will not change, the physical properties will generally change. For example, MnO_2, the catalyst for decomposition of $KClO_3$, will change from block to powder after reaction. Another example: use Pt net to catalyze ammonia oxidation, and the surface of Pt net will become rough in a few weeks.

② The catalyst has special selectivity. The selectivity of catalyst has two meanings. First, different catalysts should be selected for different types of reactions. For example, the catalysts for oxidation and dehydrogenation are different. Even for the same type of reaction, the catalysts are not necessarily the same. For example, V_2O_5 is used as catalyst for SO_2 oxidation, while Ag is used for $CH_2{=}CH_2$ oxidation. Second, for the same reactants, if different catalysts are selected, different products can be obtained. This is of great significance in industrial production. For example, the decomposition of ethanol has the following situations.

$$C_2H_5OH \xrightarrow[Cu]{473\text{-}523K} CH_3CHO + H_2$$

$$C_2H_5OH \xrightarrow[Al_2O_3]{623\text{-}633K} C_2H_4 + H_2O$$

$$2C_2H_5OH \xrightarrow[Al_2O_3]{413K} (C_2H_5)_2O + H_2O$$

$$2C_2H_5OH \xrightarrow[ZnO\cdot Cr_2O_3]{673\text{-}773K} CH_2{=}CH{-}CH{=}CH_2 + 2H_2O + H_2$$

From the thermodynamic point of view, these reactions can be spontaneous. However, a catalyst can only catalyze a specific reaction, and can not accelerate all the possible thermodynamic reactions, which is the selectivity of catalyst. Therefore, we can use this selectivity and choose different catalysts to obtain different products.

③ Catalyst activity and poisoning. The activity of catalyst refers to the catalytic capacity of catalyst, that is, the amount of product that can be obtained in unit time by using catalyst of unit mass (or unit volume) under specified conditions. In the process of using many catalysts, their activity will increase from small to large, and then gradually reach the normal level. After the activity is stable for a period of time, it starts to decline again until it is aged and can not be used. This active stabilization period is called the catalyst life. The life of the catalyst varies with the type of catalyst and service conditions. Aging catalysts can sometimes be reactivated by regeneration. During the stable period of catalyst activity, the activity of catalyst often drops immediately because of contacting a small amount of impurities, which is called catalyst poisoning. If the activity can be restored after eliminating the poisoning factors, it is called temporary poisoning, otherwise it is permanent poisoning.

The activity of solid catalyst usually depends on its surface state, which is different with different preparation methods. In other words, the physical properties of the catalyst also affect its activity. Sometimes, in order to give full play to the efficiency of the catalyst, the catalyst is often dispersed on a porous inert material with a large surface area, which is called a carrier. Common carriers include silica gel, alumina, pumice, asbestos, activated carbon, diatomite, *etc.* In specific applications, catalyst is usually a combination of principal catalyst and auxiliary catalyst. The single auxiliary catalyst has no activity or low activity, but when combined with principal catalyst, it can significantly improve the activity, selectivity and stability of the catalyst.

④ Homogeneous and heterogeneous catalysis. The dispersion state between catalyst and material in chemical industry can be divided into two types: uniform dispersion and non-uniform dispersion. The former is called homogeneous catalysis, such as hydrogen ion, transition metal ion or coordination compound in solution. The latter is called multiphase catalyst, such as various transition metal or alloy solid and metal oxide. In addition, there is a special kind of catalyst in organisms, enzyme, which has high efficiency and selectivity.

(4) Examples of catalytic applications

In the electronic industry, a kind of molecular sieve containing 0.03% palladium is sometimes used as catalyst to remove a small amount of oxygen that may be contained in hydrogen. The reaction of hydrogen and oxygen to form water can be realized rapidly at room temperature with catalyst, but this reaction can not be observed without catalyst. This is because the activation energy of the reaction is reduced and the reaction is accelerated greatly after the two gases are adsorbed on the catalyst surface. Its course is as follows:

(i) the reactants H_2 and O_2 diffuse to the catalyst surface;

(ii) oxygen molecules are adsorbed on the catalyst surface;

(iii) hydrogen molecules combine with oxygen molecules adsorbed on the catalyst surface to form water;

(iv) water molecules desorbed from the catalyst surface and diffused into the gas to complete the reaction.

Another example is the synthesis of ammonia with iron catalyst. The results show that the catalytic mechanism is that the N_2 molecule is first chemically adsorbed on the catalyst surface, thus weakening the chemical bond; then, the chemically adsorbed hydrogen atom reacts with the nitrogen atom on the surface, gradually forming ammonia molecule on the catalyst surface; finally, the ammonia molecule is desorbed from the surface to obtain the gaseous ammonia, *i.e.*

① $x\mathrm{Fe} + 0.5\mathrm{N_2} \longrightarrow \mathrm{Fe}_x\mathrm{N}$

② $\mathrm{Fe}_x\mathrm{N} + [\mathrm{H}] \longrightarrow \mathrm{Fe}_x\mathrm{NH}$

$\mathrm{Fe}_x\mathrm{NH} + [\mathrm{H}] \longrightarrow \mathrm{Fe}_x\mathrm{NH_2}$

$\mathrm{Fe}_x\mathrm{NH_2} + [\mathrm{H}] \longrightarrow \mathrm{Fe}_x\mathrm{NH_3} \longrightarrow x\mathrm{Fe} + \mathrm{NH_3}$

In the absence of catalyst, the activation energy E_a is very high, about 250-340kJ/mol. After adding iron catalyst, the reaction is divided into two steps: first, nitriding of iron, and then nitrogen hydrogenation of iron. The activation energy E_{a_1} of the first step is 125-167kJ/mol, and the activation energy E_{a_2} of the second step is very small, which is 12.6kJ/mol. Therefore, the first step is the rate control step, that is, the speed determination step. Obviously, the use of catalyst greatly reduces the activation energy of the reaction, thus greatly accelerating the reaction of ammonia synthesis.

Exercises

1. Explain the meaning of the following symbols:

$$S,\ S_{\mathrm{m}}^{\ominus},\ \Delta_{\mathrm{r}}S_{\mathrm{m}}^{\ominus}(T),\ G,\ \Delta_{\mathrm{r}}G_{\mathrm{m}}^{\ominus}(T),\ \Delta_{\mathrm{f}}G_{\mathrm{m}}^{\ominus}(T),\ Q,\ K^{\ominus}$$

2. Are the following statements correct? If not, please explain the reason.

(1) Exothermic reactions are spontaneous.

(2) The $\Delta_{\mathrm{f}}H_{\mathrm{m}}^{\ominus}(298\,\mathrm{K}), \Delta_{\mathrm{f}}G_{\mathrm{m}}^{\ominus}(298\,\mathrm{K}), S_{\mathrm{m}}^{\ominus}(298\,\mathrm{K})$ of single substance is zero.

(3) If the total number of molecules in a reaction is more than that in the reactant, the ΔS of the reaction must be positive.

(4) Both ΔH and ΔS of a reaction are positive values. When the temperature increases, ΔG will decrease.

(5) Each molecule has its own activation energy.

(6) Several main factors affecting the reaction rate have influence on the reaction rate constant.

(7) Catalyst can not only accelerate the reaction rate, but also affect the equilibrium, and also make the non spontaneous reaction can be spontaneous.

3. What is the criterion to judge whether the reaction can be spontaneous? Can we use the change of enthalpy or entropy of the reaction as the standard? Why?

4. How do you use the data of $\Delta_{\mathrm{f}}H_{\mathrm{m}}^{\ominus}(298\,\mathrm{K}), \Delta_{\mathrm{f}}G_{\mathrm{m}}^{\ominus}(298\,\mathrm{K}), S_{\mathrm{m}}^{\ominus}(298\,\mathrm{K})$ to calculate the approximate value of $\Delta_{\mathrm{r}}G_{\mathrm{m}}^{\ominus}(298\,\mathrm{K})$ and $\Delta_{\mathrm{r}}G_{\mathrm{m}}^{\ominus}(T)$ at a certain temperature T? Please illustrate the calculation method with examples.

5. How to use the data of $\Delta_{\mathrm{f}}H_{\mathrm{m}}^{\ominus}(298\,\mathrm{K}), \Delta_{\mathrm{f}}G_{\mathrm{m}}^{\ominus}(298\,\mathrm{K}), S_{\mathrm{m}}^{\ominus}(298\,\mathrm{K})$ to calculate $K_T^{\ominus}$ of the reaction? Write the relevant calculation formula.

6. $2A(g)+B(g) = 2C(g),\quad \Delta H = -x\,\mathrm{kJ/mol}$

Do you agree with the following statement?

(1) According to $K_T^{\ominus} = \dfrac{p_{\mathrm{C}}^2}{p_{\mathrm{A}}^2 p_{\mathrm{B}}}$, as the reaction proceeds, the partial pressure of C increases, the partial pressures of A and B decrease, and the equilibrium constant increases.

(2) Increasing the total pressure will increase the partial pressure of A and B and decrease the partial pressure of C, so the equilibrium moves to the right.

7. For the following equilibrium systems

$$2CO(g) + O_2(g) = 2CO_2(g),\qquad \Delta H < 0$$

(1) Write the mathematical expression of the equilibrium constant.

(2) If in the equilibrium system: ① add O_2; ② remove CO gas from the system;

③ increase the total pressure of the system; ④ reduce the temperature of the system. What will happen to the concentration of CO_2 in the system?

8. What is the meaning of chemical reaction rate? How to express the reaction rate?

9. Can we judge the order of reaction according to chemical equation? Why?

10. Please illustrate the important application of Arrhenius equation. For general chemical reactions, how many times will the reaction rate increase for every 10℃ increase in temperature?

11. In order to obtain the activation energy of the reaction, at least how many reaction rates at a certain temperature must be measured?

12. If a reaction is a single-phase reaction, what are the main factors affecting the reaction rate? How do they affect the rate constants? Why?

13. The activation energy of one reaction is 120kJ/mol, and that of the other is 78kJ/mol. Under similar conditions, which of the two reactions is faster? Why?

14. If a reaction is exothermic, the temperature rise is not conducive to the positive reaction, so the reaction will be slow at high temperature. Is that right? Why?

15. What are the similarities between the effects of total pressure and concentration on the reaction rate and equilibrium shift? What are the differences?

16. What are the similarities between the mathematical formula of temperature and equilibrium constant and that of temperature and reaction rate constant? What are the differences? Give an example and explain the meaning of each physical variable in the two formulas.

17. For multiphase reactions, what are the main factors that affect the chemical reaction rate?

18. According to the relationship between entropy and the degree of confusion, please judge: in the following changes entropy is increasing or decreasing.

(1) Salt dissolves in water;

(2) Two different gases mix;

(3) Water freezes;

(4) Adsorption of oxygen by activated carbon;

(5) Combustion of sodium metal in chlorine gas to produce sodium chloride;

(6) Ammonium nitrate decomposes by heating.

19. Instead of looking up the table, try to rank the following substances in order of $S_m^{\ominus}$ from large to small.

(1) K(s); (2) Na(s); (3) Br_2(l); (4) Br_2(g); (5) KCl(s)

20. At 353K and 101.325kPa, 1mol liquid benzene vaporizes into benzene vapor. If the vaporization heat of benzene is known to be 349.91J/g and the molar mass of benzene is 78.1g/mol, then calculate the W and $\Delta S^{\ominus}$ (353K is the normal boiling point of benzene).

21. Using the data in Appendix, calculate $\Delta_r H_m^\ominus(298\,K)$, $\Delta_r G_m^\ominus(298\,K)$, $\Delta_r S_m^\ominus(298\,K)$ for the following reaction:

$$2CuO(s) = Cu_2O(s) + \frac{1}{2}O_2(g)$$

22. According to the following data, calculate the standard generating Gibbs function variation of N_2O_4?

$$\frac{1}{2}N_2(g) + \frac{1}{2}O_2(g) = NO(g), \quad \Delta_r G_m^\ominus = 87.6\,kJ/mol$$

$$NO(g) + \frac{1}{2}O_2(g) = NO_2(g), \quad \Delta_r G_m^\ominus = -36.3\,kJ/mol$$

$$2NO_2(g) = N_2O_4(g), \quad \Delta_r G_m^\ominus = -2.8\,kJ/mol$$

23. According to $\Delta_r G_m^\ominus$ of the following reactions, calculate the standard generating Gibbs function variation of Fe_3O_4 (s) at 298K:

$$2Fe(s) + \frac{3}{2}O_2(g) = Fe_2O_3(s), \quad \Delta_r G_m^\ominus = -742.2\,kJ/mol$$

$$4Fe_2O_3(s) + Fe(s) = 3Fe_3O_4(s), \quad \Delta_r G_m^\ominus = -77.4\,kJ/mol$$

24. According to $\Delta_f H_m^\ominus$ and $S_m^\ominus$, calculate the lowest temperature at which this reaction can occur spontaneously.

$$MgCO_3(s) = MgO(s) + CO_2(g)$$

25. Calculate the lowest temperature at which $I_2(g)$ can spontaneously decompose into I(g).

26. At 100kPa and 298K, bromine evaporates from liquid to gas. Using Appendix Ⅱ data, answer the following questions.

(1) Calculate the $\Delta_r H_m^\ominus$, $\Delta_r S_m^\ominus$;

(2) From the results of (1) calculation, discuss the disorder degree of liquid and gas bromine;

(3) Calculate $\Delta_r G_m^\ominus$ of this process, and explain whether the process can be carried out automatically under this condition according to the calculation results;

(4) Try to calculate the minimum temperature of automatic evaporation.

27. The following methods can be used to prepare metallic tin (white tin) from cassiterite (SnO_2):

(1) Directly heat the ore to decompose it;

(2) Reduction of ore with carbon (heating to produce CO_2);

(3) Reduction of ore with H_2 gas (heating to produce water vapor).

It is hoped that the heating temperature will be as low as possible. Please explain which method is more suitable according to the calculation results.

28. Try to estimate the lowest decomposition temperature of $CaCO_3$ and compare it with the actual operation temperature of 900℃.

$$CaCO_3(s) \longrightarrow CaO(s) + CO_2(g)$$

29. When copper products are exposed to flowing atmosphere for a long time at room temperature, their surface will gradually be covered with a layer of black copper oxide CuO. When the product is heated above a certain temperature, the black copper oxide will be transformed into red cuprous oxide Cu_2O (artificial antique copper products). At higher temperatures, the red oxide decomposes and disappears. At 100kPa, in order to obtain Cu_2O red cover (artificial archaize), please estimate the temperature of following reaction, so as to select artificial archaize temperature.

(1) $4CuO(s) \longrightarrow 2Cu_2O(s) + O_2(g)$

(2) $2Cu_2O(s) \longrightarrow 4Cu(s) + O_2(g)$

30. Write the mathematical expression of $K_T^\ominus$ for the following reactions:

(1) $SnO_2(s) + 2CO(g) \longrightarrow Sn(s) + 2CO_2(g)$

(2) $CH_4(g) + 2O_2(g) \longrightarrow CO_2(g) + 2H_2O(l)$

(3) $Al_2(SO_4)_3(aq) + 6H_2O(l) \longrightarrow 2Al(OH)_3(s) + 3H_2SO_4(aq)$

(4) $NH_3(g) \longrightarrow \frac{1}{2}N_2(g) + \frac{3}{2}H_2(g)$

(5) $C(s) + H_2O(g) \longrightarrow CO(g) + H_2(g)$

(6) $BaCO_3(s) \longrightarrow BaO(s) + CO_2(g)$

(7) $Fe_3O_4(s) + 4H_2(g) \longrightarrow 3Fe(s) + 4H_2O(g)$

31. The standard equilibrium constants for the following reactions are known at 298K:

$$FeO(s) \longrightarrow Fe(s) + \frac{1}{2}O_2, \qquad K_1^\ominus = 1.5\times10^{-43}$$

$$CO_2(g) \longrightarrow CO(g) + \frac{1}{2}O_2, \qquad K_2^\ominus = 8.7\times10^{-46}$$

Try to calculate $K_T^\ominus$ of the following reactions at the same temperature.

$$Fe(s) + CO_2(g) \longrightarrow FeO(s) + CO(g)$$

32. The thermal decomposition reaction of phosphorus pentachloride is as follows:

$$PCl_5(g) \longrightarrow PCl_3(g) + Cl_2(g)$$

The reaction reaches equilibrium at 100kPa and a certain temperature T, and the partial pressure of PCl_5 is 20kPa. Calculate $K_T^\ominus$ of the reaction at this temperature.

33. A certain amount of N_2O_4 gas is insulated in a closed container, and the reaction reaches equilibrium. Try to calculate the following constants through the relevant data in Appendix Ⅱ:

$$N_2O_4(g) \rightleftharpoons 2NO_2(g)$$

(1) The standard equilibrium constant $K_{298}^{\ominus}$ of the reaction at 298K;

(2) The standard equilibrium constant $K_{350}^{\ominus}$ of the reaction at 350K.

34. There is the reaction as follows:

$$CuS(s) + H_2(g) \rightleftharpoons Cu(s) + H_2S(g)$$

Please calculate

(1) The standard equilibrium constant $K_{298}^{\ominus}$ of the reaction at 298K;

(2) The standard equilibrium constant $K_{798}^{\ominus}$ of the reaction at 798K.

35. A constant pressure vessel is filled with a mixture of CO_2 and H_2, and the following reversible reactions occur:

$$CO_2(g) + H_2(g) \rightleftharpoons CO(g) + H_2O(g)$$

At 100kPa, the partial pressure of CO_2 in the mixture is 25kPa. When heated to 850℃, the reaction reaches equilibrium. It is known that the standard equilibrium constant is $K^{\ominus}$ = 1.0, try to calculate and solve the following problems:

(1) The equilibrium partial pressure of each substance in the mixed gas;

(2) Percentage of CO_2 converted to CO;

(3) If the temperature remains the same, add some H_2 to the above balance system, try to determine the direction of the equilibrium shift.

36. At 763K, the $K_T^{\ominus}$ of reaction $H_2(g)+I_2(g) \rightleftharpoons 2HI(g)$ is 45.9. Which direction is the reaction going in the following two cases?

(1) $p_{H_2} = p_{I_2} = p_{HI} = 100kPa$;

(2) $p_{H_2} = 10kPa$, $p_{I_2} = 20kPa$, $p_{HI} = 100kPa$.

37. At 1073K, the $K_T^{\ominus}$ of reaction $C(s)+CO_2(g) \rightleftharpoons 2CO(g)$ is 7.5×10^{-2}. What direction is the reaction going in the following two cases?

(1) The weight of C (s) is 1kg, $p_{CO_2} = p_{CO} = 100kPa$;

(2) The weight of C (s) is still 1kg, $p_{CO_2} = 500kPa$, $p_{CO} = 5kPa$.

38. In the presence of V_2O_5 catalyst, it is known that when reaction

$$2SO_2(g)+O_2(g) \rightleftharpoons 2SO_3(g)$$

reaches equilibrium at 600℃ and 100kPa, the partial pressures of SO_2 and O_2 are 10kPa and 30kPa respectively. If the temperature remains unchanged, reduce the volume of the reaction system to the original 1/2, and explain the direction of equilibrium movement through the calculation of reaction quotient.

39. At 700℃, it is known that

$$Fe(s)+H_2O(g) \rightleftharpoons FeO(s)+H_2(g), \qquad K_T^{\ominus} = 2.35$$

If FeO is treated with H_2O and H_2 in the same amount with a total pressure of 100kPa at

700℃, will it be reduced to Fe? If the total pressure of the mixed gas of $H_2O(g)$ and $H_2(g)$ is still 100kPa and FeO is not reduced to Fe, what is the minimum partial pressure of H_2O (g)?

40. It is known that

$$Fe(s)+CO_2(g) = FeO(s)+CO(g), \quad K_{T_1}^{\ominus}$$

$$Fe(s)+H_2O(g) = FeO(s)+H_2(g), \quad K_{T_2}^{\ominus}$$

The values of $K_T^{\ominus}$ at different temperatures are as follows:

T/K	973	1073	1 173	1273
$K_{T_1}^{\ominus}$	1.47	1.81	2.15	2.48
$K_{T_2}^{\ominus}$	2.38	2.00	1.67	1.49

(1) Calculate the $K_T^{\ominus}$ of reaction $CO_2(g) + H_2(g) = CO(g) + H_2O(g)$ at two temperatures, and judge whether the positive reaction is endothermic or exothermic;

(2) Calculate the enthalpy change of the reaction.

41. At 25℃, the thermodynamic data of the following reactions are shown below:

$$CO(g)+H_2O(g) = CO_2(g)+H_2(g)$$

$K_T^{\ominus} = 3.32\times10^3$, $\Delta H = -41.2$kJ/mol, try to calculate the $K_T^{\ominus}$ at 1000K.

42. At 27℃, the thermodynamic data of the following reactions are shown below:

$$SnO_2(s)+2H_2(g) = Sn(s)+2H_2O(g)$$

$K_T^{\ominus} = 6.28\times10^{-11}$, $\Delta_r H^{\ominus} = 94.0$kJ/mol, try to calculate the $K_T^{\ominus}$ at 227℃.

43. It is known that the activation energy of the positive reaction of

$$1.5H_2+0.5N_2 = NH_3$$

is 334.7kJ/mol and the activation energy of the reverse reaction is 380.6kJ/mol. Try to calculate the reaction enthalpy change of ammonia synthesis reaction.

44. The steam of hydrogen and iodine completes the reaction at high temperature according to the following formula:

$$H_2+I_2 = 2HI$$

If the concentration of both reactants is 1mol/L, the reaction rate is 0.05mol/(L·s); Suppose the concentration of H_2 is 0.1mol/L and the concentration of I_2 is 0.5mol/L, what is the reaction rate at this time?

45. Hydrogen is produced from zinc and dilute sulfuric acid, the $\Delta_r H_m$ of the reaction is negative. The reaction rate accelerates in a period of time after the beginning of the reaction, and then slows down. Try to explain this phenomenon from the perspective of influencing factors such as concentration and temperature.

46. At 700℃, the rate constant of CH_3CHO decomposition reaction $k_1 = 0.0105s^{-1}$. If the activation energy of the reaction is 188kJ/mol, try to calculate the rate constant k_2 of the reaction at 800℃.

47. Let the activation energy of a positive reaction be 8×10^4J/mol, the activation energy of the reverse reaction is 12×10^4J/mol. If the difference of Z is ignored, try to calculate how many times the positive reaction rate and reverse reaction rate at 800K are respectively the reaction rate at 400k? According to the calculation results, for the reactions with different activation energies, which reaction rate changes greatly when the temperature increases?

48. The rate constants of formic acid decomposition on gold surface at 140℃ and 185℃ are $5.5\times10^{-4}s^{-1}$ and $9.2\times10^{-2}s^{-1}$ respectively. Try to calculate the activation energy.

49. It is known that the activation energy of a reaction is 80kJ/mol. Try to calculate:

(1) how many times the reaction rate constant increases when the temperature rises from 20℃ to 30℃;

(2) how many times the reaction rate constant increases when the temperature rises from 100℃ to 110℃.

50. Mix the solution containing 0.1mol/L Na_3AsO_3 and 0.1mol/L $Na_2S_2O_3$ with the excess dilute sulfuric acid solution to produce the following reactions:

$$2H_3AsO_3 + 9H_2S_2O_3 = As_2S_3(s) + 3SO_2(g) + 9H_2O + 3H_2S_4O_6$$

According to the experimental results, it takes 1515s from the beginning of mixing to the appearance of yellow As_2S_3 precipitation at 17℃. If the solution temperature is increased by 10℃, repeat the experiment, and the measurement takes 500s. Try to calculate the activation energy of the reaction.

51. When there is no catalyst, the decomposition reaction of H_2O_2 is as follows:

$$H_2O_2(l) = H_2O(l) + 0.5O_2(g)$$

the activation energy is 75kJ/mol. In the presence of catalyst, the activation energy of the reaction decreased to 54kJ/mol. Calculate the ratio of the two reaction rates at 298K (ignoring the difference of Z).

Chapter 7 Solution and Ion Equilibrium

Teaching contents	Learning requirements
Colligative properties of dilute solution	Master the concept and principle of the colligative properties of dilute solution. Students should be able to resolve the practical application problems by using the knowledge of the colligative properties
Electrolyte dissociation equilibrium and buffer solution	Understand the development of the acid-base theory. Master the ionization constant expressions and principles of dissociation equilibrium of weak electrolytes. Understand the common ion effect on buffer solutions, the principle of buffering effect and related applications, *etc*
Equilibrium of precipitation dissolution and rule of solubility product	Master the basic concepts and expression of solubility product and applications of solubility product rule. Understand the formation and dissolution of precipitation and the practical application of precipitation dissolution equilibrium

When winter comes, the highway management departments in North of China have to store a large amount of industrial salt (the main ingredient is sodium chloride, NaCl), and placed these salt bags on arterial roads, bridges, culverts and highways in advance. What is the relationship between industrial salt and highway maintenance? When industrial salt and ice are mixed, the freezing point of liquid water can be reduced down to −22℃. Therefore, whenever it snows, road management workers will sprinkle industrial salt on the highway in time to reduce the freezing point of water and prevent the road from freezing. So, why can the mixing of industrial salt and ice reduce the freezing point of liquid water? In this chapter, we will try to explain the science behind these phenomena and applications.

Solution usually refers to a homogeneous system (or single-phase system) formed by the dispersion of one substance in another. Many chemical reactions in the industrial process take place in the solutions, such as pickling, rust removal, electroplating, electrolytic machining, chemical etching and so on, even washing clothes and cooking in daily life. The properties of the solution can usually be divided into commonness and personality. The properties of the solution has color, conductivity, density and so on. These properties will vary with the solute after the solvent is selected, such as orange in potassium dichromate aqueous solution, purple in potassium permanganate aqueous solution and so on. The commonness of solution is vapor pressure, boiling point, freezing point, osmotic pressure and so on. These properties are not different with the type of solute after solvent selection, but only different with the amount of solute. We call these commonalities of solution as colligative properties, which have some obvious laws in non-electrolyte dilute solution.

7.1 Colligative Properties

7.1.1 Vapor Pressure Lowering

At a certain temperature, a certain amount of liquid is contained in a closed container with a definite volume, then some high energy molecules on the liquid surface overcome the inter molecular gravity and escape from the liquid level, and the gaseous molecules enter the upper space of the container, which is called evaporation [Fig.7-1(a)]. At the same time, gaseous molecules moving in the upper space of the liquid level may also encounter the liquid level and enter the liquid, called condensation. At a certain temperature, the evaporation rate of the liquid is determined, while the condensation rate is small at first [Fig.7-1(b)]. But with the increase of vapor molecules, the rate of condensation increases; when the condensation rate is equal to the evaporation rate, an equilibrium exists between the liquid and vapor [Fig.7-1(c)]. In liquid-vapor equilibrium, the number of molecules returning from the gas phase to the liquid phase per unit time is equal to the number of molecules entering the gas phase from the liquid phase, so it is a dynamic equilibrium. At a certain temperature, T_1, when the equilibrium is established, the amount of gas molecules B, n_1 and the volume, V are constants in the closed container. According to the equation of the ideal gas law, the relationship is:

$$p_1 = \frac{n_1}{V} RT_1 \tag{7-1}$$

as we know from equation (7-1), p_1 is a definite value, called the saturated vapor pressure of the substance, and vapor pressure for short. Thus, it can be seen that the saturated vapor pressure of any liquid is a constant value at a certain temperature.

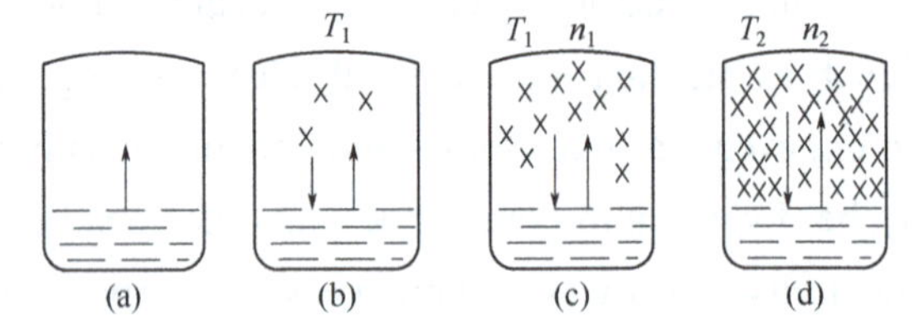

Fig.7-1 Schematic diagram of liquid evaporation and vapor condensation process

(a) Initial evaporation at temperature T_1; (b) Evaporation and condensation do not reach equilibrium at temperature T_1, the evaporation rate is greater than the condensation rate; (c) Equilibrium between evaporation and condensation at temperature T_1, the evaporation rate is equal to the condensation rate; (d) The temperature rises to T_2, and the evaporation and condensation reached equilibrium (The length of the arrow indicates the size of the evaporation rate or condensation rate)

When the temperature rises to T_2 ($>T_1$), the kinetic energy of the molecule increases. More molecules escape from unit volume per unit time, that is, the evaporation rate accelerates. As shown in Fig.7-1(d), n_2 is obviously larger than n_1; according to the ideal gas state equation,

p_2 would be greater than p_1. That is to say, the vapor pressure increases with the increase of temperature.

As can be seen from the ideal gas law, pressure is proportional to the product of n and T, and n changes with T. Hence, the curve of the relationship between vapor pressure and temperature is not a straight line.

Vapor pressure is related to the nature of the liquid, but not the amount of the liquid. For example, at 20℃ the vapor pressures are 2.339kPa, 5.847kPa and 58.932kPa for water, alcohol and ether, respectively. Usually, the higher the vapor pressure, the more volatile the substance. At a certain temperature, the value of vapor pressure for each liquid is a constant. When the temperature increase, the value of vapor pressure increases (Table 7-1).

Table 7-1 Vapor Pressure at Different Temperatures

Temperatures /℃	0	10	20	25	30	40	60
Vapor pressure /kPa	0.61	1.228	2.339	3.169	4.246	7.381	19.932
Temperatures /℃	80	100	120	150	200	375	
Vapor pressure /kPa	47.373	101.3	198.5	476.1	1554.5	22061.7	

Molecules on the solid surface can also evaporate. If you put a solid in a closed container, balance can also be reached between the solid phase and its gas phase. The vapor pressure of solid also increases with the increase of temperature. The vapor pressure of ice at different temperatures is listed in Table 7-2.

Table 7-2 Vapor Pressure of Ice at Different Temperatures

Temperature/℃	0	−1	−5	−10	−15	−20	−25
Vapor pressure /kPa	0.61	0.563	0.401	0.260	0.165	0.104	0.064

Taking vapor pressure as longitudinal coordinate and temperature as transverse coordinate, the vapor pressure curve of water and ice is drawn in the Fig.7-2.

Through a large number of experimental results, it is found that the vapor pressure of the formed solution is lower than that of the pure solvent when any non-electrolyte material is dissolved in a solvent. This is because, after the solute is dissolved into the solvent, each solute molecule binds to a number of solvent molecules to form a solvation molecule, which on the one hand reduces some high energy solvent molecules, on the other hand, occupies part of the surface of the solvent, resulting in the decrease of solvent molecules escaping per unit surface and unit time. Therefore, when the temperature is unchanged and the equilibrium is reached, the vapor pressure of nonvolatile solution is lower than that of pure solvent. At the same temperature, the difference between the vapor pressure of the pure solvent and the solution is called the decrease of the vapor pressure of the solution. Obviously, the higher the concentration of the solution, the more the vapor pressure of the solution decreases. In Fig.7-2, when any nonvolatile non-electrolyte is dissolved in the water, due to the decrease of the vapor pressure of the solution, the *aa′* line moves down to *bb′* line.

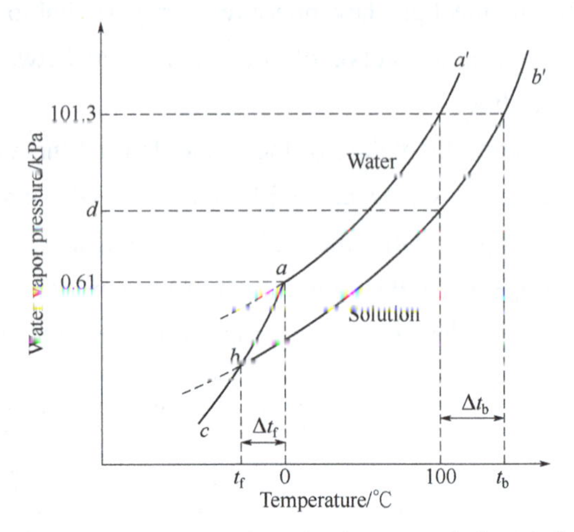

Fig.7-2 Vapor pressure lowering in a nonvolatile solution

In 1887, French physicist F. M. Raoult obtained the relationship between the vapor pressure and the amount of solute in the nonvolatile dilute solution according to the experimental results. It is

$$\Delta p = \frac{n_B}{n_B + n_A} p^* \tag{7-2}$$

Δp in formulae indicates that the vapor pressure of the solution decreases; p^* represents the vapor pressure of the pure solvent; n_A represents the amount of the solvent; n_B represents the amount of the solute. For dilute solutions, n_A is far greater than n_B, thus n_B can be ignored when calculate.

$$\Delta p \approx \frac{n_B}{n_A} p^* \tag{7-3}$$

Formula (7-2) and (7-3) can be explained as follows: at a certain temperature, the vapor pressure of the nonvolatile non-electrolyte dilute solution decreases which is proportional to the mole fraction of solute B and the vapor pressure of the solvent, but not related to the nature of the solute. This is called Raoult's Law.

Some solid substances are commonly used as desiccants, which make use of the property of vapor pressure drop. Substances such as calcium chloride, $CaCl_2$, phosphorus pentoxide electromagnetic, P_2O_5, *etc.* absorb water in the air and deliquescence. After the surface of these solid substances absorb water, the solution of the substance is formed, its vapor pressure is smaller than the partial pressure of water vapor in the air. As a result, the water vapor in the air continues to condense into the solution, so that these substances continue to hydrolyze.

Example 7-1 According to Raoult's Law, the vapor pressure of the solution decreases only with the mole fraction of solute B. At 293K, how much has the vapor pressure of the following aqueous solutions decreased? What problems do the results of calculation illustrate? (molar mass of sucrose *M*=342.3g/mol and the molar mass of urea *M*= 60.1g/mol)

(1) 17.1g sucrose solution in 100.0g water; (2) 1.5g urea is dissolved in 50.0g water.

Solution At 293K, the vapor pressure of water is 2.34kPa

(1) In sucrose aqueous solution

$$x_B = \frac{n_B}{n_A + n_B}$$
$$= \frac{17.1/342.3}{17.1/342.3 + 100.0/18.0} = \frac{0.05}{0.05 + 5.56}$$
$$= \frac{0.05}{5.61} = 8.91 \times 10^{-3}$$

According to formula (7-2) $\Delta p = \dfrac{n_B}{n_B + n_A} p^*$

There is $\Delta p = 2.34\text{kPa} \times 8.91 \times 10^{-3} = 2.08 \times 10^{-2}\text{kPa}$

(2) In urea aqueous solution

$$x_B = \frac{1.5/60.1}{50.0/18.0 + 1.5/60.1}$$
$$= 8.91 \times 10^{-3}$$

$$\Delta p = 2.34\text{kPa} \times 8.91 \times 10^{-3} = 2.08 \times 10^{-2}\text{kPa}$$

The calculation shows that the vapor pressure of both sucrose dilute solution and urea dilute solution is lower than that of pure water, and as long as the mole fraction of solute is the same, the vapor pressure reduction value of dilute solution is the same.

7.1.2 Freezing Point Depression and Boiling Point Elevation

Pure matter has a certain boiling point and freezing point. However, when the nonvolatile non electrolyte solute is added to the solvent, the boiling point and freezing point of the solution change due to the decrease of vapor pressure of the solution. The water solution in Fig.7-2 is taken as an example to illustrate. In the figure, the lines *aa'*, *ac* and *bb'* represent the relationship between vapor pressure and temperature of water, ice and aqueous solution, respectively.

We know that boiling point refers to the temperature at which the vapor pressure of the liquid is equal to the external pressure. It (*bb'* line) is always lower than the pure solvent vapor pressure of water (*aa'* line). The decrease of vapor pressure of solution is bound to the change of boiling point. Under 101.325kPa, the boiling point of pure water is 100℃. At this time, the vapor pressure of pure water is equal to 101.325kPa. Under the same temperature, the vapor pressure of the solution must be lower than 101.325kPa (That is, lower than the external atmospheric pressure). In the figure, at point *d,* the solution does not boil. In order for the solution to boil, the temperature of the solution must be raised to t_b. Therefore, the boiling point of the solution is always higher than that of the pure solvent (t_b>100℃). The difference between the boiling point of the solution and the boiling point of the pure solvent Δt_b is called the increase of the boiling point of the solution.

The freezing point of the liquid is the temperature at which the pure liquid is balanced with its solid at a certain external pressure. The freezing point of the liquid under 101.325kPa is the normal freezing point of the liquid, while the vapor pressure of the solid is equal to the vapor pressure of the liquid. For example, the freezing point of water at normal pressure is 273.15K, and at this time, the vapor pressure of liquid water is equal to that of ice, both of which are 0.61kPa located at the intersect of *aa'* line and *ac* line in the figure. Here, the freezing point of the solution is the temperature corresponding to the intersection of the line *ac* and line *bb'*.

As the vapor pressure of the solution decreases, the vapor pressure of the solution at 0℃ is less than that of ice (0.61kPa as shown in Fig.7-2). Although the vapor pressure of ice and solution decreases with the decrease of temperature, the degree of decrease of vapor pressure of ice is greater than that of solution vapor pressure, so the ratio of *ac* line is larger than the ratio of the *aa'* line. When the temperature of the system is below zero degree celsius, the vapor pressure of ice and solution is equal (less than 0.61kPa). The solid phase, ice and the liquid phase, solution reach a phase equilibrium, and the temperature is the freezing point of the solution, that is the temperature corresponding to the intersection point, t_f of line *ac* and line *bb* in Fig.7-2. It is lower than the freezing point of water (t_f<0℃). The difference between the freezing point of the solution and the freezing point of the pure solvent Δt_f is the decrease of the freezing point of the solution.

The freezing point depression and boiling point elevation caused by the decrease of vapor pressure, and the decrease of vapor pressure is related to the concentration of solution. Therefore, the freezing point depression and boiling point elevation also depend on the concentration of solution. According to the experiment, the following laws are summarized by Raoult's: the boiling point elevation and freezing point depression of nonvolatile non-electrolyte dilute solution are proportional to the mass molar concentration of the solution, which is expressed as follows:

$$\Delta p \approx \frac{x_B}{x_A} p^* = \frac{m_B}{1000/18} p^* = K\, m_B \tag{7-4}$$

$$\Delta t_b = K_b m_B \tag{7-5}$$

$$\Delta t_f = K_f m_B \tag{7-6}$$

Here, m_B is the mass molar concentration of the solution; Δt_b and Δt_f represent the freezing point depression and boiling point elevation of the solution, respectively; K_b and K_f are the boiling point elevation constant and the freezing point depression constant of the solvent, respectively.

K_b or K_f is determined by the nature of solvent, but not the nature of solute. The values of K_b and K_f varies with properties of solvents. In Table 7-3, K_b and K_f of some common solvents are given.

Table 7-3 The Values of K_b and K_f of Some Commonly Used Solvents

Solvent	Boiling point /℃	K_b/(℃·kg/mol)	Freezing point/℃	K_f/(℃·kg/mol)
Acetic acid	118.1	2.93	17	3.9
Benzene	80.2	2.53	5.4	5.12
Chloroform	61.2	3.63	−63.5	4.68
Naphthalene	—	—	80	6.8
Water	100	0.51	0	1.86

If different solute is dissolved in the same solvent, as long as the mass concentration of the solution is equal (that is, the number of solute particles in a certain amount of solvent is equal), the degree of boiling point elevation and freezing point depression must be equal respectively. For example, dissolve 0.1mol sucrose in 1000g water (the mass molar concentration of the solution is 0.1mol/kg, the freezing point depression Δt_f is 0.186℃, and dissolve 0.1mol glucose in 1000g water (the molar concentration of the solution is still 0.1mol/kg), Δt_f is also 0.186℃.

According to Raoult's Law, the relative molecular weight of solute can be calculated by measuring the freezing point depression or the boiling point elevation. Because the value of freezing point depression is easy to be measured, it is often used to determine the relative molecular weight of unknown substances.

Example 7-2 In winter, when temperature dropped to minus 10 degree celsius, in order to prevent the water from freezing in the car water tank, how many kilos of ethylene glycol can be added to 10kg water to achieve the effect of antifreeze? Please explain by calculation [K_f of water is 1.86(℃·kg/mol). Relative molecular weight of ethylene glycol is 62g/mol].

Solution According to the formula (7-6), the mass molar concentration of ethylene glycol is as follows:

$$m_B = \frac{\Delta t_f}{K_f} = \frac{0-(-10)}{1.86} = 5.3763(\text{mol/kg})$$

Hence, the mass of ethylene glycol in 10kg water is:

$$5.3763(\text{mol/kg})\times 10\text{kg}\times 62(\text{g/mol})\times 10^{-3} = 3.33\text{kg}$$

That is to say, adding 3.33kg ethylene glycol to 10kg water can make the water in the tank not freeze at −10℃, so as to achieve the effect of antifreeze.

If the concentration of the solution is increased, the influence between the solute molecules and the interaction between the solute and the solvent are greatly enhanced, then the concentration in the mathematical formula of Raoult's Law should be the effective concentration (also known as activity). For electrolyte solutions, particles are dissociated by solute (molecules and ions) with the increase of the total concentration, the value of Δp, Δt_b and Δt_f is larger than that of the non-electrolyte solution with the same concentration. For example, for 0.1mol/kg of HAc solution, due to the separation of HAc,

$$HAc \rightleftharpoons H^+ + Ac^-$$

in the solution, the total concentration of the three particles HAc, H^+ and Ac^- is bound to be greater than 0.1mol/kg. According to the formula (7-6), Δt_f is bound to be greater than the Δt_f of 0.1mol/kg sugar solution. However, the general degree of dissociation of strong electrolytes in dilute solution is always calculated at 100%. For example, in 0.1mol/kg $Ca(NO_3)_2$, the concentration of Ca^{2+} in the solution is 0.1mol/kg and the concentration of NO_3^- is 0.2mol/kg. Hence, its Δt_f is approximately three times of the same concentration of non-electrolyte solution.

We know that as the altitude rises, the atmospheric pressure decreases. The higher the altitude, the lower the boiling point. As a result, the boiling point at high altitude is lower than that at low altitude (<100℃). Then, when cooking food, it will be difficult to cooked because of the low temperature. The most commonly used pressure cooker in our daily life is to increase the pressure and raise the temperature so as to speed up the cooked food and solve the problem that it is difficult for the food to be cooked at high altitude. Another example: when we cook dumplings and other food, we add a tablespoon of salt to the water, and the dumpling or food skin tendons are easy to cook, which is the principle of boiling point elevation of the solution. It is also of great practical significance to reduce the freezing point of the solution, because when the dilute solution reaches the freezing point, the water in the solution begins to form ice and precipitates. With the precipitation of ice, the concentration of the solution increases and the freezing point decreases. Finally, when the concentration of the solution reaches the saturated solution concentration of the solute (solubility), ice and solute precipitate together (ice crystal eutectoid). At this time, although the solution continues to be cooled, the solidification temperature remains the same until the solution is completely solidified. When $CaCl_2 \cdot 2H_2O$ is mixed with ice, the temperature of the mixture can be reduced to −55℃. Hence, it can be used as a freezer. On the other hand, the freezing point of the solution can also be reduced, and some solute can be added to the solvent to prevent the solvent from solidification. For example, salt or calcium chloride is required in sand oars used in the construction industry in winter; in car radiators (water tank), alcohol or ethylene glycol is added to the water to prevent the water from freezing by lowering the freezing point of the solution.

7.1.3 Osmotic Pressure of Solution

In addition to the decrease of vapor pressure, the boiling point elevation and the freezing point depression, there is also a property of osmotic phenomenon. Infiltration must be carried out through a semi-permeable membrane that allows only solvent molecules to pass through. As shown in Fig.7-3, the solution (A) and pure solvent (B) was separated with a semi-permeable membrane. In a unit time, the number of solvent molecules entering the solution (A) from the pure solvent(B) is larger than that from the solution(A) to pure solvent(B). The process by which the solvent enters the solution through the semi-permeable membrane is called infiltration. As a result of infiltration, the volume of the solution increases gradually, and the liquid level in the vertical glass tube increases gradually. With the increase of the liquid level in the tube, the static pressure of the liquid column in the tube increases gradually, which accelerates the outward infiltration of the solvent in the tube and finally reaches the height of the liquid column. at this time, the number of solvent molecules passing through the semi-permeable membrane in two opposite directions is equal to each other, that is to say, the osmotic equilibrium is reached. The pressure above the liquid surface of the solution due to the height of the liquid column, it is the osmotic pressure of the solution. Therefore, osmotic pressure is the minimum additional pressure applied to the solution by preventing the solvent from flowing into the solution through the semi-permeable membrane.

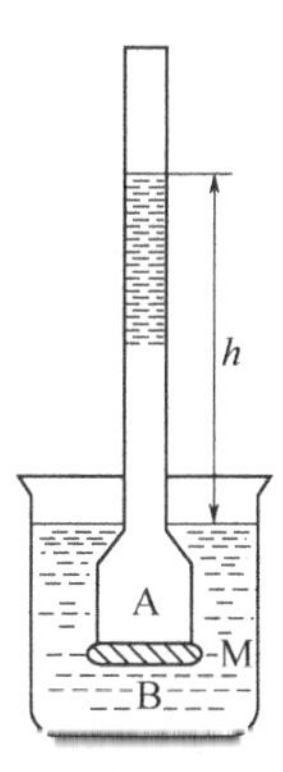

Fig.7-3 Osmotic pressure diagram

A—solution; B—pure solvent; M—semi-permeable membrane

According to the experimental results, the osmotic pressure of dilute solution is proportional to the concentration of solution when the temperature is constant, and when the concentration is constant, the osmotic pressure of dilute solution is proportional to the absolute temperature. Set Π on behalf of osmotic pressure, c represents the concentration of a substance, T represent absolute temperature, n represents the amount of substance that represents solute and V represents the volume of the solution, then

$$\Pi = cRT$$

for dilute solutions
$$c = \frac{n}{V}$$

hence,
$$\Pi = \frac{n}{V}RT$$

or
$$\Pi V = nRT \tag{7-7}$$

This equation (7-7) is called the van't Hoff osmotic pressure formula, which found by the Dutch physical chemist J. H. van't Hoff. It has a similar form with ideal gas law, but the physical significant is completely different. Gas produces pressure because its molecular motion collides with the wall of the capacitance, and the osmotic pressure of the solution is not the result of the direct movement of solute molecules. When the concentration of the actual solution is smaller, the calculation error of formula (7-7) is smaller. According to formula (7-7), it can be seen that the osmotic pressure of the solution is only related to temperature and solute concentration, but not to the type of solute. Hence, it is also a manifestation of colligative properties.

Osmotic pressure is of great significance in biology. Most of the cell membranes of organisms have the property of semi-permeable membrane, so osmotic pressure is the main force that causes water to move in animals and plants. The osmotic pressure of plant cell juice can reach 20×101.325kPa. Hence, we see that water can even be transported from the roots of plants to the top of dozens of meters high. In winter, we use salt to prevent ice on the road. If such snow is stacked on the root of the plant, it will cause the plant to wither in a large area. It is precisely because the concentration of salt outside the skin of the plant is higher than that in the epidermis that the water in the plant will continue to precipitate until it dries up and dies. The osmotic pressure of human blood is about 7.7×101.325kPa. Because the human body has the requirement to keep osmotic pressure in the normal range, when we eat too much food and perspiration strongly, we will feel thirsty because of the increase of osmotic pressure in the tissue. Drinking water can reduce the concentration of soluble matter in the tissue, thus reducing the osmotic pressure.

Example 7-3 The decrease of freezing point of human blood measured is 0.56℃, try to calculate the osmotic pressure of blood under the human body temperature of 37℃? K_f= 1.86(℃·kg/mol).

Solution　From the formula $\Delta t_f = K_f m_B$, the mass molar concentration of human blood is

$$m_B = \frac{\Delta t_f}{K_f}$$

That is,

$$m_B = \frac{0.56}{1.86} = 0.3011(\text{mol/kg})$$

For dilute solution, the mass molar concentration is approximately equal to the mass concentration of the substance c_B. Therefore, the concentration of substances in the blood c_B= 0.3011mol/L, which is equivalent to $3.011\times10^2 mol/m^3$.

At temperature 310K, by the formula $\Pi = cRT$

There is

$$\Pi = 3.011\times10^2 (mol/m^3)\times 8.314 J/(mol\cdot K)\times 310K = 776kPa$$

That is, the osmotic pressure of human blood is 776kPa under the body temperature.

To sum up, the law of dilute solution can be summarized as follows: the properties of dilute solution of nonvolatile solute (vapor pressure decrease, boiling point elevation, freezing point depression and osmotic pressure change) are proportional to the number of solute particles dissolved in a certain amount of solvent (or a certain volume of solution), but independent of the nature of solute. The law of dependence of dilute solution is finite law, the thinner the solution, the more accurate the law.

7.2 Weak Electrolyte Dissociation Equilibrium and Buffer Solution

Acid-base reaction is one of the most familiar reactions. However, it is not easy to determine whether a substance is acidic or alkaline. We need to understand the development of acid-base theory, and then clarify the definition of acid-base and acid-base dissociation equilibrium.

7.2.1 The Development of Acid-Base Theory

(1) Arrhenius ionizing theory

The ionization theory of acid and base is proposed by the Swedish chemist Arrhenius in 1884. The definition about acid and base is from Arrhenius: The substance is acid when all cations ionized in aqueous solution are H^+; the substance is alkali when all the ionized anions are OH^-. The Arrhenius acid-base theory is based on the ionization of electrolytes in aqueous solution. The essence of acid-base reaction is: the acid containing H^+ and the bases containing OH^- neutralize with each other and produce H_2O and salts. Because Arrhenius's acid-base theory is based on aqueous solution, the application of this theory is limited. There are many reactions that are not carried out in aqueous solution but they show acidity and alkalinity. For example, gaseous ammonia reacts with gaseous hydrogen chloride to form solid ammonium chloride. According to the theory of acid-base ionization, it is impossible to judge that this is a reaction between acid (HCl) and base (NH_3). In addition, there are many acids and bases that do not contain H^+ or OH^-. However, it shows acidity and alkalinity. For example: there is no OH^- in ammonia (NH_3) in the above example but it shows alkalinity. While in zinc chloride

($ZnCl_2$) there is no existence of H^+ but it is manifested as acid. Obviously, some of these problems cannot be solved in the Arrhenius acid-base theory, and then the proton theory of acid-base is developed.

(2) Brϕnsted-Lowry proton theory

In 1923, the Danish chemist J. N. Brϕnsted and the British chemist T. M. Lowry independently proposed the proton theory of acid and base almost at the same time, which is called Brϕnsted-Lowry acid-base proton theory. According to their theory, any specie that can donor proton is an acid and any specie that can accept proton is a base. For example: HCl, HS^- and $[Al(OH)(H_2O)_5]^{2+}$ are species that give protons, so they are all acids; NaOH, Ac^- and NH_3 are species that accept protons, so they are all bases. Acids and bases change through proton transfer, and we call this relationship a conjugated relationship. For conjugated acid-base pairs, the stronger the acid is, the weaker the conjugated base of the acid is. On the contrary, the stronger the base is, the weaker the conjugated acid of the base is. Because acids and bases change through proton transfer, many species have two sides, and their acidity or alkalinity depends on the species with which they react. For example:

$$\underset{\text{acid(1)}}{HCl(g)} + \underset{\text{base(2)}}{NH_3(g)} \rightleftharpoons \underset{\text{acid(2)}}{NH_4^+} + \underset{\text{base(1)}}{Cl^-}$$

From the above reaction, HCl molecule donors the proton and turns into its conjugated base Cl^- and NH_3 molecule gets protons and becomes its conjugated acid NH_4^+. We can label the HCl/Cl^- combination as "(1)" and NH_4^+/NH_3 as "(2)." Each combination is a conjugated pair. The essence of acid-base reaction is the process of proton transfer between two conjugated acid-base pairs. Let us give a few more examples:

Neutralizing reaction

$$\underset{\text{acid(1)}}{HCl} + \underset{\text{base(2)}}{NaOH} \rightleftharpoons \underset{\text{base(1)}}{NaCl} + \underset{\text{acid(2)}}{H_2O}$$

Ionization reaction

$$\underset{\text{acid(1)}}{HCl} + \underset{\text{base(2)}}{H_2O} \rightleftharpoons \underset{\text{base(1)}}{Cl^-} + \underset{\text{acid(2)}}{H_3O^+}$$

Self ionization reaction

$$\underset{\text{acid(1)}}{H_2O} + \underset{\text{base(2)}}{H_2O} \rightleftharpoons \underset{\text{base(1)}}{OH^-} + \underset{\text{acid(2)}}{H_3O^+}$$

The essence of the above reactions is proton transfer. The common conjugated relationship between acid and base is shown in Table 7-4.

Table 7-4 Conjugate Relationship of Common Acids and Bases

Acid	Formula	Conjugate base
hydroiodic acid	HI	I^-
perchloric acid	$HClO_4$	ClO_4^-
hydrochloric acid	HCl	Cl^-
sulfuric acid	H_2SO_4	HSO_4^-
hydrogen sulfate ion	HSO_4^-	SO_4^{2-}
nitric acid	HNO_3	NO_3^-
acetic acid	HAc	Ac^-
zinc hexahydrate complex	$[Zn(H_2O)_6]^{2+}$	$[Zn(OH)(H_2O)_5]^+$
ammonium ion	NH_4^+	NH_3
water	H_2O	OH^-
hydronium ion	H_3O^+	H_2O

The proton theory of acid-base extends the range of acid-base from the aqueous solution to the solution formed by various solvents, and further expands to the reaction of gas phase, liquid phase and solid phase. Therefore, the application of acid-base proton theory is more extensive than that of acid-base ionization theory. However, its application is still confined only to those reaction systems that contain protons, and it is difficult to explain these processes in which there are no protons in those reactions. For example: the concentrated $ZnCl_2$ solution is acidic, but it cannot be explained reasonably by proton theory. American chemist G. N. Lewis proposed that acid-based electronic theory can better solve this problem. You may find further details in other reference books.

7.2.2 The Self-ionization of Water and the pH Scale

(1) Self-ionization equilibrium of water

$$H_2O(l) + H_2O(l) \rightleftharpoons OH^-(aq) + H_3O^+(aq)$$

The expression of its standard equilibrium constant is:

$$K_T^\ominus = \left[\left(\frac{c(H_3O^+)}{c^\ominus}\right)\left(\frac{c(OH^-)}{c^\ominus}\right)\right] = K_w^\ominus \qquad (7\text{-}8)$$

The standard ion product constant of water in formula (7-8) is independent of concentration, but related to temperature. Usually, with the increase of temperature, the ion product constant of water increases gradually. At 0℃, the ion product constant of water is 1.15×10^{-15}, at 25℃ is 1.01×10^{-14}, at 100℃ is 5.43×10^{-13}. From 0℃ to 100℃, the ion product constant of water varies by two orders of magnitude. The ion product constant of the water we usually use is 1.0×10^{-14}. It is just a value at room temperature.

(2) pH and pOH scale

From the ionization of water, it can be seen that OH^- and H^+ exist in aqueous solution at the same time. If the concentration of OH^- and H^+ changes, the acid and alkalinity of the solution change with it. When $c(H^+) = c(OH^-) = 10^{-7}$mol/L, the solution is neutral. When the concentration $c(H^+)$ is larger than $c(OH^-)$, the solution is acidic; on the contrary, when the concentration $c(H^+)$ is smaller than $c(OH^-)$, the solution is alkaline. Thus, it can be seen that the acid and alkalinity of the solution are relative.

The acid and base solution are often represented by the negative logarithm of $c(H^+)$ or $c(OH^-)$, which is

$$pH = -\lg c(H^+)$$

or

$$pOH = -\lg c(OH^-) \tag{7-9}$$

According to the ion product constant of water, there are

$$K_w^\ominus = c_r(H_3O^+)c_r(OH^-)$$

After taking the negative logarithm on both sides at the same time, there are

$$pK_w^\ominus = pH + pOH=14 \tag{7-10}$$

that is, when $c(H^+)=10^{-7}$, the pH of the solution is 7 and the solution is neutral. When $c(H^+)>10^{-7}$, the pH of the solution is smaller than 7 and the solution is acidic. When $c(H^+)<10^{-7}$, the pH of the solution is larger than 7 and the solution is alkaline. Table 7-5 lists the pH of several common liquids.

Table 7-5 pH of Some Common Liquids

Name	pH	Name	pH
gastric juice	1.0-3.0	milk	6.5
vinegar	3.0	distilled water	7.0
orange juice	3.5	blood	7.35-7.45
urine	4.8-8.4	water exposed to air	5.5-7.0

pH can be tested by pH test-paper or pH meter. With the use of pH meter, the automatic control of industrial process can be realized quickly and accurately.

7.2.3 Single-phase Dissociation Equilibrium: The Dissociation of Weak Acid and Weak Base

(1) The dissociation equilibrium of monoprotic weak acid and weak base

In aqueous solution, the dissociation equilibrium of arbitrary HX weak electrolyte is as follows:

$$HX(aq) + H_2O(l) \rightleftharpoons H_3O^+(aq) + X^-(aq)$$

When equilibrium is reached, the expression of its standard equilibrium constant is as follows:

$$K_i^{\ominus}=\frac{[c(\mathrm{H^+})/c^{\ominus}][c(\mathrm{X^-})/c^{\ominus}]}{[c(\mathrm{HX})/c^{\ominus}]}$$

or replace $c/c^{\ominus}$ by c_r, there is a formula:

$$K_i^{\ominus}=\frac{c_r(\mathrm{H^+})c_r(\mathrm{X^-})}{c_r(\mathrm{HX})} \tag{7-11}$$

$K_i^{\ominus}$ is called the standard dissociation constant of the weak electrolyte. In general, $K_i^{\ominus}$ is independent of concentration, but related to temperature. However, the effect in aqueous solution is not great, and the temperature range of water in liquid is small, so the reaction in aqueous solution can ignore the effect of temperature on $K_i^{\ominus}$. The numerical value of $K_i^{\ominus}$ can be used to represent the degree of dissociation of arbitrary weak electrolytes, and it can also be used to represent the relative strength of arbitrary weak electrolytes. $K_i^{\ominus}$ is represented by $K_a^{\ominus}$ (acid) when it is weak acid, while it is represented by $K_b^{\ominus}$ (base) when it is a weak base. For example: for CH_3COOH (abbreviated as HAc), there are the following dissociation equilibrium in aqueous solution:

$$\mathrm{HAc} \rightleftharpoons \mathrm{H^+}+\mathrm{Ac^-}$$

When equilibrium is achieved, it can be expressed as follows:

$$K_a^{\ominus}=\frac{[c(\mathrm{H^+})/c^{\ominus}][c(\mathrm{Ac^-})/c^{\ominus}]}{[c(\mathrm{HAc})/c^{\ominus}]} \tag{7-12}$$

The standard dissociation constant of acetic acid in formula (7-12), and standard dissociation constant of some common weak electrolytes are shown in Appendix Ⅲ. For conjugate pairs, we list acids, its conjugate base can be converted by:

The dissociation of an acid is $\mathrm{HX(aq)}+\mathrm{H_2O(l)} \rightleftharpoons \mathrm{H_3O^+(aq)}+\mathrm{X^-(aq)}$

The dissociation of conjugated base is $\mathrm{X^-(aq)}+\mathrm{H_2O(l)} \rightleftharpoons \mathrm{OH^-(aq)}+\mathrm{HX(aq)}$

Hence, there is

$$K_a^{\ominus}\cdot K_b^{\ominus}=c_r(\mathrm{H_3O^+})\cdot c_r(\mathrm{OH^-})=K_w^{\ominus} \tag{7-13}$$

In other words, the $K_a^{\ominus}$ and $K_b^{\ominus}$ of conjugated acid-base pairs can be converted to each other, but the $K_a^{\ominus}$ and $K_b^{\ominus}$ of acids and bases that are not conjugated acid-base relations cannot be obtained from one ($K_a^{\ominus}$ or $K_b^{\ominus}$) to the other ($K_b^{\ominus}$ or $K_a^{\ominus}$).

The initial concentration of any weak electrolyte AB is set to be c(AB). When the dissociation degree is α, the relationship between the concentrations of each substance is as follows:

$$\begin{array}{llll} & AB & \rightleftharpoons A^+ & + \quad B^- \\ \text{initial} & c & 0 & 0 \\ \text{equilibrium} & c-c\alpha & c\alpha & c\alpha \end{array}$$

$$K_i^{\ominus}(AB) = \frac{c_r(A^+)c_r(B^-)}{c_r(AB)}$$

$$K_i^{\ominus}(AB) = \frac{(c\alpha)^2}{c(1-\alpha)} = \frac{c\alpha^2}{1-\alpha} \tag{7-14}$$

Usually, when $K_i^{\ominus} < 10^{-4}$ and $c(AB) > 0.1\text{mol/L}$, the degree of dissociation is very small, and it is often possible to ignore the dissociated part, $(1-\alpha)\approx1$. Hence, there is

$$K_i^{\ominus} \approx c\alpha^2 \quad \text{or} \quad \alpha = \sqrt{\frac{K_i^{\ominus}}{c}} \tag{7-15}$$

According to the formula (7-15), it can be seen that at a certain temperature $K_i^{\ominus}$ is a constant, so the smaller the solution concentration, the greater the degree of dissociation. This relation is also called dilution law. Particularly, the formula (7-15) can only be used in unary weak acid and weak base system, but not suitable for multivariate weak acid and weak base system, and it can be used in systems where only one component exists. Of course, the value of K_i can be used to represent the degree of dissociation of any weak electrolyte, because it is a standard equilibrium constant and the value is independent of the solution concentration.

When the initial concentration of a weak acid is c_0 and the concentration of H^+ formed by dissociation is x. The concentrations of each species in pure monoprotic weak acid and weak base are

$$\begin{array}{llll} & HX & \rightleftharpoons H^+ & + \quad X^- \\ \text{initial} & c_0 & 0 & 0 \\ \text{equilibrium} & c_0-x & x & x \end{array}$$

When the equilibrium is established,

$$K_a^{\ominus} = \frac{x^2}{c_0 - x} \tag{7-16}$$

When $K_i^{\ominus} < 10^{-4}$, and $c(HX) > 0.1\text{mol/L}$. It can be approximately considered that $(1-\alpha) \approx 1$, thus

$$x = c(H^+) = \sqrt{K_a^{\ominus} c_0} \tag{7-17}$$

At this point, the solution $pH = -\lg c(H^+)$, the degree of release is:

$$\alpha = \frac{c(H^+)}{c_0} \tag{7-18}$$

By the same reason, the correlation formula of univariate weak base can be obtained as follows:

$$K_b^\ominus = \frac{x^2}{c_0 - x} \tag{7-19}$$

Approximate calculation

$$x = c(OH^-) = \sqrt{K_b^\ominus c_0} \tag{7-20}$$

At this point, the solution pOH = $-\lg c(OH^-)$, the degree of release is

$$\alpha = \frac{c(OH^-)}{c_0} \tag{7-21}$$

If the approximate calculation conditions cannot be satisfied, the concentration of hydrogen ion in the solution must be obtained by solving the unitary quadratic equation, and then the pH of the solution must be obtained, and the degree of dissociation of the weak electrolyte.

(2) The dissociation equilibrium of polyprotic weak acids

The dissociation process of multicomponent weak acids is graded, and each stage has its standard dissociation constant, such as: H_2S the hierarchical dissociation balance is as follows

$$H_2S \rightleftharpoons H^+ + HS^- \qquad K_{a1}^\ominus = \frac{c_r(H^+)c_r(HS^-)}{c_r(H_2S)} = 8.91\times10^{-8}$$

$$HS^- \rightleftharpoons H^+ + S^{2-} \qquad K_{a2}^\ominus = \frac{c_r(H^+)c_r(S^{2-})}{c_r(HS^-)} = 1.1\times10^{-19}$$

The total dissociation equilibrium is

$$H_2S \rightleftharpoons 2H^+ + S^{2-}$$

According to the rules of multiple equilibrium, the equilibrium constant for the total equation is

$$K_a^\ominus = K_{a1}^\ominus K_{a2}^\ominus = \frac{c_r^2(H^+)c_r(S^{2-})}{c_r(H_2S)}$$

Generally speaking, it is difficult to dissociate H^+ ions from negatively charged acid roots (*e.g.*, HS^-). At the same time, the H^+ generated by first-order dissociation prevent the forward reaction in second-order dissociation equilibrium. So, the standard dissociation constant of polyprotic weak acids is significantly reduced in the order of $K_{a1}^\ominus >> K_{a2}^\ominus >> K_{a3}^\ominus$. Therefore, when comparing the acidity of polyprotic weak acids, they can be preliminarily determined by comparing the first-order dissociation constant. However, if there is no significant difference between $K_{a1}^{\ominus}$ and $K_{a2}^{\ominus}$, the second-order should be considered. In addition, the dissociation of water should also be considered if necessary.

It should also be noted that we cannot distinguish from which step the hydrogen ion dissociates. Therefore, the H^+ ion concentration in the dissociation constant formula of each level refers to the total H^+ ion concentration in the solution. In practical calculations, the concentration of first-order dissociated H^+ ions can often be approximated according to the actual situation. Polyprotic weak base has the similar situation.

Example 7-4 Calculate the concentration of H^+, OH^-, HS^- and S^{2-}, the solution pH of H_2S solution with a concentration of 0.10mol/L at 25℃.

Solution Firstly, according to the first-order dissociation equilibrium, $c(H^+)$, $c(HS^-)$ can be calculated.

Let $c(HS^-)$ is x mol/L, which are separated at the first level:

$$H_2S \rightleftharpoons H^+ + HS^-$$

Equilibrium concentration/(mol/L) $\quad 0.10-x \quad x \quad x$

$$K_{a1}^{\ominus} = \frac{c_r(H^+)c_r(HS^-)}{c_r(H_2S)} = \frac{x^2}{0.10-x} = 8.91\times10^{-8}$$

Because the value $K_{a1}^{\ominus}$, is very small, it can be calculated approximately as

$$0.10-x \approx 0.10, \qquad x^2 = 8.91\times10^{-8}$$
$$x = c(H^+) = c(HS^-) = 9.44\times10^{-5}\text{mol/L}$$

Then, according to the second order desorption equilibrium, $c(S^{2-})$, $c(OH^-)$ and solution pH can be calculated.

Let $c(S^{2-})$ is y mol/L, according to the second-order dissociation equilibrium

$$HS^- \rightleftharpoons H^+ + S^{2-}$$

Equilibrium/(mol/L) $\quad 9.44\times10^{-5}-y \quad 9.44\times10^{-5}+y \quad y$

$$K_{a2}^{\ominus} = \frac{c_r(H^+)c_r(S^{2-})}{c_r(HS^-)} = 1.0\times10^{-19}$$

Because $K_{a2}^{\ominus}$ is very small, $9.44\times10^{-5} \pm y \approx 9.44\times10^{-5}$

$$K_{a2}^{\ominus} = \frac{(9.44\times10^{-5}+y)y}{9.44\times10^{-5}-y} = 1.0\times10^{-19}$$

$$y \approx K_{a2}^{\ominus}$$

so

$$c(S^{2-}) = y = 1.0\times10^{-19}\text{mol/L}$$

and $$c_r(OH^-)=\frac{K_w^\ominus}{c_r(H^+)}=\frac{1.0\times10^{-14}}{9.44\times10^{-5}}=1.1\times10^{-10}(mol/L)$$

As can be seen from the calculation, because the value of $K_{a2}^\ominus(1.0\times10^{-19})$ is very small, the dissociation degree of HS^- ion is also very small, therefore, the concentration of H^+ and HS^- in the solution hasn't obviously changed due to the further dissociation of HS^-. The solution pH is

$$pH=-\lg(9.44\times10^{-5})=4.0$$

It can be seen from the calculation that, it is a reasonable approximate treatment to approximately replace the concentration of H^+ ions in the solution with the concentration of first order dissociated H^+ ions.

7.2.4 Buffer Solution

(1) Common ion effect

For the dissociation equilibrium of weak electrolyte, dilution increases the degree of dissociation, which is the result of the shift of dissociation equilibrium due to the change of concentration. As shown in formula (7-12), if a strong electrolyte is added to the equilibrium system of the weak electrolyte, the ion concentration in the original solution will change, and the dissociation equilibrium will also be moved. For example, in the equilibrium of acetic acid solution

$$HAc \rightleftharpoons H^+ + Ac^- \tag{A}$$

Adding strong electrolyte sodium acetate containing the same ions as weak electrolyte (NaAc), there is

$$NaAc = Na^+ + Ac^- \tag{B}$$

In original acetic acid solution, when the concentration $c(Ac^-)$ increases, the equilibrium (A) will move to the left, resulting in a decrease in the dissociation degree of acetic acid. This is because of the K_i of weak electrolytes is unchangeable, when the concentration $c(Ac^-)$ increases, the equilibrium moves to the left, the concentration $c(HAc)$ increases, and the concentration of $c(H^+)$ decreases. The result of the left shift decreases the dissociation degree of the weak electrolyte. In a word, when strong electrolytes with the same ions are added to the weak electrolyte solution, the phenomenon that the dissociation degree of the weak electrolytes can be reduced is called the common ion effect. For the HX-MX system, the calculation related the common ion effect is as follows.

After adding a strong electrolyte MX containing the same acid root ions X^-, let the concentration of H^+ dissociated from the acid is x. The equilibrium in the system can be shown as

$$HX \rightleftharpoons H^+ + X^-$$

equilibrium $\quad c_{acid}-x \quad x \quad c_{salt}+x$

$$K_a^{\ominus}=\frac{x(c_{salt}+x)}{c_{acid}-x}$$

If the approximate calculation conditions can be satisfied, then

$$K_a^{\ominus}=\frac{xc_{salt}}{c_{acid}}$$

$$c(H^+)=x=K_a^{\ominus}\cdot\frac{c_{acid}}{c_{salt}}$$

When taking the logarithm calculation on both sides of formula, we get

$$pH=pK_a^{\ominus}-\lg\frac{c_{acid}}{c_{salt}} \tag{7-22}$$

The dissociation degree of weak electrolyte is

$$a=\frac{c(H^+)}{c_{acid}}=\frac{K_a^{\ominus}}{c_{salt}} \tag{7-23}$$

In the same way, we can also take ammonia as an example to deduce and obtain similar results.

$$pOH=pK_b^{\ominus}-\lg\frac{c_{base}}{c_{salt}} \tag{7-24}$$

$$a=\frac{c(OH^-)}{c_{base}}=\frac{K_b^{\ominus}}{c_{salt}} \tag{7-25}$$

Let's take an example to see how the degree of dissociation of weak electrolytes changes with the addition of strong electrolytes containing the same ions.

Example 7-5 Try to calculate the degree of dissociation of pure HAc solution with a concentration of 0.1mol/L. If the NaAc solution with concentration of 0.2mol/L was added, then what is the degree of dissociation of HAc solution? (The volume is a constant)

Solution For the pure HAc solution, its dissociation degree can be calculated *via* formula(7-15)

$$\alpha=\sqrt{\frac{K_a^{\ominus}}{c}}=\sqrt{\frac{1.74\times10^{-5}}{0.1}}=1.32\%$$

When the NaAc solution (0.2mol/L) was added, the degree of dissociation of the solution needs to be determined by formula (7-23), there are:

$$a=\frac{K_a^{\ominus}}{c_{salt}}=\frac{1.74\times10^{-5}}{0.2}=0.0087\%$$

As can be seen from the calculation, the degree of dissociation of the pure HAc solution without NaAc is 1.32%, after the addition of NaAc solution, the degree of dissolution changed to 0.0087%. Accordingly, the solution pH changed from 2.88 to 5.06.

By using the dissociation equilibrium of weak electrolytes, a very useful solution can be obtained in chemical research and chemical production.

(2) Buffer solution

Buffer solution refers to the solution system which can resist the addition of a small amount of acid, alkali and moderate dilution, while keeping the pH of the solution basically unchanged. Such a solution is generally composed of weak acid and its salt, weak base and its salt, acid salt and secondary salt, such as HAc/NaAc, NH_3/NH_4Cl, $NaHCO_3$ and Na_2CO_3, *et al.* How does such a solution act as a buffer? What factors determine the buffering capacity? We take HAc-NaAc as an example to discuss it. In HAc-NaAc system, the concentration of HAc and NaAc are 1.0mol/L,

$$\begin{array}{ccccc} HAc & \rightleftharpoons & H^+ & + & Ac^- \\ 1.0 & & 0 & & 1.0 \end{array}$$

There's a large quantitate of HAc and Ac^- in the solution. When a small amount of acid is added, the H^+ reaction with a small number of Ac^- and produce HAc, the concentration of HAc and Ac^- don't change obviously. At the same time, the concentration of H^+ also changed a little so that pH of the solution is almost a constant. The component NaAc which can keep the pH of the solution basically unchanged is called as an acid-against ingredient. When a small amount of base is added in buffer solution, the added OH^- will react with a small number of H^+ to form H_2O. At the same time, some HAc molecule will dissociate and release H^+ ions. Therefore, the solution pH will not change obviously and the HAc is called a base-resistant ingredient.

According to the formula (7-22) and (7-24), we also can analyze the principle of buffer solution. In the buffer system of HAc-NaAc, $c_{acid} + c_{salt} = c_{total}$, the larger the buffer solution concentration c_{total}, the greater the total resistance of the buffer solution to acid and base, that is to say, the stronger the buffer capacity of the buffer solution; of course, when the ratio of c_{acid} and c_{salt} is closed to 1, the buffering ability will be the strongest. In general, the ratio of acid to alkali resistant components is between in 1∶10-10∶1. Beyond this range, it is generally believed that the solution no longer has buffer capacity. Therefore, when using the buffer solution, we should select a pH range by the value of $K_a^\ominus$ and $K_b^\ominus$. And then adjust precisely according to the ratio of the concentration of the acid or base to salt. Some common buffer pairs are shown in the Table 7-6.

Table 7-6 Some Common Buffer Solutions

Weak acid	Conjugated base	$K_a^\ominus$	pH range
Phthalic acid	Potassium hydrogen phthalate	1.3×10^{-3}	1.9-3.9
Acetic acid	Sodium acetate	1.8×10^{-5}	3.7-5.7

(Continued)

Weak acid	Conjugated base	$K_a^\ominus$	pH range
Sodium dihydrogen phosphate	Disodium hydrogen phosphate	6.2×10^{-8}	6.2-8.2
Ammonium chloride	Ammonia water	5.6×10^{-10}	8.3-10.3
Disodium hydrogen phosphate	Sodium phosphate	4.5×10^{-13}	11.3-13.3

Example 7-6 There are two solutions, one is hydrochloric acid solution (pH=5), the other is a buffer solution made up of 0.10mol/L HAc and 0.10mol/L NaAc. In each of the above two 1L solutions, 50ml 1.0mol/L of HCl solution is added separately. Please calculate the pH change of both these two solutions and make a comparison.

Solution First, for the pure hydrochloric acid solution (1L, pH = 5), after the addition of 50mL HCl solution (1.0mol/L), the volume of the HCl solution is 1.05L, in which there is about 0.05mol H^+, then

$$c(\mathrm{H}^+)=\frac{0.05}{1.05}\approx5\times10^{-2}(\mathrm{mol/L})$$

$$\mathrm{pH}=1.3$$

Compared with the original pH=5, the pH change is 3.7.

Second, for the buffer solution

$$c(\mathrm{H}^+)=K_\mathrm{a}^\ominus\frac{c(\mathrm{HAc})}{c(\mathrm{Ac}^-)}\approx1.74\times10^{-5}\times\left(\frac{0.10}{0.10}\right)=1.74\times10^{-5}(\mathrm{mol/L})$$

$$\mathrm{pH}=-\lg(1.74\times10^{-5})=4.76$$

After the addition of 50mL 1.0mol/L HCl solution into this buffer solution, nearly all the new added H^+ ions are consumed by the excessive Ac^- in the buffer solution. Therefore, the amount of HAc in solution is about 0.10+0.05=0.15(mol), and the amount of Ac^- in solution is about 0.10−0.05 = 0.05(mol)

$$\frac{c(\mathrm{HAc})}{c(\mathrm{Ac}^-)}=\frac{0.15/1.05}{0.05/1.05}=\frac{0.15}{0.05}=3$$

According to the expression equation of the standard dissociation constant of HAc, we obtain

$$c(\mathrm{H}^+)=K_\mathrm{a}^\ominus\frac{c(\mathrm{HAc})}{c(\mathrm{Ac}^-)}=3\times K_\mathrm{a}^\ominus\approx5.22\times10^{-5}$$

$$\mathrm{pH}=-\lg(5.22\times10^{-5})=5-0.72=4.28$$

Compared with the original pH value 4.76, its pH only changed 0.48.

When 50mL 1.0mol/L NaOH solution was added to the two 1L solutions respectively, similar calculations showed that the pH value of HCl solution would increase to 12.7 with a change of 7.7, while the pH value of the buffer solution increased from 4.76 to 5.24 with a change of only 0.48. The buffer effect of the buffer solution was significant.

Buffer solution is of great significance in industry, scientific research and so on. For example, the oxide on the silicon wafer (SiO_2) of semiconductor devices can usually be cleaned with the mixture of HF and NH_4F to make SiO_2 become SiF_4 gas and be removed. A buffer solution is often used to control a certain pH range in electroplating of metal devices. There are also complex and special buffers in plants and animals that keep the pH of the fluids in order to keep life going. For example, the buffer system of organic hemoglobin and plasma protein in human blood, and HCO_3^- and H_2CO_3 are the most important buffer pair, always keeps the pH of blood between 7.35 and 7.45. Beyond this range, "acidosis" or "alkalosis" can be caused in varying degrees. If the change exceeds 0.4 pH, the patient's life is in danger.

7.3 Precipitation Dissolution Equilibrium of Insoluble Strong Electrolytes

At a certain temperature, when the solid strong electrolyte is dissolved in a certain amount of solvent to form a saturated solution, there is a precipitation dissolution equilibrium between the undissolved solid strong electrolyte and the ions in the solution, which is a multiphase dissociation equilibrium. For example:

$$\underset{\text{(insoluble solid)}}{AgCl(s)} \rightleftharpoons \underset{\text{(solution)}}{Ag^+(aq) + Cl^-(aq)}$$

7.3.1 Solubility Product

(1) Solubility

When the temperature is constant, the amount of solute dissolved in a certain amount of solvent is called the solubility of the solute in that solvent. Obviously, solubility is closely related to temperature. In general, the solubility of most substances increases with the increase of temperature. Solubility is also related to the types and properties of solutes and solvents, such as the polarity of solvents, the strength of electrolytes, and so on. Insoluble strong electrolytes generally refer to the category of substances with low solubility.

(2) Solubility product

At certain temperature, the crystallization and dissolution of insoluble strong electrolytes reach equilibrium. There is a certain relationship between the ion concentration in the solution. For example, in the saturated solution of calcium carbonate, there is the following equilibrium:

$$\underset{\text{(Insoluble solid)}}{CaCO_3(s)} \rightleftharpoons \underset{\text{(Solution)}}{Ca^{2+} + CO_3^{2-}}$$

In this equilibrium of precipitation dissolution, $CaCO_3$ is a solid, and its standard equilibrium constant expression is

$$K_{sp}^{\ominus}=[c(Ca^{2+})/c^{\ominus}][c(CO_3^{2-})/c^{\ominus}]$$

That is, at a certain temperature, in the saturated solution of insoluble strong electrolyte, the product of the concentration of each ion (coefficient power) is a constant, which is called the standard solubility product constant, abbreviated as the solubility product, usually expressed as $K_{sp}^{\ominus}$. The expression of solubility product should be based on the composition of the specific compound, readers should pay more attention to the following points:

① $K_{sp}^{\ominus}$ is the characteristic constant when the insoluble strong electrolyte reaches the equilibrium of precipitation dissolution, which is independent of the concentration of each species, but the function of temperature;

② The numerical value of $K_{sp}^{\ominus}$ shows the solubility of the insoluble strong electrolyte in the solvent. In general, the larger the value is, the greater the solubility is in the solvent; the smaller the value is, the easier it is to form precipitation;

③ $\Delta_r G_m^{\ominus}$ and $K_{sp}^{\ominus}$ can be calculated *via* the equation

$$\Delta_r G_m^{\ominus}=-RT\ln K_{sp}^{\ominus}$$

Example 7-7 The saturated solubility of $CaCO_3$ at 25℃ is 9.3×10^{-5}mol/L. According to the above equation, in the solution, there is

$$c(Ca^{2+})=c(CO_3^{2-})=9.3\times10^{-5}\text{mol/L}$$

According to the expressions of the solubility product, there is:

$$\begin{aligned}K_{sp}^{\ominus}(CaCO_3)&=c_r(Ca^{2+})\cdot c_r(CO_3^{2-})\\&=(9.3\times10^{-5})\times(9.3\times10^{-5})=8.7\times10^{-9}\end{aligned}$$

(3) The relationship between solubility product and solubility

For any insoluble strong electrolyte A_xB_y, let the solubility of the electrolyte be S, the expression of the standard equilibrium constant is

$$A_xB_y(s)\rightleftharpoons xA^{y+}(aq)+yB^{x-}(aq)$$

Initial	0	0
Equilibrium	$c(A^{y+})=xS$	$c(B^{x-})=yS$

$$K_{sp}^{\ominus}(A_xB_y)=c_r^{\,x}(A^{y+})\cdot c_r^{\,y}(B^{x-}) \tag{7-26}$$

So, the conversion relationship between solubility and solubility product is

$$K_{sp}^{\ominus}(A_xB_y)=x^x y^y S^{x+y} \tag{7-27}$$

According to the formula (7-27), the relationship between solubility and solubility product of different types of insoluble strong electrolytes is as follows:

AB type	AgCl	$BaSO_4$	$K_{sp}^{\ominus} = S^2$
A_2B/AB_2 type	Ag_2CrO_4	$Mg(OH)_2$	$K_{sp}^{\ominus} = 4S^3$
AB_3 type	$Al(OH)_3$	$Fe(OH)_3$	$K_{sp}^{\ominus} = 27S^4$
A_2B_3 type	Al_2S_3	As_2S_3	$K_{sp}^{\ominus} = 108S^5$

Since the ion concentration in formula (7-26) is the concentration in the saturated solution, the solubility product can reflect the solubility. For the insoluble electrolytes of the same type [such as AgCl, AgBr, $BaSO_4$ and $BaCO_3$, they are of type AB; $Cu(OH)_2$ and $PbCl_2$ are both type AB_2], the greater the value is at the same temperature, the greater the solubility will be. However, for different types of insoluble electrolytes, one can't compare their solubility directly according to the value of solubility product.

Example 7-8 At 25℃, $K_{sp}^{\ominus}$ of AgCl and Ag_2CrO_4 are 1.77×10^{-10} and 1.12×10^{-12}, respectively. That is, $K_{sp}^{\ominus}(AgCl) > K_{sp}^{\ominus}(Ag_2CrO_4)$. However, it is calculated that the solubility of the AgCl is little (1.33×10^{-5}mol/L), but Ag_2CrO_4 has a high solubility (6.54×10^{-5}mol/L). This is because that AgCl is AB Type, $K_{sp}^{\ominus} = c_r(A^+)\cdot c_r(B^-)$, and Ag_2CrO_4 is A_2B Type, $K_{sp}^{\ominus} = c_r^2(A^+)\cdot c_r(B^{2-})$. The ion concentration index is different between the two precipitates. In this case, the solubility should be calculated according to the value $K_{sp}^{\ominus}$ and then compared. For example, the solubility of the above Ag_2CrO_4 can be calculated as follows:

$$Ag_2CrO_4 \rightleftharpoons 2Ag^+ + CrO_4^{2-}$$

Equilibrium concentration/(mol/L) $\quad 2x \quad x$

$$K_{sp}^{\ominus}(Ag_2CrO_4) = c_r^2(Ag^+)\cdot c_r(CrO_4^{2-}) = (2x)^2(x) = 4x^3 = 1.12\times10^{-12}$$

$$x = \sqrt[3]{\frac{1.12}{4}\times10^{-12}} = \sqrt[3]{0.28\times10^{-12}} = 6.54\times10^{-5}$$

Because, one mole Ag_2CrO_4 can dissociate into one mole CrO_4^{2-}, the solubility of Ag_2CrO_4 is 6.54×10^{-5}mol/L.

7.3.2 Solubility Product Rules

(1) Solubility product rules

When the crystallization and dissolution of insoluble strong electrolytes rcach equilibrium, if the conditions are changed, the multiphase dissociation equilibrium will also shift. (that is to say, precipitation or precipitation dissolution). The direction of equilibrium movement can be judged by solubility product. When the ion concentration is changed,

$$A_xB_y(s) \rightleftharpoons xA^{y+}(aq) + yB^{x-}(aq)$$

If the ion product in either state is Q, then there is:

When $Q = c_r^x(A^{y+}) \cdot c_r^y(B^{x-}) = K_{sp}^{\ominus}$, it is the saturated solution of the insoluble strong electrolyte;

When $Q = c_r^x(A^{y+}) \cdot c_r^y(B^{x-}) < K_{sp}^{\ominus}$, there is no precipitation or precipitation dissolute;

When $Q = c_r^x(A^{y+}) \cdot c_r^y(B^{x-}) > K_{sp}^{\ominus}$, precipitation generated (in principle).

These rules are called the solubility product rule.

The relationship between Q and $K_{sp}^{\ominus}$ can also be expressed based on the isotherm equation.

$$\Delta_r G_m = 2.303RT\lg\frac{Q}{K_{sp}^{\ominus}}$$

or

$$\Delta_r G_m = RT\ln\frac{Q}{K_{sp}^{\ominus}}$$

To determine the formation or dissolution of a precipitation, the following rules can be used:

When $Q < K_{sp}^{\ominus}$, there is no precipitation, it is an unsaturated solution;

When $Q = K_{sp}^{\ominus}$, the equilibrium of precipitation-dissolution is reached, which is saturated solution;

When $Q > K_{sp}^{\ominus}$, precipitation formation (in principle).

The dissolve of $CaCO_3$ in dilute hydrochloric acid solution can also be explained by solubility product rules.

$$CaCO_3(s) \rightleftharpoons Ca^{2+}(aq) + CO_3^{2-}(aq)$$

$$+$$

$$2HCl \rightleftharpoons 2Cl^- + 2H^+$$

$$\Vert$$

$$H_2CO_3 \rightleftharpoons CO_2\uparrow + H_2O$$

After adding hydrochloric acid, H_2CO_3 generated by the combination of H^+ and CO_3^{2-}. Furtherly, H_2CO_3 decompose into CO_2 and H_2O, resulting the continuous decrease of the $c(CO_3^{2-})$. Therefore, $c_r(Ca^{2+}) \cdot c_r(CO_3^{2-}) < K_{sp}^{\ominus}$ ($CaCO_3$), the precipitation-dissolution equilibrium of $CaCO_3$ is constantly moving to the right, resulting the continuous resolve of $CaCO_3$. If there's enough hydrochloric acid, $CaCO_3$ can dissolve thoroughly. On the contrary, if the Na_2CO_3 solution was added to the saturated $CaCO_3$ solution, as a result of the increase of $c(CO_3^{2-})$, there is $c(Ca^{2+}) \cdot c(CO_3^{2-}) > K_{sp}^{\ominus}$ ($CaCO_3$), the precipitation-dissolution equilibrium of $CaCO_3$ moves to the left, resulting in $CaCO_3$ precipitate until the ion product is equal to

$K_{sp}^{\ominus}$ ($CaCO_3$), and a new balance is reached.

(2) The application of solubility product rule

① Formation of precipitation. According to the solubility product rule, the conditions for the formation of precipitation are $Q > K_{sp}^{\ominus}$. The most commonly used experimental methods in practical application are as follows: adding suitable precipitator and adjusting the solution pH.

Example 7-9 The same amount of KI solution (0.01mol/L) was added to the Pb $(NO_3)_2$ solution with a concentration of 0.01mol/L, please think that whether there is PbI_2 precipitation production? If the same amount of KI solution with a concentration of 0.001mol/L was added, is there PbI_2 precipitation production? [$K_{sp}^{\ominus}(PbI_2)=8.49\times10^{-9}$]

Solution $c(Pb^{2+})$ =0.01mol/L in 0.01mol/L $Pb(NO_3)_2$ solution, after the addition of the same amount of KI solution, the amount of solution increases 1 time, then $c_r(Pb^{2+}) = 0.005$mol/L. In the same way, $c_r(I^-)$=0.005mol/L.

$$Q = c_r(Pb^{2+}) \cdot c_r^2(I^-) = 0.005\times0.005^2 = 1.25\times10^{-7} > K_{sp}^{\ominus}(PbI_2)$$

So there's PbI_2 precipitation produces in the solution.

If adding the same amount of 0.001mol/L KI solution, then $c_r(I^-) = 0.0005$mol/L.

There is

$$Q = c_r(Pb^{2+}) \cdot c_r^2(I^-) = 0.005\times0.0005^2 = 1.25\times10^{-9} < K_{sp}^{\ominus}(PbI_2)$$

Therefore, there will be no PbI_2 precipitation formed in the solution.

Example 7-10 A solution contains 0.01mol/L Fe^{3+}. Try to calculate the pH when the Fe^{3+} ions start to precipitate and the pH when the precipitation is complete.

Solution By adjusting the solution pH, Fe^{3+} can be precipitated in the $Fe(OH)_3$ form, when the precipitate process begin, that is, the equilibrium of precipitation- dissolution is established.

$$K_{sp}^{\ominus} = c_r(Fe^{3+})\cdot c_r^3(OH^-) = 2.64\times10^{-39}$$

$$c_r(OH^-) = (2.64\times10^{-39}/0.01)^{1/3} = 6.41\times10^{-13}$$

$$pH = 1.81$$

When the precipitation is complete, that is, in the solution, the concentration of Fe^{3+} is less than10^{-5}mol/L. There is

$$c_r(OH^-) = (2.64\times10^{-39}/10^{-5})^{1/3} = 6.41\times10^{-12}$$

$$pH = 2.81$$

So, when the pH of this solution is in range 1.81-2.81, Fe^{3+} will precipitate completely from the solution.

Solubility product rule is widely used in production. For example, the lime and soda method which is often used to soft hard water (generally contains more Ca^{2+} and Mg^{2+}). Why should we add Na_2CO_3 and $Ca(OH)_2$ at the same time? Can it work only use Na_2CO_3? Let's have a look at the solubility product of several substances at room temperature:

$$K_{sp}^{\ominus}[Mg(OH)_2] = 5.61\times10^{-12}$$

$$K_{sp}^{\ominus}(MgCO_3) = 6.82\times10^{-6}$$

$$K_{sp}^{\ominus}(CaCO_3) = 4.96\times10^{-9}$$

As we can see from the data, if only Na_2CO_3 is used, because $K_{sp}^{\ominus}$ $(CaCO_3)$ is very small, so Ca^{2+} can be precipitated more completely. However, on the other hand, the precipitation of Mg^{2+} is incomplete. Because $K_{sp}^{\ominus}$ $(MgCO_3)$ is larger, the Mg^{2+} ion concentration left in the solution after the reaction is still large, which does not meet the requirements of soft water. Of course, the so-called completed precipitation of Ca^{2+} doesn't mean that the Ca^{2+} is absolutely absent in water. If we want to remove boiler dirt by chemical means, we should use some matter which is easy to form inseparable substances after combining with Ca^{2+} and Mg^{2+} (For example, complex ions), and make the concentration of Ca^{2+} and Mg^{2+} in water less than the balance concentration. In this way, the boiler dirt can be removed. Of course, what kind of substance to choose should be analyzed concretely.

② Separation of multiple precipitations. In the example 7-7, the solubility product rule is used to judge whether there is PbI_2 precipitation formed after adding the precipitating agent containing I^- into the solution containing Pb^{2+}. In fact, the solution often contains a variety of ions at the same time, when a certain precipitator is added, there may be a variety of precipitation, or precipitation at the same time, or precipitation one after another. We can control the order of precipitation according to the solubility product rule. This method of sequential precipitation is called step-by-step precipitation method.

Example 7-11 In industrial wastewater, there are usually many anions and cations, and we need to adopt chemical methods to treat them. There is an industrial waste water containing 0.01mol/L Cl^- and 0.0005mol/L CrO_4^{2-}. Cl^- and CrO_4^{2-} can be removed by adding $AgNO_3$. At the beginning, white AgCl precipitate generated, and then brick-red Ag_2CrO_4 precipitation appeared. Please use the solubility product rule to explain whether this operation is reasonable?

Solution According to the solubility product rule, what is the minimum Ag^+ concentration required for the formation of AgCl and Ag_2CrO_4 precipitation? Please calculate separately (Ignore changes in volume).

The minimum concentration of Ag^+ needed for the precipitation of Cl^- is:

$$c_r(Ag^+) = \frac{K_{sp}^{\ominus}(AgCl)}{c_r(Cl^-)} = \frac{1.77\times10^{-10}}{0.01} = 1.77\times10^{-8}$$

The minimum Ag^+ concentration needed for the precipitation of CrO_4^{2-} is:

$$c_r(Ag^+) = \sqrt{\frac{K_{sp}^{\ominus}(Ag_2CrO_4)}{c_r(CrO_4^{2-})}} = \sqrt{\frac{1.12\times10^{-12}}{5.0\times10^{-4}}} = 4.73\times10^{-5}$$

From the calculated results, it can be seen that the required Ag^+ concentration to precipitate Cl^- is far lower than that for the precipitation of CrO_4^{2-}. Therefore, AgCl precipitate firstly, and then Ag_2CrO_4.

When the precipitation of Ag_2CrO_4 is first precipitated, what about the Cl^- concentration? If the change of solution amount caused by the addition of reagent is not taken into account, it can be considered that Ag^+ concentration in the solution at this time is 4.73×10^{-5}mol/L. Then the Cl^- concentration is

$$c_r(Cl^-) = \frac{K_{sp}^{\ominus}(AgCl)}{c_r(Ag^+)} = \frac{1.77\times10^{-10}}{4.73\times10^{-5}} = 3.74\times10^{-6}$$

which means that when Ag_2CrO_4 starts to precipitate, Cl^- has been completely precipitated (It is generally believed that the standard for the complete precipitation is the corresponding ion concentration is less than10^{-5}mol/L). The calculation shows that this operation is reasonable.

As can be seen from this example:

a. The greater the numerical difference of $K_{sp}^{\ominus}$ values, the better the separation effect.

b. The smaller the concentration difference of the separated ions, the easier it is to choose the appropriate precipitator.

c. When the precipitation type is the same, it can be judged directly by the value of $K_{sp}^{\ominus}$; however, if the precipitation type is not the same, it must be judged after calculation.

In analytical chemistry, according to the principle of multiple precipitations, the Cl^- content can be determined by using K_2CrO_4 solution as indicator, $AgNO_3$ solution as precipitator. With the $AgNO_3$ solution added from the titration tube to the solution, AgCl precipitation continues to form, and finally, when brick red appears, titration reaches the end.

③ Dissolution of precipitation. In practice, it is often necessary to convert insoluble solid substances into solutions. For example, the analysis of ore samples, the removal of boiler scale, the purification of materials, the fixing of impression (remove AgBr from the film) and so on, all have to dissolve the solid material. According to the solubility product rule, just make $Q < K_{sp}^{\ominus}$, in the multiphase equilibrium system of insoluble electrolytes, if a certain ion can be removed and the ion product of the insoluble electrolytes is less than its solubility product, the precipitation will dissolve.

Usually, the precipitation can be removed in the direction of dissolution by adding appropriate ions, combining with an ion in the solution to form a weak electrolyte and the precipitation equilibrium and dissolution of insoluble metal hydroxide are

$$M(OH)_n(s) \rightleftharpoons M^{n+}(aq) + nOH^-(aq)$$

After the addition of H^+ to the above balance, due to the formation of weak electrolytes H_2O, the balance will keep moving to the right, which in turn dissolves the precipitation. For example, copper hydroxide interacts with hydrochloric acid to form weak electrolyte water to dissolve copper hydroxide.

$$Cu(OH)_2(s) + 2H^+ \rightleftharpoons Cu^{2+} + 2H_2O$$

The formation of weak electrolytes H_2O result in the decrease of the OH^- concentration. $c_r(Cu^{2+}) \cdot c_r^2(OH^-) < K_{sp}^{\ominus}[Cu(OH)_2]$. The result is $Cu(OH)_2$ dissolve. For another example, $Mg(OH)_2$ dissolved in ammonium salt, the reaction is as follows:

$$Mg(OH)_2\,(s) + 2NH_4^+ \rightleftharpoons Mg^{2+} + 2NH_3 + 2H_2O$$

In the reaction NH_4^+ combines with OH^- dissociated from $Mg(OH)_2$ to form weak electrolytes H_2O and NH_3, and NH_3 escapes from a solution as gas molecular, resulting in the decrease of OH^- concentration, $c_r(Mg^{2+}) \cdot c_r^2(OH^-) < K_{sp}^{\ominus}[Mg(OH)_2]$, the result is $Mg(OH)_2$ dissolved.

The dissolution of metal sulfides is also due to the formation of weak electrolytes H_2S which causes the balance shift to the right and promotes the precipitation to dissolve. When the saturated solubility of H_2S is reached, H_2S will escape. For example, FeS dissolved in hydrochloric acid, it is H_2S formed and FeS dissolved, the reaction is as follows:

$$FeS\,(s) + 2H^+ \rightleftharpoons Fe^{2+} + H_2S\uparrow$$

By adding oxidant or reductant, the redox reaction occurs in the solution, which reduces the concentration of some ions and forward moves the equilibrium of precipitation-dissolution to the right, that is, the direction of precipitation dissolution. For example, copper sulfide interacts with oxidizing nitric acid to form a simple substance S and dissolve copper sulfide, that is,

$$3CuS(s) + 8HNO_3 \rightleftharpoons 3Cu\,(NO_3)_2 + 3S\downarrow + 2NO\uparrow + 4H_2O$$

Because of HNO_3 can oxidize the S^{2-} which dissociated from CuS to S, and reduce the S^{2-} concentration in the solution, $c_r(Cu^{2+}) \cdot c_r(S^{2-}) < K_{sp}^{\ominus}(CuS)$. The solid CuS dissolve.

The solubility product of HgS is very small [$K_{sp}^{\ominus}(HgS) = 6.44\times10^{-53}$(Black)] and HNO_3 cannot dissolve it. Nitrohydrochloric acid is used instead. In the reaction, S^{2-} was oxidized by HNO_3 in nitrohydrochloric acid. Hg^{2+} and Cl^- combine to form a ligand ion $[HgCl_4]^{2-}$. Under the dual action of oxidation and coordination, the concentrations of S^{2-} and Hg^{2+} in the solution keep decreasing, $c_r(Hg^{2+}) \cdot c_r(S^{2-}) < K_{sp}^{\ominus}(HgS)$, the result is that HgS dissolved. The reaction can be expressed by ion equation:

$$3HgS(s) + 8H^+ + 2NO_3^- + 12Cl^- \rightleftharpoons 3[HgCl_4]^{2-} + 3S\downarrow + 2NO\uparrow + 4H_2O$$

In a word, there are many dissolution methods of precipitation, which can be selected according to the needs. In order to achieve better results, two or more methods are often used in practical application at the same time.

④ Transformation of precipitation. The process of converting one precipitation into another is called the transformation of precipitation. Precipitation is always transformed to substances with smaller $K_{sp}^{\ominus}$, that is, to a more stable precipitation direction.

For example, in industrial boiler water, impurities in water often form boiler scale. The scale contains $CaSO_4$, which is both insoluble in water and insoluble in acid, is difficult to remove. However, we can try to add some kind of reagent, convert $CaSO_4$ precipitation [$K_{sp}^{\ominus}(CaSO_4)=7.1\times10^{-5}$] to $CaCO_3$ precipitation [$K_{sp}^{\ominus}(CaCO_3)=4.96\times10^{-6}$] which is loose and soluble in acid, so that facilitating the removal of boiler scale, the reaction is as follows:

$$CaSO_4(s) + Na_2CO_3 \rightleftharpoons CaCO_3(s) + 2Na^+ + SO_4^{2-}$$

As in seaport buildings, Mg^{2+} in the seawater can corrode cement which contains $Ca(OH)_2$, the principle is that $Ca(OH)_2$ precipitation {$K_{sp}^{\ominus}[Ca(OH)_2]=4.68\times10^{-6}$} is easy to convert into $Mg(OH)_2$ precipitation {$K_{sp}^{\ominus}[Mg(OH)_2]=5.61\times10^{-12}$}. The reaction is

$$Ca(OH)_2(s) + Mg^{2+} \rightleftharpoons Mg(OH)_2(s) + Ca^{2+}$$

In recent years, the principle of precipitation conversion is often used for wastewater treatment. For example, waste water including Hg^{2+} or Cu^{2+} can be treated using FeS, and the effect is very good. The reaction formula can be expressed as follows:

$$FeS(s) + Cu^{2+} \rightleftharpoons CuS(s) + Fe^{2+}$$

$$FeS(s) + Hg^{2+} \rightleftharpoons HgS(s) + Fe^{2+}$$

Obviously, the greater the difference of $K_{sp}^{\ominus}$ between the two insoluble electrolytes in the reaction, the higher the concentration of the transformed ions; and the precipitation transformation will be more complete.

Exercises

1. The boiling point of solution dissolved in 26.6g chloroform ($CHCl_3$) with 0.402g naphthalene ($C_{10}H_8$) was 0.455℃ higher than that of pure chloroform. Calculate the boiling point elevation of chloroform.

2. The vapor pressure of a dilute solution and pure water at 25℃ is 3.127kPa and 3.168kPa respectively. Determine the molality of the solution. Given that K_b = 0.51℃. Determine the boiling point of the solution.

3. How many grams of ethylene glycol ($C_2H_6O_2$) should be added to 1000g water to reduce the freezing point of the solution to −10℃?

4. Arrange the following two aqueous solutions in order of their vapor pressures from small to large.

(1) Concentrations are 0.1mol/kg: NaCl, H_2SO_4, $C_6H_{12}O_6$(glucose);

(2) Concentrations are 0.1mol/kg: CH_3COOH, NaCl, $C_6H_{12}O_6$.

5. The freezing point of a sugar aqueous solution is −0.186℃, calculate the boiling point.

6. The 5.0g solute was dissolved in 60g benzene and the freezing point of the solution was 1.38℃. Determine the relative molecular weight of the solute.

7. What is the hydrogen ion concentration of 0.1mol/kg of HCl and 1mol/kg HAc? Which is more acidic?

8. Ephedrine ($C_{10}H_{15}ON$) is a base that is used in nasal sprays to reduce congestion, $K_b^\ominus(C_{10}H_{15}ON)=1.4\times10^{-4}$.

(1) Write the ionic equation for the reaction of ephedrine with water, namely, the dissociation equation for the weak base ephedrine;

(2) Write the conjugate acid of ephedrine and calculate the value of $K_a^\ominus$.

9. Salicylic acid (*o*-hydroxybenzoic acid) $C_7H_4O_3H_2$ is a binary weak acid. Below 25℃, $K_{a1}^\ominus=1.06\times10^{-3}$, $K_{a2}^\ominus=3.6\times10^{-14}$. It is sometimes used instead of aspirin as a painkiller, but it is highly acidic and can cause stomach bleeding. Calculation of 0.065mol/kg $C_7H_4O_3H_2$ equilibrium in the solution concentration and the pH of each species.

10. 0.2mol/kg HF and 0.2mol/kg equivalent mixed NH_4F solutions, calculate the pH of solution and the dissociation degree of HF.

11. There is 2.00L of NH_3(aq) and 2.00L of HCl solution, if we want to prepare a buffer solution of pH = 9, and it's not allowed to add water, try to answer how many liters of buffer solution we can prepare. What's the $c(NH_3)$, $c(NH_4^+)$?

12. At room temperature, 100g of water can dissolve 0.0033g of Ag_2CrO_4. Try to determine the solubility product constant of Ag_2CrO_4.

13. A solution contains both Cl^- (0.1mol/kg) and CrO_4^{2-} (0.1mol/kg). If Ag^+ is slowly added to the solution, which kind of precipitation will be observed firstly in the presence of Ag^+ ion? What is the concentration of the first anion when the second anion begin to precipitate?

14. There is 0.01mol/kg Fe^{2+} in the ammonium chloride solution. How much is the pH when $Fe(OH)_2$ begins to precipitate?

15. The concentrations of Fe^{3+} and Zn^{2+} in a mixed solution are both 0.010mol/kg. Add alkali to adjust the pH, so that $Fe(OH)_3$ precipitates completely, while Zn^{2+} remains in the solution. Calculate the pH range for separating Fe^{3+} and Zn^{2+}. {$K_{sp}^\ominus$[$Fe(OH)_3$] is 2.64×10^{-39}, $K_{sp}^\ominus$[$Zn(OH)_2$] is 6.86×10^{-17}}.

16. If a solution contains 0.10mol/L Li^+ and 0.10mol/L Mg^{2+}, and NaF solution is added dropwise (ignoring the volume change), which ion will be precipitated first? When the second kind of precipitation separates out, are the ions of the first kind of precipitation completely precipitated? Is it possible to separate the two ions? [$K_{sp}^\ominus$(LiF) is 1.84×10^{-3}, $K_{sp}^\ominus$(MgF_2) is 6.86×10^{-17}].

Chapter 8 Applied Electrochemistry

Teaching contents	Learning requirements
Voltaic cells and cell diagram	Understand the basic concept of redox reaction; Master the representation of the voltaic cell diagram
Electrode potential	Understand the theory of electric double layer; Master the Nernst equation and be able to carry out the relevant calculation; Understand the application of electrode potential
Electrolysis	Understand the concepts of overpotential, overvoltage, decomposition voltage and so on; Master the method of judging electrolytic products; Understand electroplating, electric polishing, electrolytic machining and other technologies

Here is an interesting story about a rich lady, Mrs. Green, who had a pure gold false tooth. Once, in a small car accident, another tooth was knocked out. This time, the doctor installs a copper alloy tooth for her. Since then, Mrs. Green often has a headache, insomnia and irritability and many famous doctors failed to solve her problem. One day, a chemist said confidently to Mrs. Green, "I can easily solve the problem that has been bothering you for so long." "Is it true? That would be great!" Mrs. Green is about to jump with joy. Then the chemist did a small experiment for the rich lady. He found a slender piece of gold and a copper wire to connect them to the two levels of a sensitive flowmeter. Then he put the gold sheet and the other end of the copper wire in his mouth. This is, amazing thing happened! The pointer of the ammeter has deflected! Then the chemist explained what cause Mrs. Green's headache: two teeth with different materials formed the voltaic cell when they used saliva as electrolyte solution, generating a weak current that stimulated Mrs. Green's nerve endings, which caused symptoms such as headaches. So, do you know why these two teeth form the voltaic cell in the mouth? What should we do to solve Mrs. Green's headache? After learning this chapter, we will find a reasonable answer.

8.1 Voltaic Cell and Cell Diagram

8.1.1 Voltaic Cell

Voltaic cell is a device that converts chemical energy directly into electric energy by means of spontaneous redox reaction. When zinc sheets are placed in copper sulfate solution, the following redox reactions occur:

$$Zn + CuSO_4 = Cu + ZnSO_4$$

During this reaction, electrons are transferred directly from the zinc atoms to Cu^{2+} due to direct contact between the zinc and the copper sulfate solution. The flow of electrons here is disordered, and as the reaction proceeds, the temperature of the solution increases, that is, the chemical energy during the reaction is transformed into heat energy. For example, the above reaction $\Delta_r H_m^{\ominus} = -211.4 \text{kJ/mol}$. In order to use redox reaction to form the voltaic cell and convert the chemical energy into electric energy, the following three conditions must be met in order to make the charge move directed and exchange in an orderly manner.

① It must be a spontaneous redox reaction;

② Oxidation reaction and reduction reaction should be spontaneously conducted on two electrodes respectively;

③ The internal and external circuits should be properly connected.

Under the above conditions, a Cu-Zn galvanic cell can be designed according to the above overall redox reaction, as shown in Fig.8-1. A thin sheet of zinc was inserted into a beaker containing $ZnSO_4$ solution, and a thin sheet of copper was inserted into another beaker containing $CuSO_4$ solution, the two solutions are directly connected by a U-tube containing saturated KCl solution called salt bridge. The two metals are joined by wires to a voltmeter. Now, let's consider the detailed reaction process occurred in this voltaic cell shown in Fig.8-1. Zn atoms release electrons at the negative electrode and enter in to the $ZnSO_4$ solution as Zn^{2+}, the electrons migrate through the wire and the voltage meter to the positive electrode, where the Cu^{2+} in the $CuSO_4$ solution get these electrons to form metal Cu. Simultaneously, Cl^- from the salt bridge migrate to the Zn half-cell to neutralize the excess Zn^{2+}, and K^+ migrate to the Cu half-cell to neutralize the excess SO_4^{2-}. The overall reaction produce the electric current in a certain direction. This kind of electrochemical cell convert the chemical energy into

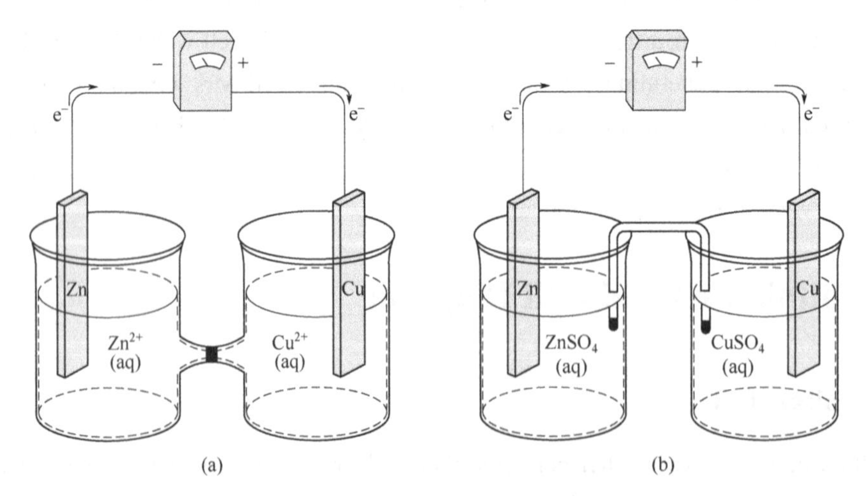

Fig.8-1 The sketch map of Daniel battery

electrical energy. Which is called Daniel battery. This battery was a universal and practical chemical power source in the 19th century.

8.1.2 Electrode, Cell Reaction and Cell Diagram

Any spontaneous oxidation-reduction reaction can be assembled into a voltaic cell by selecting an appropriate electrode (inert electrode), which causes electrons to flow in a certain direction to generate current. The electrode mentioned here is not a general electronic conductor, but an electronic conductor in contact with an electrolyte solution. It is both an electronic storage and a place where electrochemical reactions take place. Electrodes in electrochemistry are always associated with electrolyte solutions, and the characteristics of the electrodes are inseparable from the chemical reactions carried out on them. Therefore, the electrode refers to the entire system of electronic conductor and electrolyte solution. According to the nature of the electrode reaction, the electrodes can be divided into: the first type of electrode, which is composed of a metal immersed in a solution containing the metal ion, such as $Zn|Zn^{2+}$; the second type of electrode, consisting of hydrogen and oxygen, a gas electrode composed of a gas such as halogen immersed in an ionic solution containing the constituent elements of the gas, such as $Pt(H_2)|H^+$; a third type of electrode, including a metal and an insoluble salt electrode of the metal and a redox electrode, such as $Pt \mid Fe^{2+}, Fe^{3+}$.

There is a potential difference between the two electrodes of the voltaic cell. The electrode with high potential or electron inflow is the positive electrode. The electrode with low potential or electron outflow is negative. In electrochemistry, whether in the voltaic cell (spontaneous battery), electrolytic cell (non-spontaneous cell) or corrosive battery (spontaneous battery), the electrode that produces oxidation reaction is called anode, and the electrode that produces reduction reaction is called cathode. However, when the voltaic cell is transformed into an electrolytic cell (such as the recharging of the battery after discharge), their positive and negative electrode symbols remain unchanged, the original cathode becomes an anode, and the original anode becomes a cathode. This, of course, corresponds to the direction of the electrode reaction, and the direction of the electrode reaction changes, and the names of the cathode and anode change accordingly. This is why people are always willing to use positive and negative electrodes to represent the names of the two electrodes in the voltaic cell. According to this regulation, the electrode name, electrode reaction and the overall reaction in the Cu-Zn Voltaic cells are as follows:

Electrode reaction: negative electrode (zinc and zinc ion solution):

$$Zn - 2e^- \text{===} Zn^{2+} \text{ (oxidation reaction)}$$

Positive electrode (copper and copper ion solution):

$$Cu^{2+} + 2e^- \text{===} Cu \text{ (reduction reaction)}$$

Overall reaction: the overall redox reaction can be obtained by combining the reactions occurred on two electrodes:

$$Zn + Cu^{2+} \longequal Zn^{2+} + Cu \text{ (redox reaction)}$$

In order to represent the voltaic cells conveniently, the IUPAC agreement of 1953 used symbols (cell diagram) to represent the voltaic cells. The cell diagram of a voltaic cell can be written according to the following generally accepted conventions:

① Chemical formula shall be used to represent the composition of various substances in the battery, and the physical state (solid, liquid, gas, *etc.*) shall be indicated respectively. Pressure shall be indicated for gas, concentration for solution, crystal type for solid, *etc.*

② A single vertical line "|" represents the boundary between different phases. The double vertical line "||" represents the boundary between two half cells (salt bridge).

③ The negative electrode (anode) of the voltaic cell is written on the left, and the positive electrode (cathode) is written on the right. When writing cell diagram, the order of the chemical formulas and symbols should reflect the contacting order of the substances in the battery.

④ When there are multiple ions in the solution, the negative electrode is written in order of increasing oxidation state, and the positive electrode is written in order of decreasing oxidation state.

According to the above rules, the cell diagram of Cu-Zn voltaic cell can be written as:

$$(-)Zn \mid ZnSO_4(m_1) \parallel CuSO_4(m_2) \mid Cu(+)$$

Not only can the two electrodes composed of two metals and their "own" salt solution be connected by a salt bridge to form the voltaic cell, but any two different metals inserted into any electrolyte solution can also form the voltaic cell. Among them, the more active metal is the negative electrode, while the less active metal is the positive electrode. For example, volta battery:

$$(-)Zn \mid H_2SO_4 \mid Cu(+)$$

In principle, any redox reaction can be assembled into a galvanic cell as long as it is carried out spontaneously according to the rules for a galvanic cell device to generate current. For example, put a solution containing Fe^{2+} and Fe^{3+} in one beaker, another solution containing Sn^{2+} and Sn^{4+} in another beaker, insert platinum plates (or carbon rods) as electrodes, and connect them with a salt bridge. After connecting the two half cells with a wire, electrons move from the Sn^{2+} solution to the Fe^{3+} solution through the wire to generate a current. The electrochemistry reactions occurred in the voltaic cell are:

Negative electrode　$Sn^{2+}(aq) - 2e^- \longequal Sn^{4+}(aq)$ (Oxidation reaction)

Positive electrode　$Fe^{3+}(aq) + e^- \longequal Fe^{2+}(aq)$ (Reduction reaction)

Overall reaction　$Sn^{2+}(aq) + 2Fe^{3+}(aq) \longequal Sn^{4+}(aq) + 2Fe^{2+}(aq)$

The cell diagram of the cell is:

$$(-)\ Pt \mid Sn^{2+}(m_1),\ Sn^{4+}(m_2) \parallel Fe^{3+}(m_3),\ Fe^{2+}(m_4) \mid Pt(+)$$

(Oxidation reaction)　(Reduction reaction)

In this type of voltaic cell, Pt does not participate in the redox reaction and only acts as a conductor.

Each electrode reaction of the galvanic cell contains two types of substances with different oxidation numbers of the same element. Among them, those with low oxidation numbers are substances that can be used as reducing agents, called reducing substances. Those with high oxidation numbers are substances that can be used as oxidant, called oxidizing substance. For example, in the reactions of two electrodes in a Cu-Zn battery:

$$Zn-2e^- = Zn^{2+}(aq) \qquad Cu^{2+}(aq) + 2e^- = Cu$$

Reduction state Oxidation state Oxidation state Reduction state

The reduction state and the corresponding oxidation state of each electrode form a redox couple. The couple can be represented by the symbol "oxidation state/reduction state". For example, the couples of zinc electrode and copper electrode are Zn^{2+}/Zn and Cu^{2+}/Cu, respectively. Not only the metal and its ions can form a couple, but also the ions of different oxidation states of the same metal or the non-metallic element and its corresponding ions can form a couple. For example, Fe^{3+}/Fe^{2+}, Sn^{4+}/Sn^{2+}, H^+/H_2, O_2/OH^- and Cl_2/Cl^- *etc.* But in these couples, since they are not metal conductors, an inert electrode that can conduct electricity without participating in electrode reactions must be added. Platinum or graphite is usually used as an inert electrode. The electrodes composed of these couples can be expressed as $Pt \mid Fe^{3+}, Fe^{2+}$; $Pt \mid Sn^{4+}, Sn^{2+}$ (redox electrode); $Pt(H_2) \mid H^+$; $Pt(O_2) \mid OH^-$ and $Pt(Cl_2) \mid Cl^-$ (non-metallic electrode).

Example 8-1 Try to write the cell diagram, electrode reactions, electric couples and electrodes of the voltaic cells in the following redox reactions:

$$2MnO_4^- + 10Cl^- + 16H^+ = 2Mn^{2+} + 5Cl_2 + 8H_2O$$

Solution According to the change of the oxidation number of each substance in the equation, find out that the oxidizing couple is MnO_4^-/Mn^{2+}, and the reducing couple is Cl_2/Cl^-. Then write the cell diagram as:

$$(-)Pt(Cl_2, p_1) \mid Cl^-(m_1) \parallel MnO_4^-(m_2), Mn^{2+}(m_3), H^+(m_4) \mid Pt(+)$$

Negative electrode reaction $2Cl^- - 2e^- = Cl_2$

Positive electrode reaction $MnO_4^- + 8H^+ + 5e^- = Mn^{2+} + 4H_2O$

Electric couples $Cl_2 \mid Cl^-$, $MnO_4^- \mid Mn^{2+}$

Electrodes $Pt(Cl_2) \mid Cl^-$, $Pt \mid MnO_4^-, Mn^{2+}$

8.2 Generation, Determination, Influencing Factors and Application of Electrode Potential

In the voltaic cell, the two electrodes are connected by a wire, and the current passes through the wire, which indicates that there is a potential difference between the two electrodes. It means that the two electrodes that make up the voltaic cell have different electrode potentials. That is to say, the generation of the voltaic cell current is caused by the difference of the electrode potential of the two electrodes. So, how does the electrode potential come from?

8.2.1 Generation of Electrode Potential: Electric Double Layer Theory

When a new interface is formed between the electrode and the solution, the free charge or dipole from the solution is rearranged on the interface to form a double electric layer, and there is a potential difference between the two electric layers, as shown in Fig.8-2. The formation of the electric double layer can be explained from the Gibbs function combined with the internal structure of the metal. If the electrode is a metal, the metal consists of free ions and "free electrons". In general, the Gibbs free energy of metal ions in the metal phase and the Gibbs free energy of the same kind of ion in the solution phase are not equal when they are not in contact. Therefore, when the metal is in contact with the solution, the transfer of metal ions between the two phases will occur. For example, if the Zn electrode is inserted into a $ZnCl_2$ solution, the Gibbs free energy of the Zn^{2+} ion in the metal zinc is higher than that in a $ZnCl_2$ solution at certain concentration, the Zn^{2+} on the zinc metal will spontaneously be transferred into the solution, and the zinc oxidation reaction will occur. After the Zn^{2+} ions on the metal are transferred into the solution, the electrons stay on the metal and the surface of the metal is negatively charged. It will attract the positive charge (such as Zn^{2+} ions) in the solution with Coulomb force, leaving it near the electrode surface, so a potential difference appears at the interface of the two phases. This potential difference prevents the Zn^{2+} ions from continuing to enter the solution. On the contrary, it can cause the Zn^{2+} ions in the solution to return to the metal. As the number of Zn^{2+} ions entering the solution on the metal increases and the potential difference becomes larger, the rate of Zn^{2+} ions entering the solution gradually decreases, and the rate of Zn^{2+} ions returning to the metal in the solution increases continuously. Finally, under the influence of the potential difference, a state of equal speed in two directions is established, that is, a solution-deposition balance is reached. At this time, electric double layer where metal zinc was negatively charged and the solution is positively charged is formed between the two-phase interface, as shown in Fig.8-2(a). Hence, a certain electrode potential was produced.

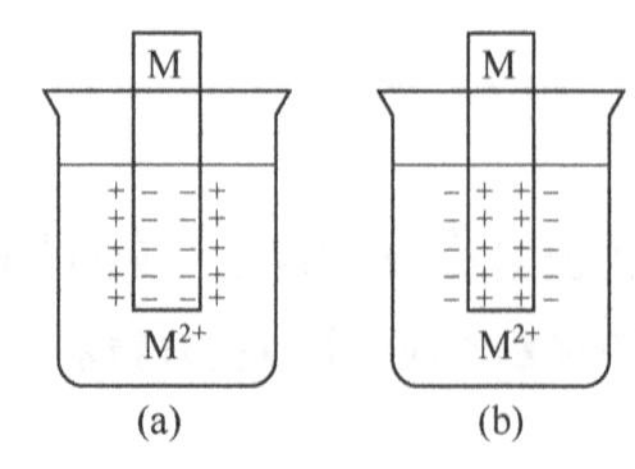

Fig.8-2 Schematic diagram of electric double layer

If the Gibbs free energy of positive ions (such as Cu^{2+} ions) on the metal is lower than that in the solution, the positive ions in the solution will spontaneously deposit on the metal, making the metal surface positively charged. This transfer of positive ions to the metal also destroys the electrical neutrality of the solution. The excess negative ions in the solution are attracted near the surface by the positive charge of the metal surface, forming a positively charged metal surface, and a negatively charged ion double layer of the solution as shown in Fig.8-2(b).

The spontaneous formation of an electric double layer is very rapid and can generally be completed in an instant of 1/1000000 seconds.

In some cases, the ionic electric double layer cannot be spontaneously formed even the metal is in contact with the solution. For example: pure mercury is placed in a KCl solution, because mercury is relatively stable, it is not easy to be oxidized, and K^+ ions are also difficult to be reduced. Therefore, it cannot spontaneously form an ionic double layer.

8.2.2 Measurement of Standard Electrode Potential

The potential of metal electrode reflects the trend of electron gain or loss of metal in its salt solution. If the electrode potential can be measured quantitatively, it will help us to judge the relative strength of oxidant and reductant. However, up to now, the absolute value of electrode potential of metal in its salt solution cannot be measured. Usually, the electrode potential of one particular electrode is set to zero, and the potential of another electrode is compared with this electrode, and then their electrode potential is determined. This approach is the same as setting the sea level at zero as the standard for altitude. At present, the standard electrode is hydrogen electrode, which is called standard hydrogen electrode.

The assembling of the standard hydrogen electrode is to insert a platinum sheet coated with a sponge-like fluffy platinum black into a sulfuric acid solution of $m^{\ominus}(H^+)=1$mol/kg, and continuously feed pure hydrogen with a pressure of 100kPa at 298.15K, and the hydrogen is adsorbed by platinum black. The hydrogen adsorbed by platinum black and the hydrogen ions in the solution form the electric couple H^+/H_2, and the electrode reaction is as follows:

$$\frac{1}{2}H_2(100\text{kPa})-e^- = H^+(aq)\ (1\text{mol/kg})$$

During the determination, a reference standard state is specified for the state of the substance. For gases, the standard state is its partial pressure of 100kPa; for solutions, the standard state is the concentration of the solution at standard pressure of 1mol/kg (denoted by $m^{\ominus}$); for liquids and solids, the standard state is pure substance under standard pressure. Since all the substances in the electrode reaction are in the standard state, the device becomes the standard hydrogen electrode, as shown in Fig.8-3. The potential it has is called the standard electrode potential of a standard hydrogen electrode, and its symbol is $\varphi^{\ominus}(H^+/H_2)$. The standard hydrogen electrode is used as a reference datum. It is artificially specified that the standard electrode potential at 298.15K is zero volts, that is:

$$\varphi^{\ominus}(H^+/H_2) = 0.000V$$

To determine the electrode potential of an electrode, the standard electrode and the standard hydrogen electrode of the electrode to be measured can be combined into a voltaic cells, as shown in Fig.8-4. The standard battery electromotive force (E) of the voltaic cells is equal to the potential difference between the two electrodes that make up the voltaic cells. In 1953, IUPAC identified the reduction potential as the electrode potential. The so-called "reduction potential" is the electrode potential measured by the reduction reaction when the electrode to be measured is used as the positive electrode when forming the voltaic cells for measurement. The general formula of the electrode reaction can be written as follows:

$$a\,(\text{oxidized state}) + ne^- = b\,(\text{reduced state})$$

Standard battery EMF(electromotive force)

$$E^{\ominus} = \varphi^{\ominus}_{+\text{To be tested}} - \varphi^{\ominus}_{-\text{Hydrogen electrode}} \tag{8-1}$$

In the formula, $\varphi^{\ominus}_{+}$ and $\varphi^{\ominus}_{-}$ indicate the standard electrode potential of the positive electrode and the negative electrode, respectively. Since the electrode potential of the standard hydrogen electrode is zero, the value of the electrode potential of the electrode to be measured can be determined by measuring the value of the electromotive force of the voltaic cells. Since the electrode potential is not only determined by the nature of the substance, but also related to temperature, concentration, *etc.*, for comparison, it is used at a temperature of 298.15K, when the concentration of the relevant ions in the electrode is 1mol/L, and the pressure of the relevant gas is 100kPa, the measured electrode potential is the standard electrode potential, expressed as $\varphi^{\ominus}$.

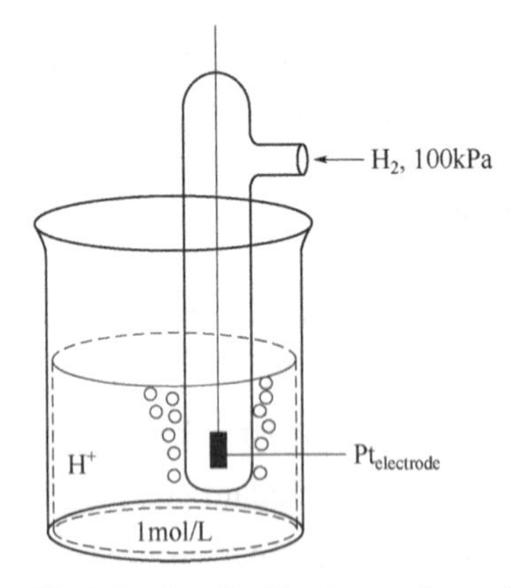

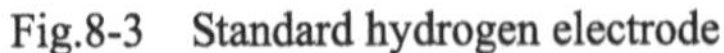

Fig.8-3　Standard hydrogen electrode

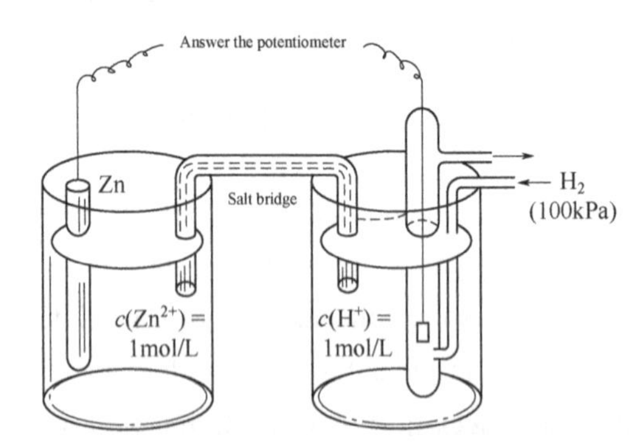

Fig.8-4　Device for measuring standard electrode potential

If the electrode to be tested is a zinc electrode, the electric potential of the voltaic cell is −0.7628V, which is equal to the difference between the potential of the electrode to be measured and the potential of the standard hydrogen electrode, as shown in Fig.8-4.

Electromotive force $E^{\ominus}=\varphi_{+}^{\ominus}(Zn^{2+}/Zn)-\varphi_{-}^{\ominus}(H^{+}/H_2)=-0.7628V$

Because $\varphi_{-}^{\ominus}(H^{+}/H_2)=0.0000V$

$$E^{\ominus}=\varphi^{\ominus}(Zn^{2+}/Zn)=-0.7628V$$

"–" in the formula indicates that the electrode potential is lower than that of the standard hydrogen electrode, and Zn is more susceptible to electron loss than H_2. It also indicates that when the electrode and the standard hydrogen electrode form a galvanic cell, the electrode should actually be a negative electrode. The electrode reaction is as follows:

Negative pole $Zn-2e^{-} = Zn^{2+}$

Positive pole $2H^{+}+2e^{-} = H_2$

Battery reaction $2H^{+}+Zn = H_2+Zn^{2+}$

If the zinc electrode is replaced with a copper electrode, the EMF of the voltaic cell is measured to be 0.337V.

Electromotive force $E^{\ominus}=\varphi^{\ominus}(Cu^{2+}/Cu)-\varphi^{\ominus}(H^{+}/H_2)=0.337V$

Because $\varphi^{\ominus}(H^{+}/H_2)=0.0000V$

$$E^{\ominus}=\varphi^{\ominus}(Cu^{2+}/Cu)=0.337V$$

The "+" sign indicates that the electrode potential is higher than the standard hydrogen electrode potential, and H_2 is more susceptible to electron loss than Cu. It also indicates that when the electrode and the standard hydrogen electrode form a voltaic cell, the electrode should actually be a positive electrode. The electrode reaction is as follows:

Negative pole $H_2-2e^{-} = 2H^{+}$

Positive pole $Cu^{2+}+2e^{-} = Cu$

Battery reaction $H_2+Cu^{2+} = 2H^{+}+Cu$

By using a similar method, the standard electrode potential values of electron pairs composed of various substances can be measured. The standard electrode potential of some substances cannot be measured at present, but can be calculated by indirect method (see calculation below). Some of the standard electrode potentials are arranged in to Table 8-1.

Table 8-1 Standard Electrode Potential (298.15K)

Electron pair (oxidation state/reduction state)	Electrode reaction (oxidation state + ne^{-} = b reduction state)	$\varphi^{\ominus}$/V
K^{+}/K	$K^{+}+e^{-} = K$	−2.931
Ca^{2+}/Ca	$Ca^{2+}+2e^{-} = Ca$	−2.868
Na^{+}/Na	$Na^{+}+e^{-} = Na$	−2.714
Mg^{2+}/Mg	$Mg^{2+}+2e^{-} = Mg$	−2.372
Al^{3+}/Al	$Al^{3+}+3e^{-} = Al$	−1.662
Mn^{2+}/Mn	$Mn^{2+}+2e^{-} = Mn$	−1.185
H_2O/H_2	$2H_2O+2e^{-} = H_2+2OH^{-}$	−0.828(碱性)
Zn^{2+}/Zn	$Zn^{2+}+2e^{-} = Zn$	−0.7618
Fe^{2+}/Fe	$Fe^{2+}+2e^{-} = Fe$	−0.447

(Continued)

Electron pair (oxidation state/reduction state)	Electrode reaction (oxidation state + ne^- = b reduction state)	$\varphi^\ominus$ /V
Cd^{2+}/Cd	$Cd^{2+} + 2e^- = Cd$	−0.4030
PbI_2/Pb	$PbI_2 + 2e^- = Pb + 2I^-$	−0.365
Pb^{2+}/Pb	$PbSO_4 + 2e^- = Pb + SO_4^{2-}$	−0.3588
$PbCl_2/Pb$	$PbCl_2 + 2e^- = Pb + 2Cl^-$	−0.2675
Co^{2+}/Co	$Co^{2+} + 2e^- = Co$	−0.28
Ni^{2+}/Ni	$Ni^{2+} + 2e^- = Ni$	−0.257
Sn^{2+}/Sn	$Sn^{2+} + 2e^- = Sn$	−0.1375
Pb^{2+}/Pb	$Pb^{2+} + 2e^- = Pb$	−0.1262
Fe^{3+}/Fe	$Fe^{3+} + 3e^- = Fe$	−0.037
H^+/H_2	$H^+ + e^- = \frac{1}{2} H_2$	0
$S_4O_6^{2-}/S_2O_3^{2-}$	$S_4O_6^{2-} + 2e^- = 2S_2O_3^{2-}$	+0.08
S/H_2S	$S + 2H^+ + 2e^- = H_2S$	+0.142
Sn^{4+}/Sn^{2+}	$Sn^{4+} + 2e^- = Sn^{2+}$	+0.151
SO_4^{2-}/H_2SO_3	$SO_4^{2-} + 4H^+ + 2e^- = H_2SO_3 + H_2O$	+0.172
$AgCl/Ag$	$AgCl + e^- = Ag + Cl^-$	+0.22233
Hg_2Cl_2/Hg	$Hg_2Cl_2 + 2e^- = 2Hg + 2Cl^-$	+0.26808
Cu^{2+}/Cu	$Cu^{2+} + 2e^- = Cu$	+0.347
O_2/OH^-	$\frac{1}{2} O_2 + H_2O + 2e^- = 2OH^-$	+0.401(Alkaline)
Cu^+/Cu	$Cu^+ + e^- = Cu$	+0.521
I_2/I^-	$I_2 + 2e^- = 2I^-$	+0.5355
$I_3^-/3I^-$	$I_3^- + 2e^- = 3I^-$	+0.536
O_2/H_2O	$\frac{1}{2} O_2 + 2H^+ + 2e^- = H_2O$	+0.695
Fe^{3+}/Fe^{2+}	$Fe^{3+} + e^- = Fe^{2+}$	+0.771
Hg_2^{2+}/Hg	$\frac{1}{2} Hg_2^{2+} + e^- = Hg$	+0.7973
Ag^+/Ag	$Ag^+ + e^- = Ag$	+0.7996
Hg^{2+}/Hg	$Hg^{2+} + 2e^- = Hg$	+0.851
NO_3^-/NO	$NO_3^- + 4H^+ + 3e^- = NO + 2H_2O$	+0.957
HNO_2/NO	$HNO_2 + H^+ + e^- = NO + H_2O$	+0.983
Br_2/Br^-	$Br_2 + 2e^- = 2Br^-$	+1.0873
MnO_2/Mn^{2+}	$MnO_2 + 4H^+ + 2e^- = Mn^{2+} + 2H_2O$	+1.224
O_2/H_2O	$O_2 + 4H^+ + 4e^- = 2H_2O$	+1.229
$Cr_2O_7^{2-}/Cr^{3+}$	$Cr_2O_7^{2-} + 14H^+ + 6e^- = 2Cr^{3+} + 7H_2O$	+1.33
Cl_2/Cl^-	$Cl_2 + 2e^- = 2Cl^-$	+1.35827
PbO_2/Pb^{2+}	$PbO_2 + 4H^+ + 2e^- = Pb^{2+} + 2H_2O$	+1.455
MnO_4^-/Mn^{2+}	$MnO_4^- + 8H^+ + 5e^- = Mn^{2+} + 4H_2O$	+1.507
MnO_4^-/MnO_2	$MnO_4^- + 4H^+ + 3e^- = MnO_2 + 2H_2O$	+1.679
H_2O_2/H_2O	$H_2O_2 + 2H^+ + 2e^- = 2H_2O$	+1.776
$S_2O_8^{2-}/SO_4^{2-}$	$S_2O_8^{2-} + 2e^- = 2SO_4^{2-}$	+2.01
F_2/F^-	$F_2 + 2e^- = 2F^-$	+2.866

Note: Since the pH of the solution affects the electrode potential of many pairs, the standard electrode potential meter is generally divided into an acid meter (marked as A) and an alkali meter (marked as B). In addition to the electrode potentials of the O_2/OH^- and H_2O/H_2 pairs, the standard electrode potentials in the table are all hydrogen standard electrode potentials in acidic solutions. Data recorded from Haynes W.M., CRC Handbook of Chemistry and Physics, 95th Edition, CRC Press, 2014-2015.

It can be seen that the electrode potential often used in actual work does not refer to the potential difference on a single electrode, but refers to the voltaic cells composed of the electrode and the standard hydrogen electrode, and the electrode is a positive electrode and the standard hydrogen electrode is a negative electrode. The potential difference between the two end points, namely the electromotive force, is usually called the hydrogen standard electrode potential.

The standard electrode potential, $\varphi^{\ominus}$, can be a criterion of the oxidation or reduction capacity of metal ions. A small algebraic value of $\varphi^{\ominus}$ implicates a pair of weak oxidant and strong reductant. On the contrary, a large value implicates a pair of strong oxidant and weak reductant. The following points should be noted in the use of standard electrode potentials.

The main results are as follows:

① The standard electrode potential of the same substance in different media is different, and the redox capacity is also different. Such as $KMnO_4$:

In acidic medium $MnO_4^- + 8H^+ + 5e^- = Mn^{2+} + 4H_2O$, $\varphi^{\ominus}(MnO_4^-/Mn^{2+}) = 1.507V$

In neutral media $MnO_4^- + 2H_2O - 3e^- = MnO_2 + 4OH^-$ $\varphi^{\ominus}(MnO_4^-/MnO_2) = 1.679V$

In a strongly alkaline medium $MnO_4^- + e^- = MnO_4^{2-}$ $\varphi^{\ominus}(MnO_4^-/MnO_4^{2-}) = 0.558V$

② For the same electron pair in the same medium, the stoichiometric coefficient in the equilibrium equation has no effect on the value of the standard electrode potential. For example:

$$Zn^{2+} + 2e^- = Zn \qquad \varphi^{\ominus}(Zn^{2+}/Zn) = -0.7618V$$

$$2Zn^{2+} + 4e^- = 2Zn \qquad \varphi^{\ominus}(Zn^{2+}/Zn) = -0.7618V$$

③ There is no summation of the standard electrode potential. For example:

$$Fe^{2+} + 2e^- = Fe, \qquad \varphi^{\ominus}(Fe^{2+}/Fe) = -0.447V$$

$$+ \quad Fe^{3+} + e^- = Fe^{2+} \qquad \varphi^{\ominus}(Fe^{3+}/Fe^{2+}) = 0.771V$$

$$Fe^{3+} + 3e^- = Fe, \qquad \varphi^{\ominus}(Fe^{3+}/Fe) \neq 0.324V$$

And $\varphi^{\ominus}(Fe^{3+}/Fe) = -0.037V$

④ The value of the standard electrode potential is independent of the positive and negative electrode of the electron pair as the voltaic cell. For example, the standard electrode potential of copper:

$$\varphi^{\ominus}(Cu^{2+}/Cu) = 0.347V$$

It acts as positive electrode when it is formed with zinc standard electrode, and the reaction of electrode is as follows:

$$Cu^{2+} - 2e^- = Cu$$

When the original cell is formed with the silver standard electrode, the copper is the negative electrode, and the reaction of the electrode is as follows:

$$Cu - 2e^{-} \longequal Cu^{2+}$$

Whether as a positive electrode or a negative electrode, its standard electrode potential is

$$\varphi^{\ominus}(Cu^{2+}/Cu) = 0.347V$$

8.2.3 Nernst Equation: Effect of Concentration on Electrode Potential

(1) Nernst equation

The standard electrode potential is that the electrode is in equilibrium and is measured under a thermodynamic standard state (pure substance, each gas pressure is 100kPa, ion concentration is 1mol/L). Electric potential, its value reflects the nature of matter about the difficulty of gaining and losing electrons in the oxidized and reduced states of the electricity pair.

In practical applications, it is not always in the thermodynamic standard state, so what happens to the electrode potential under the non-standard state? According to the electric double layer theory [refer to Fig.8-2(a)], it can be seen that if the concentration of cations (oxidized state) is high, the rate at which it is deposited on the electrode surface increases. At equilibrium, there will be more positive charges on the electrode surface, and the electrode potential algebraic value will increase; if the ions in the solution are reduced species (such as Cl^{-} in the electron pair Cl_2/Cl^{-}), then the greater the ion concentration, the smaller the electric potential algebraic value of the electrode. In addition, the electrode potential is also related to temperature (generally treated as 298.15K when conditions are not specified).

In this chapter, the relationship between electrode potential and concentration is discussed, and the relationship between electrode potential and temperature is not involved for the time being. The relationship between electrode potential and concentration is based on the Nernst equation. If the electrode reaction is:

$$a\,(\text{oxidation state}) + ne^{-} \longequal b\,(\text{reduction state})$$

Then the electrode potential of the electrode is

$$\varphi = \varphi^{\ominus} + \frac{RT}{nF}\ln\frac{m_{\mathrm{r}}^{a}(\text{oxidation state})}{m_{\mathrm{r}}^{b}(\text{reduced state})} \tag{8-2}$$

In the formula: φ—electrode potential at any concentration;

$\varphi^{\ominus}$—the standard electrode potential of the electrode;

m_{r}(oxidized state)—the relative concentration of oxidized substances;

m_{r}(reduced state)—the relative concentration of reduced substances;

a, b—their stoichiometric coefficient in the electrode reaction formula;

n—the number of electrons in the electrode reaction;

T—absolute temperature, K;

R—gas constant, $R = 8.3145\text{J/(mol·K)}$;

F—Faraday constant, $F = 96485\text{C/mol}$.

The formula (8-2) is called the Nernst equation. At T= 298.15K, the above values are substituted into equation (8-2) and become the common logarithm, then

$$\varphi=\varphi^{\ominus}+\frac{8.314\times 298.15\times 2.303}{n\times 96485}\lg\frac{m_{\mathrm{r}}^{a}(\text{oxidation state})}{m_{\mathrm{r}}^{b}(\text{reduced state})}$$

That is

$$\varphi=\varphi^{\ominus}+\frac{0.0592}{n}\lg\frac{m_{\mathrm{r}}^{a}(\text{oxidation state})}{m_{\mathrm{r}}^{b}(\text{reduced state})} \tag{8-3}$$

The Nernst equation can be used to calculate and discuss the electrode potential of the electrode and different concentrations and normal temperature (T= 298.15K).

(2) Effect of concentration on electrode potential

According to the Nernst equation, the effect of concentration on the electrode potential at room temperature (T= 298.15K) can be calculated or discussed. Pay attention to the following points when applying the Nernst equation:

① If the substance that makes up the electrode is a solid or pure liquid (its concentration is specified as 1), it is not included in the Nernst equation. In the case of a gas, the relative partial pressure of the gas ($p_{\mathrm{i}} / p^{\ominus}$) is calculated and, in the case of a solution, the relative concentration ($c_{\mathrm{i}} / c^{\ominus}$) is for calculation.

② If the stoichiometric coefficient before oxidizing and reducing substances in the electrode reaction formula is not equal to 1, the concentrations of oxidized and reduced substances should be indexed by their respective stoichiometric coefficient.

③ If H^+ or OH^- participate in the reaction in the electrode reaction, the concentration of these ions should also be written in the Nernst equation according to the balanced electrode reaction formula (the reason will be described later), but H_2O is not written (it is a pure liquid, concentration is 1).

④ Scope of application: calculate the balance (that is, the current of the outer conductor tends to zero) M^{n+}/M electrode potential.

Example 8-2 Calculates 25℃, $c(Zn^{2+})$= 0.001mol/L Zn^{2+}/Zn electrode potential.

Solution

$$Zn^{2+} + 2e^- \longrightarrow Zn$$

$$\varphi=\varphi^{\ominus}+\frac{0.0592}{n}\lg\frac{c_{\mathrm{r}}^{a}(\text{oxidation state})}{c_{\mathrm{r}}^{b}(\text{reduced state})}=\varphi^{\ominus}+\frac{0.0592}{2}\lg\frac{c_{\mathrm{r}}(Zn^{2+})}{1}$$

$$=-0.7628+\frac{0.0592}{2}\lg 0.001=-0.8516(\text{V})$$

That is, when $c(Zn^{2+})$=0.001mol/L, the electrode potential of Zn^{2+}/Zn is −0.8516V.

Example 8-3 Calculate the electrode potential of the O_2/OH^- electrode at 25℃, $p(O_2)$=100kPa, $c(OH^-)$=10^{-7}mol/L.

Solution $$O_2 + 2H_2O + 4e^- \longrightarrow 4OH^-$$

$$\varphi = \varphi^{\ominus} + \frac{0.0592}{n}\lg\frac{c_r^a(\text{oxidation state})}{c_r^b(\text{reduced state})} = \varphi^{\ominus} + \frac{0.0592}{4}\lg\frac{\frac{100}{100}\times 1}{10^{-28}}$$

$$= 0.401 + \frac{0.0592}{4}\lg(10^7)^4 = 0.8154(\text{V})$$

That is, when $p(O_2)$=100kPa and $c(OH^-)$=10^{-7}mol/L, the electrode potential of O_2/OH^- electrode is 0.8154V.

As can be seen from the above two examples:

① Ion concentration has an effect on the electrode potential, but it has little effect on the electrode potential. As in example 8-2, when the concentration of metal ions is reduced from 1 to 0.001mol/L, the electrode potential changed by only 0.0888V.

② When the concentration of metal (or hydrogen) ion (oxidized state) is reduced, the corresponding electrode potential algebra value is reduced, and the metal (or hydrogen) will easily lose electrons to transform into ions and enter the solution, that is to say, the reducibility of metal (or hydrogen) will be enhanced. On the contrary, the reducibility become weak.

③ For non-metallic negative ions, when the concentration of ions (reduced state) decreases, the corresponding electrode potential value increases, that is to say, the oxidation of non-metals is enhanced. On the contrary, the oxidization decreased.

If a battery reacts as $$a\text{A} + b\text{B} \longrightarrow g\text{G} + d\text{D}$$

Then the relationship between the electromotive force of the battery and the concentration of each substance can be obtained according to the relationship between the thermodynamic function and the EMF, as well as the thermodynamic isotherm equation (6-10). Because:

$$\Delta_r G_m = -nFE$$

$$\Delta_r G_m^{\ominus} = -nFE^{\ominus} \tag{8-4}$$

$$\Delta_r G_m = \Delta_r G_m^{\ominus} + RT\ln\frac{c_r^g(\text{G})c_r^d(\text{D})}{c_r^a(\text{A})c_r^b(\text{B})}$$

So

$$-nFE = -nFE^{\ominus} + RT\ln\frac{c_r^g(\text{G})c_r^d(\text{D})}{c_r^a(\text{A})c_r^b(\text{B})}$$

$$E = E^{\ominus} - \frac{RT}{nF}\ln\frac{c_r^g(\text{G})c_r^d(\text{D})}{c_r^a(\text{A})c_r^b(\text{B})}$$

After substituting each constant, there is:

$$E = E^{\ominus} - \frac{0.0592}{n}\lg\frac{c_r^g(\text{G})c_r^d(\text{D})}{c_r^a(\text{A})c_r^b(\text{B})} \tag{8-5}$$

In the formula, n is the number of gain and loss electrons after battery reactive balancing.

(3) Effect of pH on electrode potential

Example 8-4 For the half cell reaction $Cr_2O_7^{2-} + 14H^+ + 6e^- = 2Cr^{3+} + 7H_2O$, $\varphi^\ominus = 1.33V$, using the Nernst equation, what are the values when $c(H^+) = 10mol/L$ and $c(H^+) = 1.0\times10^{-3}mol/L$, and the other ion concentrations are standard concentrations? According to the calculation results, the influence of acidity on the strength of $Cr_2O_7^{2-}$ redox is compared.

Solution According to the electrode reaction,

$$\varphi = \varphi^\ominus + \frac{0.0592}{6}\lg\frac{c_r(Cr_2O_7^{2-})}{c_r^2(Cr^{3+})}\times\frac{c_r^{14}(H^+)}{1}$$

When $$c(H^+) = c(Cr^{3+}) = c(Cr_2O_7^{2-}) = 1mol/L$$

$$\varphi = \varphi^\ominus = +1.33V$$

When $$c(H^+) = 10mol/L \text{ and } c(Cr^{3+}) = c(Cr_2O_7^{2-}) = 1mol/L$$

$$\varphi = 1.33 + \frac{0.0592}{6}\lg\frac{10^{14}}{1} = 1.4682(V)$$

When $$c(H^+) = 1\times10^{-3}mol/L \text{ and } c(Cr^{3+}) = c(Cr_2O_7^{2-}) = 1mol/L$$

$$\varphi = 1.33 + \frac{0.0592}{6}\lg\frac{(1\times10^{-3})^{14}}{1} = +0.9156(V)$$

The calculation results of ($Cr_2O_7^{2-}/Cr^{3+}$) above: when $c(H^+)$=10mol/kg, φ = 1.4682V; when $c(H^+)$=1mol/L, φ = +1.33V; when $c(H^+) = 1.0\times10^{-3}mol/L$, φ = +0.9156V. It can be seen from the above that the oxidation capacity of $Cr_2O_7^{2-}$ is obviously weakened as the acidity decreases. Therefore, where there is a redox reaction in which H^+ and OH^- participate, and the H^+ and OH^- are measured in a larger amount in the reaction formula, the acidity has a greater influence on the electrode potential. Therefore, when calculating the electrode potential at any concentration, you must first write the balanced electrode reaction.

8.2.4 Application of Electrode Potential and EMF

(1) The judgment of positive and negative electrodes and the calculation of EMF

In galvanic cells, the electrode with higher potential is the positive electrode and the lower one is the negative electrode. The electromotive force of the galvanic cell $E = \varphi_+ - \varphi_- > 0$.

Example 8-5 The spontaneous battery was composed of zinc electrode Zn^{2+} (0.1mol/L) | Zn and copper electrode Cu^{2+}(0.01mol/L) | Cu. The positive and negative electrodes of the battery were judged, and the electromotive force of the battery was calculated.

Solution Check Table 8-1, $\varphi^\ominus(Zn^{2+}/Zn) = -0.7618V$, $\varphi^\ominus(Cu^{2+}/Cu) = 0.347V$,

$$\varphi(Zn^{2+}/Zn) = \varphi^\ominus(Zn^{2+}/Zn) + \frac{0.059}{2}\lg m(Zn^{2+})$$

$$=-0.7618+\frac{0.059}{2}\lg(0.1)=-0.7913(\text{V})$$

$$\varphi(\text{Cu}^{2+}/\text{Cu})=\varphi^{\ominus}(\text{Cu}^{2+}/\text{Cu})+\frac{0.059}{2}\lg m(\text{Cu}^{2+})$$

$$=0.347+\frac{0.059}{2}\lg(0.01)=0.288(\text{V})$$

From the above calculation results, it can be seen that: $\varphi(Cu^{2+}/Cu) > \varphi(Zn^{2+}/Zn)$. Therefore, in this cell, Cu^{2+}(0.01mol/L) | Cu is positive electrode and Zn^{2+}(0.1mol/L) | Zn is negative electrode.

Electromotive force $E=\varphi(\text{Cu}^{2+}/\text{Cu})-\varphi(\text{Zn}^{2+}/\text{Zn})=0.288-(-0.7913)=1.079\text{V}$

(2) The strength and selection of oxidants and reductants

① Electrode potential and the strength of oxidant and reductant. It is known that $\varphi^{\ominus}(Zn^{2+}/Zn)= -0.7618V$, and copper electrode $\varphi^{\ominus}(Cu^{2+}/Cu) = 0.347V$. As mentioned above, after the two electrodes are connected, zinc in anode loses electrons and the copper ion in cathode accepts electrons, which indicates that the zinc has stronger reducibility than copper and Cu^{2+} has stronger oxidizability than Zn^{2+} in their standard states. It should be noted that the strength of reducibility (lose electrons) or oxidizability (accept electrons) is relative. For example, the reducibility of copper is weaker than zinc. However, if compared with silver, copper is a stronger reductant.

In Table 8-1, the values of standard electrode potential increase from top to bottom, and the oxidizatility of oxidants increases in the same order. Correspondingly, the reducibility of reductants decrease from top to bottom in the table. When the value of the potential is very close or the electrode is not in standard state, it needs to be calculated and determined according to the Nernst equation.

② Selection of oxidant and reducing agent. After judging the strength of the oxidant/ reducing agent by the magnitude of the electrode potential, it can also be used in the selection of oxidant/reducing agent in specific reactions in practice. For example: in a mixed system, if only one component is desired to be oxidized or reduced, the other components do not change. In this case, it is necessary to select the appropriate oxidant or reducing agent, and this can be achieved by comparing the electrode potential.

Example 8-6 In a mixed solution containing Br^-, I^-, under standard conditions, I^- want to be oxidized to I_2, without Br^- oxidized to Br_2, ask which oxidant of $Fe_2(SO_4)_3$ and $KMnO_4$ can meet the requirements.

Solution Analysis: to make I^- to I_2 without Br^- to Br_2, then the oxidizing agent you choose should be greater than I_2 and less than Br_2. Therefore, the electrode potential of the corresponding electron pair should be greater than I_2/I^- and less than Br_2/Br^-.

Lookup table, $\varphi^{\ominus}(Br_2/Br^-) = 1.0873V$, $\varphi^{\ominus}(I_2/I^-) = 0.5355V$,

$$\varphi^{\ominus}(Fe^{3+}/Fe^{2+})=0.771V,\quad \varphi^{\ominus}(MnO_4^-/Mn^{2+}) = 1.507V$$

Obviously, the electrode potential between 0.5355V and 1.0873V should be selected as the oxidant, that is, $Fe_2(SO_4)_3$ should be selected.

(3) Judgment of reaction direction

Whether a chemical reaction can be carried out can be judged by the change of Gibbs function, that is:

$\Delta_r G_m > 0$ Positive reactions cannot proceed spontaneously

$\Delta_r G_m < 0$ Forward reaction can proceed spontaneously

$\Delta_r G_m = 0$ The reaction is in equilibrium

After the voltaic cell assembled by the spontaneous oxidation-reduction reaction generates current, the voltaic cell does work on the environment (outside circuit). This work is called electrical work W, which is equal to the amount of charge transferred from one pole to the other (q). The product of the electromotive force (E) and the battery's work on the environment is negative, that is

$$W_{max} = -qE \tag{8-6}$$

If there is a certain amount of material reaction at the electrode and there is 1mol of electron transfer, 96485 C of electricity, which is a Faraday electricity (F), will be generated. If there is n mol electron transfer in the reaction, there is n×96485 C of electricity, so

$$W_{max} = -n\times 96485\times E = -nFE$$

The electrical work is similar to other work. The reduction of Gibbs free energy of the galvanic cell reaction under the reversible conditions of constant temperature and constant pressure must be equal to the electrical work done by the system to the environment, that is:

$$\Delta_r G_m = W_{max} = -nFE \tag{8-7}$$

In the above formula, n and F are both positive integers. By formula (8-7), the $\Delta_r G_m$ criterion for judging the reaction direction can be successfully converted into the electromotive force criterion. Then according to $E = \varphi_+ - \varphi_-$, there are:

$E > 0$ or $\varphi_+ > \varphi_-$ Forward reaction can proceed spontaneously

$E < 0$ or $\varphi_+ < \varphi_-$ Positive reactions cannot proceed spontaneously

$E = 0$ or $\varphi_+ = \varphi_-$ The reaction is in equilibrium

Here are two points to note:

① Why is the electromotive force here negative? This is because we arbitrarily specified the oxidant and reductant before determining the spontaneous direction of this redox reaction.

② When $E < 0$, $\Delta_r G_m > 0$, the inverse reaction $\Delta_r G_m < 0$ and the inverse reaction is spontaneous. Therefore, $E < 0$ does not mean that the battery does not exist, it just indicates that the direction of the battery reaction is the opposite of the direction of the original judgment (or hypothesis).

Example 8-7 Try to determine the direction of the following redox reactions:

$$2Fe^{2+} + I_2 \longrightarrow 2Fe^{3+} + 2I^-$$

if the concentration of all ions in the solution is 1mol/L.

Solution　It can be seen from the reaction formula that if the reaction proceeds in the forward direction, the electrode corresponding to the pair Fe^{3+}/Fe^{2+} should be negative, and the electrode corresponding to the pair I_2/I^- should be positive.　At this time:

$$\varphi_+ = \varphi^\ominus(I_2/I^-) = 0.5355V$$

$$\varphi_- = \varphi^\ominus(Fe^{3+}/Fe^{2+}) = 0.771V$$

That is　$$E = \varphi_+ - \varphi_- = 0.5355-0.771 = -0.2355(V)$$

$$E < 0$$

Therefore, this reaction cannot proceed to the right spontaneously, and its reverse reaction must be $E > 0$ and can proceed spontaneously. If E is negative, it means that an external voltage is required to make a positive response.

Example 8-8 Try to determine the direction of the following concentration battery reactions:

$$Cu + Cu^{2+}(1mol/L) \longrightarrow Cu^{2+}(1.0\times10^{-4}mol/L) + Cu$$

Solution　Assuming that the reaction follows the direction of the positive reaction, then $Cu^{2+}(1.0\times10^{-4}mol/L)$ | Cu should be a negative electrode, Cu^{2+} (1mol/L) | Cu should be a positive electrode.

$$\varphi_+ = \varphi^\ominus(Cu^{2+}/Cu) = 0.347V$$

$$\varphi_- = \varphi^\ominus(Cu^{2+}/Cu) + \frac{0.059}{2}\lg m(Cu^{2+}) = 0.347 + \frac{0.059}{2}\lg(10^{-4}) = 0.229(V)$$

Electromotive force　$$E = \varphi_+ - \varphi_- = 0.347 - 0.229 = 0.118(V)$$

Because $E > 0$, the reaction proceeds spontaneously to the right.

When judging the direction of the oxidation-reduction reaction, a standard electromotive force can usually be used to make a rough judgment. This is because in general, the ion concentration has little effect on the electrode potential. However, if the standard electrode potentials of the two pairs that make up the battery are very different, and E or $\Delta_r G_m$ is close to zero, the change in ion concentration may cause the redox reaction to proceed in the opposite direction.

For example, in the redox reaction　$Pb^{2+} + Sn \longrightarrow Pb + Sn^{2+}$

When　$$m(Pb^{2+}) = m(Sn^{2+}) = 1mol/kg$$

$$\varphi^\ominus_{Ox} = (Pb^{2+}/Pb) = -0.1262V$$

$$\varphi^\ominus_{Red} = (Sn^{2+}/Sn) = -0.1375V$$

$$E^\ominus = \varphi^\ominus_{Ox} - \varphi^\ominus_{Red} = -0.1262-(-0.1375) = 0.0113(V)$$

Although E is close to zero, the reaction can proceed spontaneously to the right. The redox property of the substance is: oxidizability, Pb^{2+}>Sn^{2+}; reducibility, Sn>Pb.

If $c(Sn^{2+})$ = 1mol/L, and $c(Pb^{2+})$ = 0.1mol/L. Then, it can not be judged directly by the standard electrode potential, but should be calculated separately.

$$\varphi(Pb^{2+}/Pb) = \varphi^{\ominus}(Pb^{2+}/Pb) + \frac{0.059}{2}\lg(0.1) = -0.1262 - 0.0295 = -0.1557(V)$$

$$\varphi(Sn^{2+}/Sn) = \varphi^{\ominus}(Sn^{2+}/Sn) = -0.1375V$$

$$E = \varphi(Pb^{2+}/Pb) - \varphi(Sn^{2+}/Sn) = -0.1557 - (-0.1375) = -0.0182(V)$$

That is, $E < 0$.

Conclusion is opposite to the above, the above reaction cannot be carried out spontaneously to the right. The spontaneous direction of the reaction is:

$$Pb + Sn^{2+} = Pb^{2+} + Sn$$

In the reverse reaction of the above reaction, the redox ability of the substance has changed. The result is: oxidizability, $Sn^{2+} > Pb^{2+}$; reducibility, Pb > Sn.

Therefore, when the electrode potential is used to discuss the problem, if the difference between the standard electrode potentials of the two pairs is very small (generally less than 0.3V), the ion concentration is not 1mol/L. The correct conclusion can only be drawn after the Nernst equation is calculated.

In addition, the direction of redox reaction can be determined qualitatively without calculation by using the strength of oxidant and reductant, which is convenient in many cases.

In the upper right of Table 8-1, the reduction state in the pair with a smaller $\varphi^{\ominus}$— generation value is a strong reductant. In the lower left of the table, the oxidation state in the pair with larger $\varphi^{\ominus}$ generation value is a stronger oxidant. The direction of the oxidation-reduction reaction is that a stronger oxidant and a stronger reductant act to produce a weaker oxidant and a weaker reductant, that is:

$$(\text{strong oxidant})_1 + (\text{strong reductant})_2 \longrightarrow (\text{weak reductant})_1 + (\text{weak oxidant})_2$$

For example $Sn^{4+} + 2e^- = Sn^{2+}$, $\varphi^{\ominus}(Sn^{4+}/Sn^{2+}) = 0.151V$

$Fe^{3+} + e^- = Fe^{2+}$, $\varphi^{\ominus}(Fe^{3+}/Fe^{2+}) = 0.771V$

Available $2Fe^{3+} + Sn^{2+} = 2Fe^{2+} + Sn^{4+}$

(strong) (strong) (weak) (weak)

It can be seen that the reduction state in the upper right of Table 8-1 is used as reductant, and the oxidation state in the lower left is used as oxidant, and the reaction can be carried out spontaneously. This kind of diagonal direction reaction rule is commonly referred to as the "diagonal rule".

Of course, when the difference between the two electron pairs $\varphi^{\ominus}$ involved in the reaction is small, and under non-standard conditions, it is inaccurate to use $\varphi^{\ominus}$ to determine the direction of the reaction. It is necessary to obtain the electromotive force E after calculating φ by the Nernst equation.

(4) Limits of redox reactions

The standard equilibrium constant of redox reaction K represents the limit of redox reaction. The standard equilibrium constant can be calculated by the standard EMF of the redox reaction as

$$\Delta_r G_m^\ominus = -2.303RT\lg K^\ominus$$

$$\Delta_r G_m^\ominus = -nFE^\ominus$$

So $$nFE^\ominus = 2.303RT\lg K^\ominus$$

If you substitute, $F = 96485\text{C/mol}$, $R = 8.314\text{J/(K·mol)}$, $T = 298.15\text{K}$, get:

$$E^\ominus = \frac{2.303 \times 8.314 \times 298.15}{n \times 96485}\lg K^\ominus = \frac{0.059}{n}\lg K^\ominus$$

$$\lg K^\ominus = \frac{nE^\ominus}{0.059} \tag{8-8}$$

Therefore, as long as the standard electromotive force of the voltaic cells composed of the redox reaction is known, the standard equilibrium constant of the redox reaction can be calculated, so that the extent of the reaction can be judged. However, it should be noted that n in the formula is the number of electron transfers after the total reaction is balanced.

Example 8-9 Determine the extent to which the following reactions took place:

$$Cu + 2Ag^+ = Cu^{2+} + 2Ag$$

Solution Assuming that the above reaction proceeds in the forward direction, the negative electrode is $Cu \mid Cu^{2+}$ and the positive electrode is $Ag^+ \mid Ag$.

The reaction corresponds to the standard EMF of the voltaic cell.

$$E^\ominus = \varphi^\ominus(Ag^+/Ag) - \varphi^\ominus(Cu^{2+}/Cu) = 0.7996 - 0.347 = 0.4526(V)$$

$$\lg K^\ominus = \frac{nE^\ominus}{0.059} = \frac{2 \times 0.4526}{0.059} = 15.342$$

$$K^\ominus = 2.2 \times 10^{15}$$

The standard equilibrium constant of 2.2×10^{15} is very large, so this reaction proceeds very thoroughly in the forward direction.

8.3 Electrolytic Cell

8.3.1 Composition and Electrode Reaction of Electrolytic Cells

The process of redox of current through electrolytic cell solution (or molten salt) is called electrolysis, this process is a non-spontaneous process, with the help of an external power supply to make the $\Delta_r G_m > 0$ process in which the redox reaction is carried out. In order to complete this process, the device that converts electricity into chemical energy is called the electrolytic cell (non-spontaneous battery). In an electrolytic cell, the electrode connected to the positive electrode of the power supply is called anode, and the electrode connected to the negative electrode of the power supply is called cathode. Electrons flow from the negative pole of the power supply to the cathode of the electrolytic cell along the wire. On the other hand, electrons leave the anode of the electrolytic cell and flow back to the positive pole of the power supply along the wire. Therefore, the oxidation state ion in the electrolyte moves to the cathode, and the electron is obtained on the cathode for reduction reaction; the reduced ion moves to the anode and loses the electron on the anode for oxidation reaction. The process in which electrons are lost or accepted in half-cell is called discharge.

It should be noted that in the electrolytic cell, the electrode name, electrode reaction and the direction of electron flow are different from the original battery, and cannot be confused with each other.

8.3.2 Main Factors Affecting Electrode Reaction

When electrolyzing an aqueous salt solution, in addition to the ions of the electrolyte, there are H^+ ions and OH^- ions separated by hydrolysis in the electrolyte solution. Therefore, there may be at least two oxidized species ions discharged at the cathode, usually metal ions and H^+ ions; there may also be at least two reduced state ions that may be discharged at the anode, namely acid ions and OH^- ions. It is necessary to analyze the electrode potential and overpotential in order to judge which substance is discharged first.

(1) Electrode potential

In the electrolytic cell, the anode is performing an oxidation reaction, and the cathode is performing a reduction reaction. In the anode, anions (reductant) usually aggregate by electrostatic attraction. The reductant with the smallest φ value discharges firstly because of the strongest reducibility. In the cathode, cations (oxidant) with the largest φ value discharge firstly because of the strongest oxidizability.

It is known in Section 8.2 that φ is related to the nature ($\varphi^{\ominus}$) and ion concentration of the substance. It can be calculated using the Nernst equation. We call it the theoretical precipitation potential φ_{theory} (Nernst potential). In theory, as long as the calculation based on the theoretical value of φ_{theory} of each substance that may be discharged at the two poles, according to the above principles, it can be determined which substance is discharged first.

For example, electrolysis of 1mol/L $CuCl_2$ aqueous solution (the gas produced is 100kPa), H^+ and Cu^{2+} ions tend to the cathode, the electrode reaction is $Cu^{2+} + 2e^- = Cu$, $2H^+ + 2e^- = H_2$, H^+ ion $\varphi^{\ominus} = 0$, concentration 10^{-7}mol/L, and Cu^{2+} ion $\varphi^{\ominus} = 0.347V$, the concentration is 1mol/L, according to this calculation

$$\varphi(Cu^{2+}/Cu) = \varphi^{\ominus}(Cu^{2+}/Cu) = 0.347V$$

$$\varphi(H^+/H_2) = \varphi^{\ominus}(H^+/H_2) + \frac{0.059}{2}\lg\frac{\left(\frac{10^{-7}}{1}\right)^2}{(100/100)} = -0.413(V)$$

The theoretical precipitation potential of Cu^{2+} is greater than that of H^+, that is:

$$\varphi(Cu^{2+}/Cu)_{theory} > \varphi(H^+/H_2)_{theory}$$

So, in the cathode is Cu^{2+} ions discharge first.

OH^- and Cl^- may be discharged at the anode. The electrode reaction is: $2Cl^- - 2e^- = Cl_2$, $4OH^- - 4e^- = O_2 + 2H_2O$. When the concentration of OH^- ion is 10^{-7}mol/L

$$\varphi(O_2/OH^-) = \varphi^{\ominus}(O_2/OH^-) + \frac{0.059}{4}\lg\frac{(100/100)}{\left(\frac{10^{-7}}{1}\right)^4} = 0.401 + 0.413 = 0.814(V)$$

$$\varphi(Cl_2/Cl^-) = \varphi^{\ominus}(Cl_2/Cl^-) + \frac{0.059}{2}\lg\frac{\left(\frac{100}{100}\right)}{(2/1)^2} = 1.358 - 0.0177 = 1.3417(V)$$

The theoretical precipitation potential of OH^- is 0.814V, which is much smaller than $\varphi(Cl_2/Cl^-)$ (1.3417V). According to the above principle, it seems that OH^- discharges firstly. But in fact, the product of anode is Cl_2. There must be other factors that determine the order of ion discharge.

(2) Polarization of the electrode

During electrolysis, a DC power source must be added to pass current. In the process of the oxidized and reduced substances moving toward the cathode and anode respectively and discharging, it is affected by many factors. The actual potential values of the discharge ion at the electrode often deviates from the theoretical values. This phenomenon of electrode potential deviating from its equilibrium value φ_{theory} when current passes through the electrode is called electrode polarization. The common causes of electrode polarization inclued the existence of concentration gradient (nonequilibrium concentration), electrode material and shape, electrolytic products, and so on.

① Concentration polarization. When the electrode is in equilibrium, the distribution of electrolytes in the solution is uniform. After the current flows, the situation changes. With the progress of the electrode reaction, the reactants on and near the electrode surface have been consumed, and the products are constantly formed. In order to maintain the stability of the current, it is ideal that the reactants on the electrode surface can be supplemented by the reactants in the depths of the solution in time, and the products can leave immediately. However, the actual situation is that the diffusion and migration rates of reactants and products cannot keep up with the reaction rate, resulting in the change of electrolyte concentration near the electrode, thus forming a concentration gradient in the solution. For the cathode, the concentration of oxidized substances in the electrode surface solution is smaller, while the concentration of reduced substances is relatively larger, if still calculated by Nernst formula. Obviously, the actual electrode potential will decrease at this time, while for the anode, on the contrary, the real-time potential will increase. This polarization phenomenon caused by the difference between the ion concentration near the electrode surface and the ion concentration in equilibrium is called concentration polarization. It can be seen that the current is controlled by the speed of ion movement when the concentration polarization is carried out.

② Electrochemical polarization. Electrode reaction is heterogeneous chemical reaction on the surface of the electrode. The reaction is naturally constrained by kinetic factors, so we have to consider the reaction rate. Usually, each electrode reaction is composed of several basic continuous steps (such as ion discharge, atomic binding into molecules, bubble formation and escape, *etc.*). And one of them may be the slowest step in the electrode process, which has the highest activation energy and therefore the slowest rate. In order to make the electrode reaction go on continuously, the external power supply needs to increase a certain voltage to overcome the activation energy of the reaction. The polarization caused by the slow reaction rate of the electrode is called electrochemical polarization (also known as kinetic polarization or activation energy polarization). In the case of electrochemical polarization, the current flowing through the electrode is controlled by the reaction rate of the electrode.

③ The IR loss of the battery. For the battery of the electrochemical system, whether it is an electrolytic cell or a primary battery, there is another polarization factor besides the concentration polarization and the electrochemical polarization. This is the IR drop of the battery (R is also called ohmic internal resistance). This is because when the current flows through the electrolyte solution, the oxidized and reduced ions each migrate to the two poles. Because the battery itself has a certain internal resistance R, the movement of the ions is subject to a certain "resistance". In order to overcome the internal resistance, an additional voltage must be added to "push" the ion forward. The voltage required to overcome the internal resistance of the battery is equal to the product of the current I and the internal resistance R of the battery, that is, IR drop. It is usually converted to the environment in the form of heat. This extra power loss is I^2R.

(3) Over potential

The polarization of the electrode is discussed above. In order to measure the degree of electrode polarization, a new concept, over potential, needs to be introduced.

Due to the polarization phenomenon on the electrode, the actual potential of the electrode deviates from the equilibrium potential. This deviation value is called the over potential, and is represented by the symbol η. It should be noted that when polarization occurs, the anode potential φ positive increases and the cathode potential φ negative decreases. But traditionally η takes positive values, with $\eta_{cathode}$ and η_{anode} representing the over potentials of cathode and anode respectively, $\varphi_{\text{cathode (theory)}}$ and $\varphi_{\text{anode (theory)}}$ representing the equilibrium potentials of cathode and anode (also called the theoretical potential); $\varphi_{\text{cathode (actual)}}$ and $\varphi_{\text{anode (actual)}}$ represent the actual precipitation potential of the cathode and anode poles, respectively.

$$\varphi_{\text{cathode (actual)}} = \varphi_{\text{cathode (theory)}} - \eta_{\text{cathode}}, \quad \eta_{\text{cathode}} = \varphi_{\text{cathode (theory)}} - \varphi_{\text{cathode (actual)}} \tag{8-9}$$

$$\varphi_{\text{anode (actual)}} = \varphi_{\text{anode (theory)}} + \eta_{\text{anode}}, \quad \eta_{\text{anode}} = \varphi_{\text{anode (actual)}} - \varphi_{\text{anode (theory)}} \tag{8-10}$$

This is consistent with the polarization mentioned earlier that the cathode potential decreases and the anode potential increases.

According to several reasons for polarization, the total over potential η for a single electrode should be the sum of concentration over potential $\eta_{\text{concentration difference}}$, electrochemical over potential $\eta_{\text{electrification}}$, ohmic voltage drop η_{ohm}, *etc.*

$$\eta = \eta_{\text{concentration difference}} + \eta_{\text{electrification}} + \eta_{\text{ohm}} + \cdots \tag{8-11}$$

At present, the value of over potential cannot be calculated theoretically, the difficulty is that the influencing factors include some factors that cannot be predicted and controlled, but the over potential can be measured by experiment. It can be seen from the experiment that the over potential is not constant for the same substance, it is related to the following factors:

① The values of over potential are different when the electrolytic products are different. The over potential of metal is generally small, but the over potential of iron, cobalt and nickel is larger. For gas products, especially hydrogen and oxygen, the over potential is larger, while the over potential of halogen is smaller, as shown in Table 8-2 and Table 8-3.

② The electrode material and surface state are different, even if the electrolytic product is the same material, its over potential is different. On tin, lead, zinc, silver, mercury and other "soft metal" electrodes η is significant, particular the mercury electrode, as shown in Table 8-2.

③ With the higher current density, the greater the over potential, as shown in Table 8-3 and Table 8-4.

④ With the higher temperature, or by stirring, the over potential will decrease.

Table 8-2　Over Potential of Hydrogen and Oxygen on Different Metals at Room Temperature

Electrode material	Over potential/V	
	Hydrogen	Oxygen
Pt (platinum-plated black)	0.00	0.25
Pd	0.00	0.43
Au	0.02	0.53
Fe	0.08	0.25
Pt(smooth)	0.09	0.45
Ag	0.15	0.41
Ni	0.21	0.06
Cu	0.23	—
Cd	0.48	0.43

(Continued)

Electrode material	Over potential/V	
	Hydrogen	Oxygen
Sn	0.53	—
Pb	0.64	0.31
Zn	0.70	—
Hg	0.78	—
graphite	0.90	1.09

Note: measured under the condition of current density at the beginning of the onset of a significant bubble.

Table 8-3 Over Potential of Chlorine Precipitating on Graphite Electrode in Saturated NaCl Solution at 25℃

Current density/(A/M²)	400	700	1000	2000	5000	10000
Over potential/V	0.186	0.193	0.251	0.298	0.417	0.495

Table 8-4 Over Potential of Oxygen Precipitated on Graphite Electrode in 1mol/kg KOH Solution at 25℃

Current density/(A/m²)	100	200	500	1000	2000	5000
Over potential/V	0.869	0.963	—	1.091	1.142	1.186

8.3.3 Decomposition Voltage and over Voltage

During electrolysis, a certain voltage must be applied to the two poles of the electrolytic cell, so that the reaction on the electrode can proceed smoothly. How much voltage should be applied? This is related to the over potential. Now take platinum as the electrode, electrolysis c (NaOH) = 0.1mol/L aqueous solution as an example (the gas generated is 100kPa).

When electrolyzing NaOH aqueous solution, hydrogen is precipitated at the cathode and oxygen is precipitated at the anode, and part of the hydrogen and oxygen are adsorbed on the surface of the platinum sheet, thus forming the following primary battery:

$$(-)(\mathrm{Pt})\mathrm{H_2} \mid \mathrm{NaOH}(c = 0.1\mathrm{mol/L}) \mid \mathrm{O_2}(\mathrm{Pt})(+)$$

Its EMF is the difference between the electrode potential of the positive electrode (oxygen electrode) and the electrode potential of the negative electrode (hydrogen electrode). The values can be calculated as follows:

In an aqueous solution with $c(\mathrm{NaOH}) = 0.1\mathrm{mol/L}$, $c(\mathrm{OH^-}) = 0.1\mathrm{mol/L}$, then

$$c(\mathrm{H^+}) = \frac{10^{-14}}{10^{-1}} = 10^{-13}(\mathrm{mol/L})$$

Positive reaction $\mathrm{O_2} + 4\,\mathrm{H_2O} + \mathrm{e^-} == 4\mathrm{OH^-}$ $\varphi^{\ominus}(\mathrm{O_2/OH^-}) = 0.401\mathrm{V}$

Positive potential $\varphi = \varphi^{\ominus} + \dfrac{0.059}{4}\lg\dfrac{(p_{\mathrm{O_2}}/p^{\ominus})}{m_{\mathrm{r}}^{4}(\mathrm{OH^-})} = 0.401 + \dfrac{0.059}{4}\lg(0.1)^{-4} = 0.459(\mathrm{V})$

Negative electrode reaction $2\mathrm{OH^-} + \mathrm{H_2} - 2\mathrm{e^-} = 2\mathrm{H_2O}$ $\varphi^{\ominus}(\mathrm{H_2O/H_2}) = -0.828(\mathrm{V})$

Negative potential $\varphi = \varphi^{\ominus} + \frac{0.059}{2}\lg\frac{1}{(p_{H_2}/p^{\ominus})m_r^2(OH^-)} = -0.828 + \frac{0.059}{2}\lg(0.1)^{-2}$

$$= -0.769(V)$$

The EMF of the hydrogen-oxygen battery is:

$$E = \varphi_+ - \varphi_- = 0.459-(-0.769) = 1.228(V)$$

The direction of the current in the battery is opposite to the direction of the external DC power supply. According to this, in theory, when the applied voltage is equal to the electromotive force of the oxyhydrogen cell, the electrode reaction is in equilibrium. As long as the applied voltage slightly exceeds the electromotive force (1.228V), electrolysis seems to be able to proceed, but the experimental results and theoretical calculations are quite different, that is, the voltage is not 1.228V but 1.769V (Fig.8-5, the difference between A and C). That is to say, when the applied voltage reaches 1.769V, there are obvious bubbles on the two poles (at this time, the current should be the indication value of point B), the current increases rapidly, and the electrolysis can proceed smoothly. The lowest voltage is the actual decomposition voltage. The actual decomposition voltage of various substances is determined by experiments, such as 1mol/L HCl decomposition voltage is 1.31V; 1mol/L HBr decomposition voltage is 0.94V; 1mol/L HI has a decomposition voltage of 0.54V; electrolytic salt solution (diaphragm method) has a decomposition voltage of 3.4V.

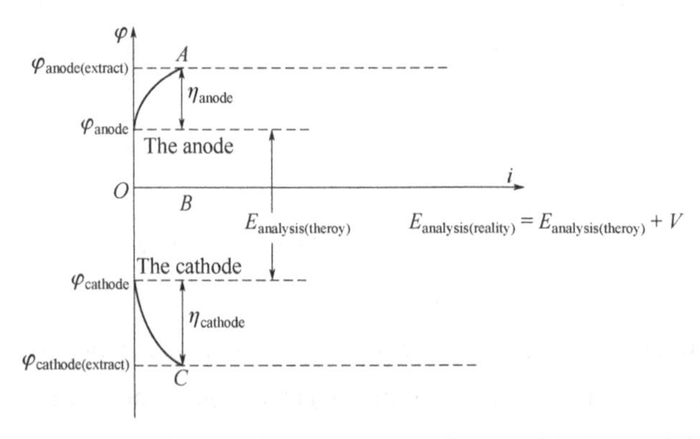

Fig.8-5 Schematic diagram of cathode and anode potentials during electrolysis

The decomposition voltage of the electrolyte is related to the electrode reaction. As shown in Table 8-5, the decomposition voltages of the NaOH, KOH, and KNO_3 solutions are very similar. This is because the electrode reaction products of these solutions are H_2 and O_2.

Table 8-5 Decomposition Voltage of Several Electrolyte Solutions (c = 1mol/L) (Room Temperature, Platinum Electrode)

Electrolyte	HCl	KNO_3	KOH	NaOH
Decomposition voltage/V	1.31	1.69	1.67	1.69

Why is there a difference between the actual decomposition voltage and the theoretical decomposition voltage? One reason is that both the solution and the wire have resistance, and there will be a voltage drop (IR) when energized. However, in general electrolysis, if the current I and the resistance R are not large, the value of IR is not large[1].

The other main reason is that the over potential is produced by the polarization of the electrode, and the overvoltage is caused by the over potential. Therefore, the actual decomposition voltage($V_{\text{decomposition}}$) is often larger than the theoretical decomposition voltage (V_{theory}).

The actual decomposition voltage equals to the theoretical decomposition voltage plus overvoltage, and the relationship between them is as follows:

$$V_{\text{decomposition}} = \varphi_{\text{anode (actual)}} - \varphi_{\text{cathode (actual)}} = [\varphi_{\text{anode (theory)}} + \eta_{\text{anode}}] - [\varphi_{\text{cathode (theory)}} - \eta_{\text{cathode}}]$$

$$= [\varphi_{\text{anode (theory)}} - \varphi_{\text{cathode (theory)}}] + (\eta_{\text{anode}} + \eta_{\text{cathode}})$$

$$= \text{Theoretical decomposition voltage} + \text{over voltage} \quad (8\text{-}12)$$

From the above formula, it can be seen that the sum of the polar over potential is the over voltage of the electrolytic cell, while the actual decomposition voltage is mainly the sum of the theoretical decomposition potential and the over potential.

$$\underbrace{\underset{\varphi_{O_2\text{(theory)}}}{0.4602} - \underset{\varphi_{H_2\text{(theory)}}}{(-0.7688)}}_{V_{\text{(theory)}}} + \underbrace{\underset{(\eta_{O_2})}{0.45} + \underset{(\eta_{H_2})}{0.09}}_{V_{\text{(over)}}} = \underset{V_{\text{decomposition}}}{1.769(\text{V})}$$

The above relationship is shown in Fig.8-5.

The existence of the over potential causes more electricity to be consumed during electrolysis, which is disadvantageous. In general electrolysis, we always hope to reduce the over potential to save electric energy and increase productivity. For example, when electrolyzing water (NaOH solution) in industry, nickel is used as the anode[2] and iron as the cathode, this is because the over potential of oxygen on nickel is small, and the over potential of hydrogen on iron is also small. But on the other hand, over potential is of great significance in production. For example, because H_2 has a large over potential, when electrolyzing a more active metal salt solution, more active metals, such as zinc, are only likely to precipitate at the cathode. This principle used to make zinc by electroplating zinc in weakly acidic (pH = 5) zinc salt solution is called electrolytic method. In electroplating and electrolytic processing, the rational use of polarization can improve product quality.

---

[1] In some processes (such as electrolytic processing and high-speed electroplating), the current I is very large, and the influence of the IR value should be considered.

[2] It can be seen from the electrode potential that Ni is more likely to lose electrons than OH^- ions, but here, nickel does not dissolve because nickel is passivated in alkaline solution.

8.3.4 General Law of Electrolytic Products

After understanding the factors affecting the electrode reaction and the concepts of decomposition voltage and over voltage, the general law of electrolytic products can be further discussed. Taking the preparation of caustic soda from electrolytic brine as an example, the product of the electrode and the required decomposition voltage are judged from the electrode potential, concentration and overvoltage.

The concentration of NaCl used in electrolytic saturated brine is generally not less than 315g/kg. The pH of the solution is controlled at about 8. Graphite is used as the anode and iron is used as the cathode. The gas produced is 100kPa. After the NaCl solution is energized, Na^+ and H^+ ions move to the cathode, and Cl^- and OH^- ions move to the anode. The first ion discharged on the electrode depends on the actual precipitation potential of various substances.

At the cathode $\varphi^\ominus(H^+/H_2) = 0.000V$ $\quad\varphi^\ominus(Na^+/Na) = -2.714V$

Electrode reaction $2H^+ + 2e^- \longrightarrow H_2$ $\quad Na^+ + e^- \longrightarrow Na$

Because the solution pH = 8, when the power is turned on, $m(H^+) = 10^{-8}$mol/kg, then the theoretical electrode potential of H_2 is calculated as

$$\varphi_{H_2(theory)} = \varphi^\ominus(H_2/H^+) + \frac{0.059}{2}\lg m_r^2(H^+) = 0 + 0.059\lg 10^{-8} = -0.472(V)$$

Looking up the table, the over potential of H_2 on the iron is 0.08V, therefore, the actual precipitation potential of H_2 is:

$$\varphi_{H_2(analysis)} = \varphi_{H_2(theory)} - \eta_{H_2} = -0.4736 - 0.08 = -0.5536(V)$$

This value is much larger than the standard electrode potential of sodium (−2.71V). Even if the concentration of Na^+ ions in a saturated solution of NaCl is large, the electrode potential will increase a little, and it is impossible to be as large as −0.5536V. Therefore, H^+ ion discharge is at the cathode, that is:

$$2H^+ + 2e^- \longrightarrow H_2$$

With the discharge of H^+ ions, the alkalinity of the solution in the cathode region gradually increased, and finally the NaOH concentration was about 10% (2.7mol/kg). At this time, it can be calculated that $\varphi_{H_2(theory)} = -0.85V$ and $\varphi_{H_2(analysis)} = -0.93V$, which is still much larger than the electrode potential of sodium, so when electrolyzing NaCl solution, the cathode always gets hydrogen.

At the anode $\quad\varphi^\ominus(Cl_2/Cl^-) = 1.358V$ $\quad\varphi^\ominus(O_2/OH^-) = 0.401V$

Electrode reaction $\quad Cl_2 + 2e^- \longrightarrow 2Cl^-$ $\quad 2H_2O + O_2 + 4e^- \longrightarrow 4OH^-$

In electrolytic salt water, the concentration of NaCl is not less than 315g/kg, which is 5.38mol/kg, $m(Cl^-) = 5.38$mol/kg, When chlorine gas is evolved, its partial pressure is 100kPa, then the theoretical electrode potential of chlorine is:

$$\varphi_{Cl_2(theory)} = +1.358 + \frac{0.059}{2}\lg\frac{100/100}{(5.38/1)^2} = 1.315(V)$$

While the over potential of chlorine on graphite is 0.25V, the actual precipitation potential of chlorine is:

$$\varphi_{Cl_2(analysis)} = \varphi_{Cl_2(theory)} + \eta_{Cl_2} = 1.315 + 0.25 = 1.57(V)$$

When pH = 8 $m(OH^-) = 10^{-6}$mol/kg

$$\varphi_{O_2(theory)} = 0.401 + \frac{0.059}{4}\lg\frac{1}{(10^{-6})^4} = 0.754(V)$$

The over potential of oxygen on graphite is 1.09V. Then the actual precipitation potential of oxygen is:

$$\varphi_{O_2(analysis)} = \varphi_{O_2(theory)} + \eta_{O_2} = 0.754 + 1.09 = 1.844(V)$$

Therefore, the anode should be a substance that actually precipitates in the reduced state with a small electrode potential value, that is, Cl^- ion discharge and chlorine gas is precipitated:

$$2Cl^- - 2e^- = Cl_2$$

Theoretical decomposition voltage = $\varphi_+ - \varphi_- = 1.315 - (-0.472) = 1.787(V)$

Actual decomposition voltage= $\varphi_{anode\ (theory)} - \varphi_{cathode\ (theory)} + \eta_{anode} + \eta_{cathode} = 1.787 + 0.33 = 2.117(V)$

That is, when the applied voltage be greater than 2.5V, electrolysis can proceed smoothly. The voltage used in actual production is even greater to overcome the voltage loss of the electrolyte and separator.

8.3.5 Application of Electrolysis

Electrolysis is widely used in mechanical industry and electronic industry. Electrolytic method is widely used for machining and surface treatment in mechanical industry and electronic industry, such as electroplating, electro polishing, anodizing and electrolytic machining, *etc.*

(1) Electroplating

Electroplating is the process of plating one metal (or non-metal) on the surface of another metal (or non-metallic) part by electrolysis. Take galvanization as an example. When galvanizing, the galvanized parts are used as cathode and metal zinc as anode. Electroplating solution cannot be used directly with simple zinc ion salt solution. If zinc sulfate is used as electroplating solution, because of the high concentration of zinc ion, the coating is rough, the thickness is not uniform, and the adhesion between the coating and the matrix metal is poor. For example, using alkaline zincate zinc plating, the coating is fine and smooth, this electroplating solution is prepared by zinc oxide, sodium hydroxide and additives. Zinc oxide is mainly formed in sodium hydroxide solution $Na_2[Zn(OH)_4]$ (habitually written as sodium zincate Na_2ZnO_2):

$$ZnO + 2NaOH + H_2O = Na_2[Zn(OH)_4]$$

$[Zn(OH)_4]^{2-}$ ions are in the following equilibrium in solution:

$$[Zn(OH)_4]^{2-} = Zn^{2+} + 4OH^-$$

Due to the formation of $[Zn(OH)_4]^{2-}$ ions, the ion concentration of Zn^{2+} is reduced, and the rate of crystal nucleation during the precipitation of metal crystals on the plating is reduced, which is beneficial to the formation of new crystal nucleus and can obtain crystallized and smooth coating. As the electroplating progresses, the Zn^{2+} ions are continuously discharged, and at the same time the above mentioned balance is continuously moved to the right, thereby ensuring that the concentration of Zn^{2+} ions in the plating solution is basically stable. The main reaction of the two poles is:

Cathode $\quad Zn^{2+} + 2e^- = Zn$

Anode $\quad Zn-2e^- = Zn^{2+}$

After electroplating, the plating parts are passivated in chromic acid solution to increase the beauty and corrosion resistance of the coating.

(2) Electric polishing

Electric polishing is one of the finishing methods of metal surface. Electric polishing can obtain smooth and glossy surface. The principle of electric polishing is that in the process of electrolysis, the dissolution rate of the protruding part on the metal surface is greater than that of the concave part on the metal surface, so that the surface is smooth and bright.

When electro polishing, the work piece (steel) is used as the anode, the lead plate is used as the cathode, and the electrolyte is electrolyzed with phosphoric acid, sulfuric acid, and chromic anhydride (CrO_3). At this time, the anode iron of the work piece is oxidized and dissolved.

Anode reaction $\quad Fe - 2e^- = Fe^{2+}$

Then Fe^{2+} ions and $Cr_2O_7^{2-}$ ions in the solution (chromic anhydride forms $Cr_2O_7^{2-}$ ions in an acidic medium) redox reaction, that is

$$6Fe^{2+} + Cr_2O_7^{2-} + 14H^+ = 6Fe^{3+} + 2Cr^{3+} + 7H_2O$$

Fe^{3+} further forms hydrogen phosphate [$Fe_2(HPO_4)_3$, *etc.*] and sulfate [$Fe_2(SO_4)_3$] with hydrogen phosphate in solution.

The cathode is mainly the reduction reaction of H^+ ions and $Cr_2O_7^{2-}$ ions:

$$2H^+ + 2e^- = H_2\uparrow$$

$$Cr_2O_7^{2-} + 14\,H^+ + 6e^- = 2Cr^{3+} + 7H_2O$$

(3) Electrochemical machining

Electrolytic machining is based on the principle that metal can dissolve in the electrolyte, and the work piece is formed. The principle is the same as that of electro polishing. In the process of electrolytic processing, the choice of electrolyte is closely related to the material to be processed. The commonly used electrolyte is a 2.7-3.7mol/kg sodium chloride solution, which is suitable for the electrolytic processing of most ferrous metals or alloys. The following uses steel processing as an example to illustrate the electrode reaction of the electrolytic process:

Anode reaction $Fe-2e^- = Fe^{2+}$

Cathode reaction $2H^+ + 2e^- = H_2\uparrow$

The reaction product Fe^{2+} ion combines with the OH^- ion in the solution to form $Fe(OH)_2$, and can be oxidized by oxygen dissolved in the electrolyte to produce $Fe(OH)_3$. Electrolytic machining is widely used, and can process high hardness metal or alloy, as well as complex profile work piece, and the processing quality is good. It also saves the tool. However, this method can only process metal materials that can be electrolyzed, and the precision can only meet the general requirements.

(4) Anodic oxidation

Some metals can form oxide protective films in the air, so that the internal metals are generally protected from corrosion. For example, a layer of uniform and dense oxide film (Al_2O_3) is formed after aluminum metal is in contact with air to protect it. However, this naturally formed oxide film (only 0.02-1μm) cannot meet the requirements of protecting the work piece. Anodizing is to use the metal as the anode in the electrolysis process to obtain an oxide film with a thickness of 3-250μm. The anodization of aluminum and aluminum alloys is now taken as an example.

After degreasing the surface, lead plate is used as cathode, aluminum parts as anode and dilute sulfuric acid (or chromic acid) solution as electrolyte. After electrification, a layer of aluminum oxide film can be formed on the aluminum parts of anode by properly controlling current and voltage conditions. However, because alumina can dissolve in sulfuric acid solution, it is necessary to control the concentration of sulfuric acid, voltage, current density and so on, so that the rate of alumina produced by anodizing aluminum is faster than that of sulfuric acid. The reaction is as follows:

Anode $2Al + 3H_2O - 6e^- = Al_2O_3 + 6H^+$

$$H_2O - 2e^- = \frac{1}{2}O_2 + 2H^+$$

Cathode $2H^+ + 2e^- = H_2\uparrow$

The oxide film obtained by anodizing is very firmly bonded to metal, so the corrosion resistance of aluminum and alloy is greatly improved. In addition, alumina protective film is also porous and has a good adsorption capacity, which can adsorb a variety of pigments. A variety of colors of aluminum products are usually filled with dyes to fill the pores of the oxidation film. So many optical instruments and instruments need to reduce the reflective properties of some aluminum parts, are often filled with black pigments.

Finally, it should be pointed out that in the application of electrolysis, the solution used or its products may cause environmental pollution, which should be properly solved.

8.4 Chemical Power Supply

Chemical power supply is one of the main power sources commonly used in modern production and life. The voltaic cells discussed earlier is a kind of chemical power supply. In addition, there are batteries (secondary batteries), fuel cells and reserve cells. A reserve battery is a battery that separates the electrolyte (liquid) from the battery stack during storage and can be discharged by adding electrolyte and water when use. At present, solid dielectric batteries have also been developed to provide microampere current for pacemakers, such as Li-I_2, Ag-R_4NI_3, *etc.* There are many kinds of chemical power supply, and new types of power supply continue to appear. This section only introduces several kinds of chemical power supply with large requirements. In order to have a general understanding of chemical power supply, we first introduce the classification of chemical power supply.

8.4.1 Classification of Chemical Power Sources

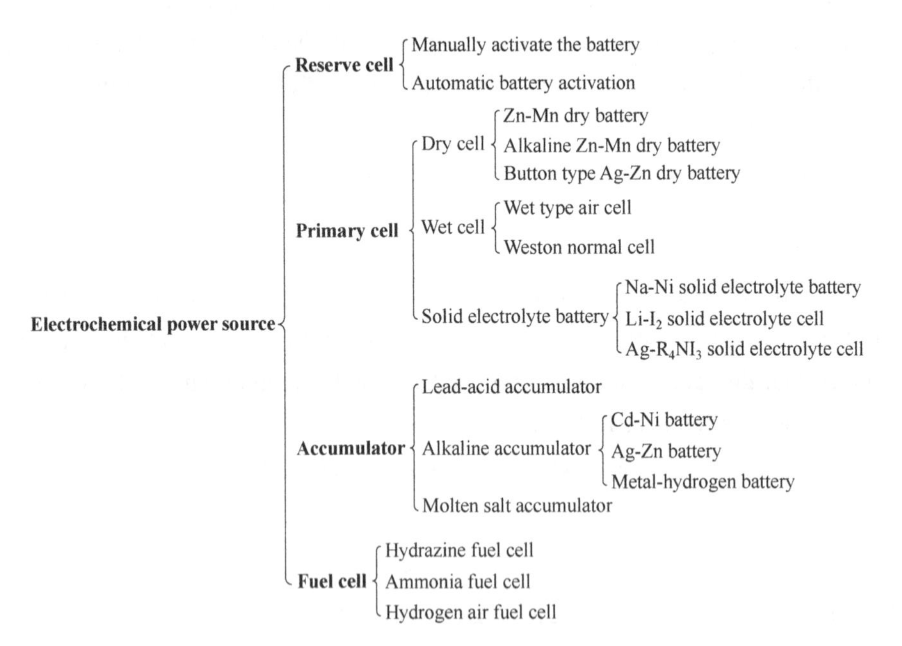

8.4.2 Electromotive Force, Open Circuit Voltage and Working Voltage of Chemical Power Supply

Chemical power supply is a device that converts chemical energy into electrical energy. According to thermodynamic knowledge, the reduction of the Gibbs function of the system is equal to the maximum useful work done by the system under the isothermal and equal pressure reversible process. Formulated as:

$$\Delta_r G_m = -nFE$$

Where E represents the reversible battery electromotive force. The reversible battery must be reacted on two electrodes, and can be carried out in both directions. The discharge process is carried out in a reversible manner, that is, the current charged or discharged is very small, and the battery works in a state close to equilibrium. The E value can be calculated based on the thermodynamic data in the battery reaction.

The open circuit voltage of a battery refers to the terminal voltage of a "new" battery where the battery is fully charged. Only the open circuit voltage of the reversible battery is its electromotive force. The open circuit voltage of a general battery is only close to its electromotive force. The primary batteries in chemical energy are all irreversible batteries, and the open circuit voltage of the voltaic cells is less than its electromotive force. The open circuit voltage of the secondary battery and fuel cell is equal to its electromotive force.

The working voltage of the battery refers to the discharge voltage when the battery is connected to the load, that is, the terminal voltage when the battery does not pass current. It varies with the size of output current, discharge depth and temperature. There are three polarizations (electrochemical polarization, concentration polarization and ohmic polarization) when the battery passes current, making the discharge voltage of the battery lower than the open circuit voltage. If the battery is a reversible battery (battery), its terminal voltage is lower than the electromotive force when the battery is discharged, its terminal voltage is higher than the electromotive force when charging. The operating voltage V_i is:

Operating voltage during discharging $V_i = E - \eta_{anode} - \eta_{cathode} - IR$

Working voltage during charging $V_i = E + \eta_{anode} + \eta_{cathode} + IR$

In the formula, η_{anode} is the over potential of the anode; $\eta_{cathode}$ is the over potential of the cathode; IR is the ohmic voltage drop of the battery during charging and discharging.

8.4.3 Voltaic Cells

A voltaic cell is a kind of battery that should not be recharged after discharge, such as zinc-manganese dry battery, zinc-mercury battery and so on. In order to carry and use conveniently, the electrolyte is absorbed in gel or paste without free flow in dry battery. The performance and working principle of this kind of battery are described below.

(1) Acid zinc manganese dry battery

This type of battery uses zinc as a negative electrode, a mixture of MnO_2 and activated carbon powder as a positive electrode, an aqueous solution of NH_4Cl and $ZnCl_2$ as an electrolyte, and a starch paste to condense the electrolyte without flowing. The upper opening is closed with some sealing materials to protect the moisture inside the battery. The cell diagram is

$$(-)Zn\,|\,NH_4Cl(3.37mol/kg),ZnCl_2(1mol/kg)\,|\,MnO_2\,|\,C(+)$$

When the battery is discharged, Zn is oxidized and MnO_2 reduced, and the open circuit voltage is 1.55 to 1.70V.

Because the electrolyte is acidic, the reactions between the two poles of the battery are as follows:

Positive pole (+) $2MnO_2 + 2H_2O + 2e^- \!=\!=\!= 2MnOOH(s) + 2OH^-$

Negative pole(−) $Zn - 2e^- \!=\!=\!= Zn^{2+}$

$$Zn^{2+} + 4NH_4Cl \!=\!=\!= (NH_4)_2ZnCl_4 + 2NH_4^+$$

The battery reaction is

$$Zn + 2MnO_2 + H_2O + 4NH_4Cl \!=\!=\!= (NH_4)_2ZnCl_4 + Mn_2O_3 + 2NH_4OH$$

Since the electrolyte in the battery is acidic (pH=5), there are no zinc-ammonia coordination ions formed by Zn^{2+} and NH_3 in the battery reaction product, while Cl^- ions and Zn^{2+} form $(ZnCl_4)^{2-}$ coordination ions.

3.37mol/kg NH_4Cl solution also freezes at −20℃ and precipitates NH_4Cl crystal. Therefore, the most suitable temperature of the battery is 15-35℃. When the temperature is lower than −20℃, the battery cannot work. Alkaline zinc-manganese dry batteries can be used in alpine regions.

(2) Alkaline zinc manganese dry battery

Alkaline zinc manganese batteries are sometimes called alkaline manganese batteries. The main difference between it and the acidic zinc manganese battery is that the electrolyte is an aqueous solution of KOH, the negative electrode is amalgamated Zn powder (not a Zn cylinder), and the positive electrode is a mixture of MnO_2 powder and carbon powder contained in a steel case. It can discharge continuously with large current, and the battery capacity at high rate discharge is 3 to 4 times that of acid zinc manganese battery. This battery has good low temperature discharge performance, and can still discharge at −40℃. The reaction during discharge is:

Positive pole (+) $2MnO_2 + 2H_2O + 2e^- \!=\!=\!= 2MnOOH + 2OH^-$

MnOOH has a certain solubility in alkaline solution

$$2MnOOH + 6OH^- + 2H_2O \!=\!=\!= 2Mn(OH)_6^{3-}$$

Negative pole (−) $Zn + 2OH^- \!=\!=\!= Zn(OH)_2 + 2e^-$

$$Zn(OH)_2 + 2OH^- \!=\!=\!= [Zn(OH)_4]^{2-}$$

Battery reaction $2MnO_2 + Zn + 4H_2O + 8KOH \!=\!=\!= 2K_3[Mn(OH)_6] + K_2[Zn(OH)_4]$

The positive electrode reaction of the battery is not a solid phase reaction, and the product of the negative electrode is soluble $[Zn(OH)_4]^{2-}$, so it can be discharged with a large current and can also be used in cold areas. The disadvantage is that the problem of "alkali climbing" cannot be solved.

The zinc manganese battery is an irreversible battery. Its open circuit voltage is around 1.5V, and its operating voltage is very unstable. Another disadvantage of it is that it has serious self-discharge, so it has poor storage performance and can only be stored for 6 months.

8.4.4 Secondary Battery

This kind of battery is a kind of energy memory, and the battery reaction can be carried out along the forward and reverse direction. The battery is a spontaneous battery when discharged and an electrolytic cell (non-spontaneous battery) when charged. After charging, the capacity of the battery can be restored, and the number of charge and discharge can reach thousands of times. The following only introduces the principle, characteristics and maintenance methods of several commonly used batteries.

(1) Lead-acid battery

The negative electrode is sponge lead, the positive electrode is PbO_2 (attached to the lead plate), and the electrolyte is a sulfuric acid solution with a density of 1.25-1.28g/cm^3. The reaction during discharge is:

Positive pole (+)

$$PbO_2 + H_2SO_4 + 2H^+ + 2e^- = PbSO_4 + 2H_2O$$

Negative pole (−)

$$Pb + H_2SO_4 - 2e^- = PbSO_4 + 2H^+$$

Battery reaction

$$Pb + PbO_2 + 2H_2SO_4 \underset{\text{charge}}{=} 2PbSO_4 + 2H_2O$$

During discharge, the active materials of both poles are gradually converted into $PbSO_4$ by the action of sulfuric acid, H_2SO_4 in the electrolyte gradually decreases, and the density gradually decreases. When the surface of the active material on the two poles is covered with non-conductive $PbSO_4$, the discharge voltage drops quickly. The open circuit voltage of the battery can be calculated with the Nernst formula:

$$\varphi_- = \varphi^{\ominus}_{Pb^{2+}/Pb} + \frac{RT}{2F}\ln\frac{m_r^2(H^+)}{m_r^2(H^+)m_r(SO_4^{2-})} = -0.385 + \frac{RT}{2F}\ln\frac{1}{m_r(SO_4^{2-})}$$

$$\varphi_+ = \varphi^{\ominus}_{PbO_2/Pb^{2+}} + \frac{RT}{2F}\ln m_r^4(H^+)m_r(SO_4^{2-}) = 1.455 + \frac{RT}{2F}\ln m_r^4(H^+)m_r(SO_4^{2-})$$

$$E = \varphi_+ - \varphi_- = 1.79 + \frac{RT}{2F}\ln m_r^4(H^+)m_r^2(SO_4^{2-}) = 1.79 + \frac{RT}{F}\ln m_r^2(H^+)m_r(SO_4^{2-})$$

The open circuit voltage of this battery (that is, the electromotive force of the battery) varies slightly with temperature and the concentration of H_2SO_4, which is generally 2.05-2.1V. The terminal voltage of the storage battery varies with the discharge rate. When the discharge rate is large, the degree of polarization is large, and the terminal voltage drops rapidly. On the contrary, when the battery discharge rate is small, the degree of polarization is also small, and the terminal voltage decreases slowly. Therefore, the cut-off voltage of the battery discharge also varies with the discharge rate. It must be charged immediately after the discharge is over.

(2) Cadmium-nickel battery

According to the different fabrication methods of the plate, the battery can be divided into two types: sintered type and plate box. The active material of the positive electrode is nickel hydroxyl oxide, in order to increase the conductivity graphite is added to the nickel hydroxyl oxide. The negative material is sponge metal cadmium, which is packed in a nickel plate box with holes or sintered on the substrate. Electrolyte is 1.16-1.19g/cm^3 KOH solution. The reaction at discharge is as follows.

Positive electrode reaction

$$2NiO(OH) + 2H_2O + 2e^- = 2Ni(OH)_2 + 2OH^-$$

Negative electrode reaction

$$Cd + 2OH^- - 2e^- = Cd(OH)_2$$

Overall reaction

$$2NiO(OH) + Cd + 2H_2O \underset{\text{charge}}{\overset{\text{electric}}{=\!=\!=}} 2Ni(OH)_2 + Cd(OH)_2$$

The open circuit voltage of this battery is 1.38V, and the charge is cut off from 1.40 to 1.45V. This type of dry battery requires no maintenance and is easy to carry and use. Currently it is mainly used in calculators, miniature electronic instruments, satellites, and space probes. Long service life is one of the advantages of this battery.

8.4.5 Fuel Cells

(1) The principle and significance of fuel cell

The device where the fuel is directly oxidized in the battery to generate electricity is called a fuel cell. This kind of chemical power supply is different from the general battery. The general battery stores all the active materials in the battery body, while the fuel cell continuously feeds the negative electrode as the active material, and sends oxygen or air to the positive electrode as the oxidant, and the products are continuously discharged. Positive and negative electrodes do not contain active materials, just a catalytic conversion element. Therefore, fuel cells are veritable “energy conversion machines” that convert chemical energy into electrical energy. The use of general fuel must first convert chemical energy into heat energy, and then convert the heat energy into electrical energy through the heat engine, so it is limited by the "heat engine efficiency". The highest energy utilization rate (diesel engine) after heat conversion is no more than 40%, and the energy utilization rate of the steam locomotive is less than 10%. Most of the energy is dissipated into the environment, causing environmental pollution and waste of energy. The fuel cell directly oxidizes the fuel, which can be regarded as a constant temperature energy conversion device. It is not limited by the efficiency of the heat engine. The energy utilization rate can be as high as 80% or more, and there is no exhaust gas emission, no pollution to the environment. In addition, in the development of new energy sources fuel cells also play an important role. The future energy will be mainly atomic energy and solar energy. Using atomic energy to generate electricity, electrolyzed water produces a large amount of hydrogen, and the hydrogen is sent to users (factories and homes) through pipelines, or liquefied and transported to remote areas, and hydrogen-oxygen fuel cells are used to generate electrical energy for people to use. Solar cells can also be used for electrolysis hydrogen that can be stored. When there is no solar energy, the hydrogen is passed through a hydrogen-oxygen fuel cell to generate electricity. In this way, the influence of time and climate change on solar energy is overcome.

The acidic hydrogen-oxygen fuel cell is taken as an example to illustrate the principle of the fuel cell. Hydrogen flows through the electrode and dissociates into atoms. The hydrogen atoms release electrons on the electrode to form hydrogen ions. The electrons flow through the external circuit to promote the load and flow to the electrode that passes oxygen. The oxygen combines with H^+ ions from the other electrode in the solution. Water is generated on the oxygen electrode. The reaction is:

Negative (−) $$H_2 - 2e^- = 2H^+$$

Positive (+) $$\frac{1}{2}O_2 + 2H^+ + 2e^- = H_2O$$

Overall reaction $$H_2 + \frac{1}{2}O_2 = H_2O$$

(2) Types of fuel cells

There are many kinds of fuel cells, which can be divided into the following categories.

① Hydrogen-oxygen fuel cell. Hydrogen-oxygen fuel cell is the most important fuel cell at present. According to the different electrolyte properties, it can be divided into acid, alkaline and molten salt and other types of fuel cells. Take alkaline fuel cells as an example:

Alkaline fuel cell (AFC) is a fuel cell with potassium hydroxide solution as electrolyte. The mass fraction of potassium hydroxide is generally 30% to 45%, up to 85%. Redox reactions are easier in alkaline electrolytes than in acidic electrolytes. AFC is a fuel cell that was vigorously researched and developed in the 1960s and successfully applied in manned spaceflight. It can provide power and water for spaceflight, and has high specific power and specific energy.

The oxidation of hydrogen on the anode is as follows:

$$H_2 + 2OH^- = 2H_2O + 2e^- (\varphi_1 = -0.828V)$$

The reduction of oxygen on the cathode is as follows:

$$\frac{1}{2}O_2 + H_2O + 2e^- = 2OH^- (\varphi_2 = 0.401V)$$

The overall reaction is as follows:

$$H_2 + \frac{1}{2}O_2 = H_2O + \text{electric energy} + \text{heat} \quad (\varphi_0 = \varphi_2 - \varphi_1 = 1.229V)$$

When it comes to alkaline batteries, we must mention the American Apollo moon landing plan. In the 1960s and 1970s, space exploration was the focus of competition in several developed countries. Due to the urgent need for high power density and high energy density in manned space flight, there has been an upsurge of AFC research in the world. Different from general civil projects, there is no need to consider too much cost in the choice of power supply, but only a strict inspection of performance. Through comparison with various chemical batteries, solar cells and even nuclear energy, fuel cells are the most suitable for spacecraft.

The Apollo system uses pure hydrogen as fuel and pure oxygen as oxidant. The anode is a nickel electrode with a double-hole structure, and the cathode is nickel oxide with a double-hole structure, and platinum is added to improve the catalytic reaction activity of the electrode.

With the funding of NASA, the asbestos membrane alkaline fuel cell system for the space shuttle was successfully developed. The battery pack is composed of 96 single cells, with the size of 35.6cm×38.1cm×114.3cm, mass 118kg, output voltage 28V, average output power 12kW up to 16kW, system efficiency 70%. And first in April 1981 it has been used in space flight and has accumulated 113 flights so far, with a running time of about 90264 hours. The battery system was overhauled every 13 flights (with an operating time of approximately 2600 hours), and the overhaul interval was later extended to 5000 hours. The successful application of AFC in aerospace flight not only proves that the alkaline fuel cell has higher weight/volume power density and energy conversion efficiency (50%-70%), but also fully proves that this power supply has high stability and reliability.

Phosphoric acid fuel cell (PAFC) is a fuel cell with phosphoric acid as the electrolyte. The anode is fed with a reformed gas rich in hydrogen and containing carbon dioxide, and the cathode is fed with air. The operating temperature is around 200℃. PAFC is suitable for installation in residential areas or user-intensive areas. Its main features are high efficiency, compactness, no pollution, easy availability of phosphoric acid, and mild reaction. It is currently the most mature and commercialized fuel cell.

The concept of the molten carbonate fuel cell (MCFC) first appeared in the 1940s. In the 1950s, Broes *et al.* demonstrated the world's first molten carbonate fuel cell, and the pressurized molten carbonate fuel cell in the 1980s started operation. It is expected that it will enter the commercial stage after the first-generation phosphate fuel cell, so it is usually called the second-generation fuel cell. Molten carbonate fuel cell is a kind of high temperature battery, which can use many fuels, such as hydrogen, coal gas, natural gas and biofuel. The battery construction material is cheap and the electrode catalytic material is not precious metals. The battery pack is easy to assemble and has the advantages of high working efficiency (more than 40%), low noise, no pollution and high utilization rate of waste heat. It can be widely used in green power stations.

Solid oxide fuel cell (SOFC) is an ideal fuel cell suitable for large power plants and industrial applications. SOFC has the advantages of high efficiency and environmental friendliness similar to other fuel cells. SOFC has developed rapidly in recent years. Since 2003, SOFC has become a representative of high-temperature fuel cells. If the waste heat power generation is included, the conversion rate of SOFC fuel to electrical energy is as high as 60%. Recently, scientists have discovered that SOFC can work at a relatively low temperature (600℃), which greatly broadens the choice of battery materials, simplifies the manufacturing process of battery stacks and materials, and reduces the cost of battery systems.

Proton exchange membrane fuel cell (PEMFC), also known as polymer electrolyte membrane fuel cell, was first developed by General Electric Company for NASA. In addition to the general advantages of fuel cells, proton exchange membrane fuel cells also have outstanding features such as rapid startup at room temperature, no electrolyte loss and corrosion problems, easy water drainage, long life, high specific power and high specific energy. Therefore, the proton exchange membrane fuel cell can be used not only for the construction of decentralized power stations, but also particularly suitable for use as a portable power source.

Substantial progress has been made in the research and development of proton exchange membrane fuel cells. Following the successful demonstration of the PEMFC electric bus by the Canadian Ballard Power Company in 1993, internationally renowned car companies have attached great importance to PEMFC, successively launched their respective concept cars and put into demonstration operation. On November 16, 2004, Japanese Honda Corporation announced that it will lease two 2005 Honda FCX cars to New York State for a full-year demonstration operation. The 2005FCX electric car uses high-pressure hydrogen as fuel, the battery pack power is 86kW, and the engine power is 80kW. It can be started below 0℃, the maximum speed of the car is 150km/h, and it can drive 306km in one hydrogenation.

Another huge market for PEMFC is the submarine power source. Nuclear-powered submarines are expensive to manufacture, and difficult to handle when retired. The submarines powered by diesel engines are noisy, high heat, and with poor concealment. Therefore, the German Siemens company has successively built four hybrid-driven submarines powered by 300kW PEMFC, which are planned to be used as the power source of the navy's 212 new submarine. With the improvement of PEMFC technology and the continuous reduction of costs, new application markets are bound to emerge.

② Organic compound oxygen fuel cell. Direct methanol fuel cell (DMFC): This is a low-temperature organic fuel cell. Both positive and negative poles can be made of porous platinum, and other materials can also be used as electrodes. For example, the negative electrode uses a small number of precious metals as a nickel electrode for the catalyst, and the positive electrode uses silver or a catalyst-supported activated carbon electrode. The electrolyte can be H_2SO_4 solution or KOH aqueous solution. The fuel is methanol. Methanol is dissolved in the electrolyte, and it is brought to the electrode for reaction through the circulating flow of the electrolyte. Oxygen or air are oxidants. The specific reactions are as follows:

Negative pole (−)　　$CH_3OH + H_2O - 6e^- = CO_2 + 6H^+$

Positive pole (+)　　$\frac{3}{2}O_2 + 6H^+ + 6e^- = 3H_2O$

Battery reaction　　$CH_3OH + \frac{3}{2}O_2 = CO_2 + 2H_2O$

The electromotive force of the battery is 1.20V, and the starting voltage is 0.8-0.9V; the operating voltage is 0.4-0.7V, and the operating temperature is 60℃. Methanol is a liquid fuel, which is very convenient in storage and transportation. It is easy to dissolve in the electrolyte, it is very superior to gas fuel, and it is also a more active organic fuel in electrochemical reaction. However, in addition to electrochemical oxidation at the electrode, methanol also undergoes chemical oxidation, so the open circuit voltage of the battery is only 65% of the electromotive force.

Although the direct methanol fuel cell (DMFC) started late, it has developed rapidly in recent years. Due to its simple structure, small size, convenience and flexibility, abundant fuel sources, cheap price, easiness for carrying and storing, it has become one of the hotspots of international fuel cell research and development. The theoretical energy density of a direct methanol fuel cell is about 10 times that of a lithium-ion battery, and it has obvious advantages compared with various conventional batteries in terms of specific energy density.

③ Metal-oxygen fuel cell

Various metal-oxygen fuel cells, such as magnesium-oxygen, lead-oxygen, zinc-oxygen and other fuel cells, are currently being studied. The advantage of metal fuel cell is that it is very safe and easy to use. The disadvantage is that it is easy to release hydrogen due to the autolysis of metals, and the price of metal as fuel is high. There are also many kinds of fuel cells, such as hydrazide-oxygen fuel cells, regenerated fuel cells and so on, which are no longer introduced here.

8.4.6 Environmental Pollution Caused by Chemical Power Supply and Its Treatment Measures

With the rapid development of Chinese economy and the continuous updating of electronic industry technology, the use of various chemical batteries by the people has risen sharply, mainly used in various digital products, electric motorcycles, various small electronic devices and other fields. According to statistics, the annual output of batteries in the world is about 25 billion, of which China accounts for about 1/2 of the total, and is growing at a rate of 20% per year. The use of chemical batteries has brought great convenience to people's lives. However, due to the current low awareness of the recycling of waste batteries, it has caused great pollution to the environment.

The harmful substances in the battery mainly include heavy metal substances such as mercury, cadmium, and lead. If these substances slowly penetrate the soil or water body through the discarded battery, they will cause great pollution to the soil and water body. Relevant information shows that a 1# battery slowly corrodes and rots in the soil, which will permanently lose the use value of $1m^2$ of soil. The heavy metal in a coin battery can pollute 600t of water, which is equivalent to the amount of water a person can drink for a lifetime. If heavy metals that have penetrated into the soil or water are transferred into the human body through the food chain, it will cause great harm to human health. If mercury enters the body's central nervous system, it will cause symptoms such as neurasthenia syndrome, neurological dysfunction, and mental retardation. If cadmium enters the rice through irrigation water, people who use this cadmium-containing rice for a long time will suffer from pain, and the disease manifests as joint pain in the waist, hands, and feet. After the disease lasts for several years, neuralgia and bone pain will occur in various parts of the patient's body. It is difficult to relieve, and even breathing will bring unbearable pain. In the later stages of the disease, the patient's bones softened or shrank, the limbs were bent, the spine was deformed, and the bones were brittle. Even coughing could cause fractures. Lead can have adverse effects on human organs and systems such as the chest, kidneys, reproductive and cardiovascular, which manifested as decreased intelligence, kidney damage, infertility and high blood pressure. It can be seen from the above that if used batteries are discarded without effective recycling or disposal, they will not only cause a lot of waste of resources, but also cause great harm to the environment and human health, and even harm next generations. Therefore, this year's battery recycling and utilization have become a subject of increasing concern.

The recycling of batteries can not only alleviate environmental pollution problems, but also generate renewable secondary resources. For example: 100kg waste lead batteries can recover 50-60kg lead, and 100kg waste batteries containing cadmium can recover about 20kg of metal cadmium. There are roughly three kinds of disposal methods for waste batteries in the world: ①solidification and deep burial; ②storage in mines; ③recycling. There are mainly two metallurgical treatment methods for the recycling of waste dry batteries: wet method and fire method. At present, developed countries have accumulated more successful experience in the recycling of used batteries. For example, Denmark is the first country in Europe to recycle used batteries. Germany was the first to legally determine the obligatory subject for recycling used batteries. A non-profit organization, GRS (global recycle standard), strictly operates the entire system. After the collection and transportation of waste batteries, they are strictly classified, disposed, and recycled. The recovery rate of Japanese secondary batteries has also reached 84%. The United States is the country with the most detailed legislation on waste battery pollution management. It not only has established a complete waste battery recycling system, but also established a number of waste battery treatment plants. In comparison, Chinese prevention and control of used batteries started late. In order to standardize the management of waste batteries and strengthen the prevention and control of waste battery pollution, the State Environmental Protection Administration issued the Technical Policy on the Prevention and Control of Waste Battery Pollution in 2003, which is currently the only special provision for waste battery management in China. However, the policy does not formulate detailed rules for battery recycling. There are no rewards and penalties for recycling and non-recycling, and lack of operability.

Thus it can be seen that there is still a big gap between China and developed countries in the recovery and utilization of waste batteries, which requires us to start from ourselves, enhance awareness of environmental protection, and vigorously publicize the harm caused by waste batteries to the environment and human health, and minimize the use of batteries. At the same time, government departments should also give great importance to legislation, and establish the corresponding waste battery recovery and treatment institutions. Promote the recovery and recycling of waste batteries to form industrialization, and realize the reduction, resource utilization and innocuity of waste batteries.

Exercises

1. Please representing the following redox reactions by cell diagram.

(1) $Zn + CdSO_4 = ZnSO_4 + Cd$

(2) $Fe^{2+} + Ag^+ = Fe^{3+} + Ag$

2. There are three kinds of oxidants available: H_2O_2, CrO_7^{2-}, Fe^{3+}. From the standard electrode potential analysis, if the I^- in the mixed solution containing I^-, Br^- and Cl^- is oxidized to I_2, but Br^- and Cl^- without changing, which oxidant is suitable?

3. Given that:

$$MnO_4^- + 8H^+ + 5e^- = Mn^{2+} + 4H_2O$$

$\Delta_f G_m^\ominus$/(kJ/mol) −447.2 0 −228.1 −237.1

Determine the standard electrode potential $\varphi_{MnO_4^-/Mn^{2+}}$ for this reaction.

4. The metal sheet of tin and lead is inserted into the salt solution containing the metal ion to form the voltaic cell, respectively.

(1) $m(Sn^{2+}) = 1mol/kg$, $m(Pb^{2+}) = 1mol/kg$

(2) $m(Sn^{2+}) = 1mol/kg$, $m(Pb^{2+}) = 0.01mol/kg$

Calculate their electromotive force and write the cell diagrams, the electrode reactions and the total reaction equation.

5. A voltaic cell composed of a standard hydrogen electrode and a nickel electrode. When the concentration of Ni^{2+} is 0.01mol/kg, the electromotive force is 0.316V. Ni is the negative electrode. Please calculate the standard electrode potential of the nickel electrode.

6. Given that $\varphi^{\ominus}_{Ag^+/Ag} = 0.7996V$, determine the electrode potential of Ag^+/Ag when $m(Ag^+) = 0.1mol/kg$, 0.001mol/kg respectively.

7. The standard electrode potential is used to judge and explain:

(1) When the iron piece is put into solution $CuSO_4$, will Fe be oxidized to Fe^{2+} or Fe^{3+}?

(2) What is the product of the reaction between iron and excess chlorine?

(3) Which of the following substances is the strongest oxidant? Which is the strongest reductant?

MnO_4^- $Cr_2O_7^{2-}$ I^- Cl^- Na^+ HNO_3

8. The voltaic cells is composed of a standard cobalt electrode and a standard hydrogen electrode, and its electromotive force is measured to be 1.6365V. At this time, the cobalt electrode is used as a negative electrode. It is known that the standard electrode potential of chlorine is +1.3595V. Please answer the questions below.

(1) What is the direction of the reaction of this battery?

(2) What is the electrode potential of cobalt standard electrode?

(3) When the pressure of chlorine increases or decreases, how will the battery's electromotive force change? Explain the reason.

(4) When the Co^{2+} ion concentration is reduced to 0.01mol/kg, how will the battery's electromotive force change? What is the change value?

9. In the copper-zinc voltaic cells, when $m(Zn^{2+}) = m(Cu^{2+}) = 1mol/kg$, the electromotive force of the battery is 1.1037V.

(1) Calculate the value of $\Delta_r G_m^{\ominus}$ for this reaction.

(2) From $E^{\ominus}$ and $\Delta_r G_m^{\ominus}$, calculate the standard equilibrium constant of the reaction.

10. (1) Using the standard electrode potential of the half-cell reaction, calculate the standard equilibrium constant of the following reaction and the electromotive force of the composed battery.

(2) After the equal amounts of 2mol/kg of Fe^{2+} and 2mol/kg of I^- are mixed, does the electromotive force and standard equilibrium constant change? Why? (With the help of the Nernst equation, there is no need to calculate. Note that $m(Fe^{2+}) \neq 1$ in the solution, but its concentration is very small):

$$Fe^{3+} + I^- = Fe^{2+} \frac{1}{2} I_2$$

11. A $ZnSO_4$ solution contains m (Mn^{2+}) =0.1mol/kg of Mn^{2+}. Under acidic conditions (pH=5), $KMnO_4$ can be added to oxidize Mn^{2+} to MnO_2 and the precipitate is removed. At the same time, $KMnO_4$ itself is also reduced to MnO_2 and finally, the concentration of the excess MnO_4^- [$m(MnO_4^-)$] equals to 10^{-3}mol/kg. Please answer the following question by calculation: What is the concentration of the remaining Mn^{2+} in the solution when equilibrium is reached?

12. Insert the Ag electrode into the $AgNO_3$ solution, the copper electrode into the $Cu(NO_3)_2$ solution of $m[Cu(NO_3)_2]$ = 0.1mol/kg, the two half-cells are connected, add excessive HBr to the Ag half-cell to produce AgBr precipitation, and make $m(Br^-)$ = 0.1mol/kg in the AgBr saturated solution, then the battery electromotive force is measured to be 0.21V, and Ag is the negative electrode. Try to calculate the solubility product constant of AgBr.

13. Given that $\varphi^{\ominus}_{Fe^{3+}/Fe^{2+}} = 0.771V$, $\varphi^{\ominus}_{Ag^+/Ag} = 0.7991V$. Use it to make a voltaic cells. If ammonia water is added to Ag half-cell to $m(NH_3) = m[Ag(NH_3)_2^+] = 1mol/kg$, the electromotive force is larger or smaller than $E^{\ominus}$? Why? What is $\varphi_{Ag^+/Ag}$ at this time? [Given $\lg K_f^{\ominus}$ of $Ag(NH_3)_2^+$ is 7].

14. A solution contains $CdSO_4$ of $m(CdSO_4)$ = 10^{-2}mol/kg, $ZnSO_4$ of $m(ZnSO_4)$ = 10^{-2}mol/kg, put the solution between two platinum electrodes for electrolysis, ask:

(1) Which metal is first deposited on the cathode?

(2) When another metal begins to deposit, what is the remaining concentration of the metal ions that precipitated first in the solution?

15. At 25℃, the solution pH=7, the over potential of H_2 on Pt is 0.09V, the over potentials of O_2 and Cl_2 on graphite are 1.09V and 0.25V, respectively, when $p_{Cl_2} = p_{O_2} = p_{H_2} = 100kPa$, the applied voltage causes electrolysis of the following electrolytic cell:

$$\text{cathode Pt}\begin{cases} m(CdCl_2) = 1mol/kg\ CdCl_2 \\ m(NiSO_4) = 1mol/kg\ NiSO_4 \end{cases} |\text{(graphite)anode}$$

When the applied voltage increases gradually, what is the first reaction on the electrode? At this point, what is the applied voltage (considering over potential)?

16. In the mixed solution where $m(CuSO_4)$ = 0.05mol/kg and $m(H_2SO_4)$ = 0.01mol/kg H_2SO_4, plate Cu on the platinum electrode. If the over potential of H_2 on Cu is 0.23V, ask when the applied voltage increases to the point where H_2 precipitates on the electrode. What is the concentration of remaining Cu^{2+}?

17. When 25℃ $p_{Cl_2} = p_{O_2} = p_{H_2} = 100kPa$ and pH=7, Pt is used as the cathode and graphite is used as the anode. Electrolysis contains a mixed aqueous solution of $FeCl_2$ [$m(FeCl_2)$ = 0.01mol/kg] and $CuCl_2$[$m(CuCl_2)$ = 0.02mol/kg]. If the over potentials of Cl_2 and

O_2 on graphite are 0.25V and 1.09V, ask:

(1) What kind of metal precipitates first?

(2) How much voltage has to add when the second metal precipitates?

(3) When the second metal precipitates, what is the concentration of the first metal ion?

18. A solution contains three kinds of cations, and their concentrations are $m(Fe^{2+})$ = 0.01mol/kg, $m(Ni^{2+})$ = 0.1mol/kg, $m(H^+)$ = 0.001mol/kg, $p(H_2)$ = 100kPa. It is known that the super potential of H_2 on Ni is 0.21V. Try to describe the discharge order of the three ions through calculation when Ni is used as a cathode to electrolyze the above solution.

Chapter 9 Corrosion and Material Protection

Teaching contents	Learning requirements
Principle and rate of metal corrosion	Understand the principle of electrochemical corrosion and the factors that affect the corrosion rate
Protection of metal corrosion	Understand several common methods to prevent metal corrosion
Protection of polymer materials	Understand the concepts of photooxidative aging, thermal oxidative aging, chemical reagent aging and so on Understand the concepts of light stabilizer, antioxidant, flame retardant and their protection principle to polymer materials

9.1 Principle and Rate of Metal Corrosion

When the metal comes into contact with the surrounding media (air, CO_2, H_2O, acid, base, salt, *etc.*), various degrees of damage will occur. After this phenomenon, the shape, color and mechanical properties of the metal itself have changed. This kind of metal is damaged by the action of the surrounding medium, which is called corrosion of metal.

The loss of metal due to corrosion is serious, which not only causes great harm to the national economy, but also causes the loss of metal structure (such as machinery, equipment and instruments, *etc.*). The consequences of product quality reduction, environmental pollution, aircraft crash, ship leakage, power outage, water outage and explosion cannot be calculated by the amount of lost metal. Therefore, engineers and technicians should understand the basic principle of corrosion, reduce or avoid corrosion factors as far as possible in construction and design, or take effective protective measures, which is of great significance for increasing production and saving, safety in production.

According to the mechanism of metal corrosion, corrosion can be divided into chemical corrosion and electrochemical corrosion. Chemical corrosion is caused by chemical interaction between metal surface and dry gas or non-electrolyte. It is not easy to occur at room temperature and atmospheric pressure. At the same time, this kind of corrosion often only occurs on the metal surface, and the harm is generally less than electrochemical corrosion. Electrochemical corrosion refers to the corrosion caused by the dissolution of metal as anode when the metal surface and electrolyte solution form the original cell. It can occur at room temperature and atmospheric pressure, and can penetrate into the metal interior. Compared with chemical corrosion, it is more harmful and more common. Therefore, the electrochemical corrosion of metals is discussed below.

9.1.1 Principle of Electrochemical Corrosion

The electrochemical corrosion of metals is the damage caused by the electrochemical interaction between metal and medium. The electrochemical effect mentioned here is essentially the result of the formation of the voltaic cell because of the different electrode potential on the metal surface. The voltaic cell is called corrosion battery (corrosion microcell). In the corrosion battery, the oxidation reaction on the negative electrode, often called anode, which dissolves and is corroded; the reduction reaction on the positive electrode is often called cathode, which only plays the role of electron transfer and is not corroded.

To explain the cause of the electrochemical corrosion of metals, the atmospheric corrosion that occurs at room temperature when two metals are in contact is taken as an example for analysis, as shown in Fig. 9-1. Because the air contains water vapor, CO_2 and SO_2 and other gases, the water vapor is adsorbed on the metal surface, and the metal surface is covered with a thin water film. Iron and copper seem to be immersed in H^+, OH^-, HSO_3^-, HCO_3^- as in the plasma solution, Cu-Fe corrosion battery is formed, which leads to electrochemical corrosion. Because the electrode potential of iron is lower than that of copper, iron is the anode and copper is the cathode. The polar reactions are as follows:

Anode (iron) $\quad Fe - 2e^- = Fe^{2+}$ (oxidation reaction)

$$Fe^{2+} + 2OH^- = Fe(OH)_2 \downarrow$$

Cathode (copper) $\quad 2H^+ + 2e^- = H_2 \uparrow$ (reduction reaction)

$$O_2 + 4e^- + 2H_2O = 4OH^-$$

Corrosion battery reaction

$$Fe+2H_2O = Fe(OH)_2 \downarrow +H_2 \uparrow$$

$$2Fe+O_2+2H_2O = 2Fe(OH)_2 \downarrow$$

Then $Fe(OH)_2$ is oxidized to $Fe(OH)_3$ (or $Fe_2O_3{\cdot}nH_2O$) by oxygen in the air, and partially dehydrated $Fe(OH)_3$ (or $Fe_2O_3{\cdot}nH_2O$) is called rust.

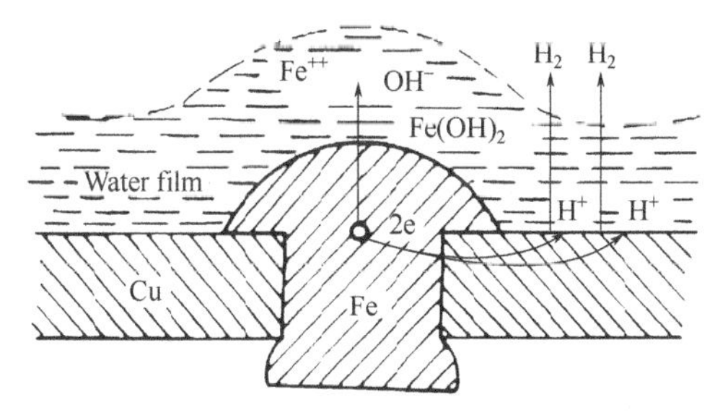

Fig.9-1 Corrosion of copper and iron contact

From the above example, it can be seen that this is the electrochemical corrosion of two different metals in contact with electrolyte solution, which can be seen by the naked eye, so it is called macro battery corrosion.

Electrochemical corrosion can also occur if one metal is not in contact with other metals and placed in electrolyte solution. Because, in general, pure metals in industry often contain impurities. For example, industrial zinc iron impurities FeZn, Fe_3C in steel, graphite of cast iron, *etc.*, because of the high potential of these components, when they are in contact with electrolyte solution, many micro cathodes can be formed on the surface of metal, and the metal with lower potential can be used as anode to form countless micro battery (micro cell), and cause metal corrosion. We call this corrosion microcell corrosion, as shown in Fig.9-2.

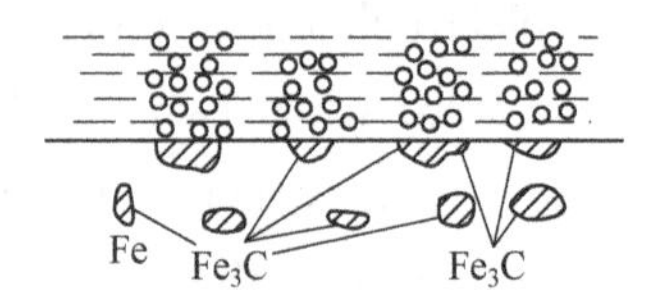

Fig.9-2 Corrosion of iron and steel

To sum up, it is not difficult to see that the necessary conditions for causing electrochemical corrosion of metals are:

(i) Areas with different electrode potentials on the metal surface;

(ii) Electrolyte solution is present.

The corrosion process can be regarded as composed of three links:

① On the anode, the metal dissolves into ions and transfers into the solution, and the oxidation reaction occurs, that is $M - ne^- \longequal M^{n+}$;

② Electrons flow from anode to cathode;

③ On the cathode, electrons are accepted by substances in the solution that can bind to electrons, and the reduction reaction occurs. In most cases, it's H^+or O_2 in the solution, that is,

$$2H^+ + 2e^- = H_2\uparrow \qquad \text{hydrogen corrosion}$$

$$O_2 + 4e^- + 2H_2O = 4OH^- \qquad \text{oxygen corrosion}$$

The former often occurs in acidic solution, while the latter occurs in neutral or alkaline solution. These three links are interrelated and indispensable, otherwise the whole corrosion process will stop.

If we understand the causes and conditions of electrochemical corrosion, we can distinguish the possibility of corrosion of metals under some conditions. However, in order to understand the reality of corrosion, it is also necessary to know the speed of corrosion, so what are the main factors that will affect the corrosion rate?

9.1.2 Polarization of Corrosion Battery and Factors Affecting Corrosion Rate

In a corroded battery, the metal of the anode dissolves and is corroded by the loss of electrons. Obviously, the more electrons a metal loses, the more electrons flow out of the anode, and the more metals dissolve and corrode. The relationship between the amount of metal dissolution and corrosion and the amount of electricity can be expressed by Faraday's law:

$$W = \frac{QA}{nF} = \frac{ItA}{nF} \tag{9-1}$$

Where W—Metal corrosion;

Q—Amount of energy flowing through (in t seconds);

F—Faraday constant;

n—Oxidation number of metals;

A—Relative atomic mass of metal;

I—Current intensity (potential is v).

Since corrosion rate (v) refers to the weight [g/(m^2·h)] that metal loses per unit area in per unit time, it can be expressed as follows:

$$v = QA/nF = 3600\ IA/SnF \tag{9-2}$$

As can be seen from the formula, the greater the current intensity (I) of the corrosion battery, the greater the metal corrosion rate.Therefore, the value of current intensity can be used to measure the corrosion rate.

According to Ohm's law, the relationship between I and the polarization potential difference and the resistance of the battery is:

$$I_{\text{corrosion beginning}} = \frac{\phi_{\text{cathode beginning}} - \phi_{\text{anode beginning}}}{R_{\text{beginning}}} = \frac{E_{\text{beginning}}}{R_{\text{beginning}}} \tag{9-3}$$

Where $\varphi_{\text{cathode beginning}}$ and $\varphi_{\text{anode beginning}}$ are the potential of the cathode and anode at the beginning of corrosion; $R_{\text{beginning}}$ is the battery resistance at the beginning of the corrosion.

(1) Polarization of corroded batteries

It is obvious from the above formula that $I_{\text{corrosion reality}}$ is affected by two factors: one is the potential difference between the two poles, the other is the battery resistance.

The electrode polarization will also occur in the corrosion battery, which results in the increase of the anode potential and the decrease of the cathode potential, resulting in a decrease in the potential difference between the two poles, as shown in Fig.9-3.

$$E_{\text{reality}} = (\phi_{\text{cathode beginning}} - \eta_{\text{cathode}}) - (\phi_{\text{anode beginning}} + \eta_{\text{anode}}) = \phi_{\text{cathode}} - \phi_{\text{anode}} \tag{9-4}$$

$$I_{\text{corrosion reality}} = \frac{\phi_{\text{cathode}} - \phi_{\text{anode}}}{R_{\text{reality}}} = \frac{E_{\text{reality}}}{R_{\text{reality}}} \tag{9-5}$$

From $E_{\text{beginning}}$ to E_{reality}, it is caused by the polarization of the corrosion battery electrode.

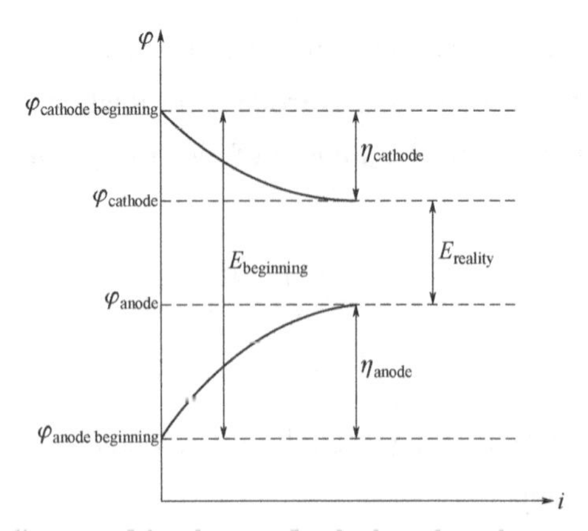

Fig.9-3 Schematic diagram of the change of cathode and anode potential during corrosion

$R_{reality}$ is the actual resistance of the corrosion battery, and it is actually not a constant. For example, the distance between the poles near the cathode and anode interface is very close, $R_{reality}$ is very small, and the distance from the interface is larger; with the anode being corroded, the impurities (cathode) originally wrapped in the interior will be exposed again, and the anode area will constantly change, all of which will change with it. At present, there is no good way to calculate the distribution of corrosion current.

The corrosion rate of metal depends on the current intensity of corrosion battery. Therefore, all the factors that affect the $I_{corrosion\ reality}$ will affect the corrosion rate.

(2)The main factors affecting the corrosion rate

① The electrode potential of metal: it can be seen from formula (9-3) that the greater the initial potential, the larger the $I_{corrosion\ reality}$. Therefore, the greater the initial potential difference between the metal component and the different metal or impurity in wet air or in aqueous solution, the faster the corrosion of the metal that forms the two poles. There are stress and deformation parts in metal components; the larger the potential difference at the grain boundary is, the more likely it is to be corroded.

② The polarization of the electrode and the properties of the medium: it can be seen from formula (9-5) that if the other conditions are the same, the smaller the polarization degree of the electrode, the greater the $I_{corrosion\ reality}$. When the polarization degree of different metals is different, the corrosion rate will be different.

In addition, the overpotential is related to the current density, so in the corrosion battery, the electrode area of the cathode also has an impact on the corrosion rate. The smaller the cathode area is, the higher the current density is; the larger the overpotential of hydrogen is, and the smaller the corrosion rate is. On the contrary, the corrosion rate is accelerated, so from the point of view of preventing corrosion we should avoid connecting to a very small anode with a very large cathode.

It can be seen that the ion or some additives in the solution can enhance the polarization or increase the resistance of the battery that can slow down the corrosion. On the contrary, the corrosion will be accelerated. For example, ions that can be dissolved in anodic metal can be formed into ion coordinating agents, such as NH_3, CN^-, Cl^-, Br^-, I^- and other active ions, which can accelerate the corrosion of iron and steel. So it is necessary to clean the metal parts after molten salt quenching or electroplating to avoid Cl^- accelerate corrosion.

When the dissolved oxygen or oxidant in the solution can make some metal (such as Al, Cr, Ni) surfaces form a dense oxide film, which increases the resistance value, causes the change of electrode potential, and slows down the corrosion. Al, Cr, Ni and other electrodes have a negative electric potential, but they are not easily corroded in the medium containing oxidant, because these metal surfaces form an oxide film, which is tightly and firmly covered on the metal surface, thus the metal is no longer corroded, which is called metal passivation.

In order to play a protective role, the oxide film must be continuous, that is to say, the volume of the oxide must be larger than the volume of the simple substance consumed. If V_{oxide} is used to represent the volume of oxide generated by oxidation of elemental substances, $V_{simple\ substance}$ is used to represent the volume of elemental substances consumed by oxidation. Only when the $V_{oxide}/V_{simple\ substance}$ is more than 1 can the oxide form a continuous surface film, which covers the surface of the metal and has a protective effect. If $V_{oxide}/V_{simple\ substance}<1$, since the oxide film cannot be continuous and cannot cover the metal, it has no protective effect. Table 9-1 lists the volume ratios of some oxides to simple substances.

Table 9-1 Volume Ratio of Some Oxidizing Substances to Simple Substances

Simple substances	Oxide	$V_{oxide}/V_{simple\ substances}$	Simple substances	Oxide	$V_{oxide}/V_{simple\ substances}$
K	K_2O	0.45	Zr	ZrO_2	1.35
Na	Na_2O	0.55	Zn	ZnO	1.57
Ca	CaO	0.64	Ni	NiO	1.60
Ba	BaO	0.67	Be	BeO	1.70
Mg	MgO	0.81	Cu	Cu_2O	1.70
Cd	CdO	1.21	Si	SiO_2	1.88
Ge	GeO	1.23	U	UO_2	1.94
Al	Al_2O_3	1.28	Cr	Cr_2O_3	2.07
Pb	PbO	1.29	Fe	Fe_2O_3	2.14
Sn	SnO_2	1.31	W	WO_3	3.35
Th	ThO_2	1.32	Mo	MoO_3	3.45

As can be seen from Table 9-1, the oxide film of metals in the s region (except Be) is impossible to be continuous and has no protective effect on the oxidation of metals in the air. The oxide films of aluminum, chromium, nickel, copper, silicon and so on may be continuous and may form protective films. But, $V_{oxide}/V_{simple\ substance}>1$,is the necessary condition for the protection of the surface film, not the only condition. If the stability of the oxide is poor, or the thermal expansion coefficient between the film and the simple substance is quite different, or $V_{oxide}/V_{simple\ substance}\geqslant 1$, but the film is brittle, and easy to break and become a discrete structure, losing its protective effect. For example, the oxide of MoO_3, when the temperature exceeds 520℃, it begins to volatilize and loses its protective effect. Another example is V_{WO_3}/V_W=3.35(⩾1). However, WO_3 film is brittle and easy to break, and the protective effect of this membrane is poor. The reason why aluminum, chromium, silicon and so on are quite stable in air and can be used as high temperature heat resistant (antioxidant) alloy elements is related not only to the continuous structure of oxide film, but also to that oxide (Al_2O_3, Cr_2O_3, SiO_2) has a high degree of thermal stability. Iron can form a dense oxide protective film (such as blackened Fe_3O_4) under certain conditions. However, the composition of oxide scale or rust usually varies with temperature, and the structure is loose and the protection performance is poor, which plays an important role in accelerating corrosion in electrochemical corrosion.

③ Temperature and humidity: increasing the temperature can accelerate most chemical reactions, decrease the resistance value of the battery and decrease the polarization of the electrode, so it can also accelerate the corrosion rate. However, things often have two sides, for oxygen absorption corrosion, because of the increase of temperature, dissolved oxygen decreases, so sometimes the corrosion rate slows down. When the metal is corroded, the considerable humidity of the atmosphere has a great influence on the corrosion rate. Because of the high humidity, the water film on the metal surface is thick, the solution resistance is small, and the corrosion is quick.

The purpose of the above analysis about the main factors affecting the corrosion rate of metals is to master the methods to control the corrosion rate and prevent electrochemical corrosion.

9.2 Main Methods to Prevent Metal Corrosion

It is not difficult to see from formula (9-5) that all measures that can reduce $I_{\text{corrosion reality}}$ or make $I_{\text{corrosion reality}} = 0$ can effectively prevent electrochemical corrosion.

First of all, the potential difference between $\varphi_{\text{cathode beginning}}$ and $\varphi_{\text{anode beginning}}$ can be minimized by various measures to reduce the possibility of corrosion. Secondly, the resistance of corrosion battery and the polarization of battery electrode can be increased. In production practice, in order to prevent metal corrosion, it is often necessary to consider the above factors to analyze and choose the best scheme.

9.2.1 Improve the Anticorrosion Performance of Metal

Removing or reducing harmful impurities in metal as much as possible, reducing the possibility of forming corrosive batteries, or adding some components that can increase battery resistance and electrode polarization to reduce corrosion rate and improve the anticorrosion performance of metals, such as adding 18% Cr, 8% Ni and a small amount of titanium to iron to make stainless steel. In addition, reducing the roughness of metal surface can also improve its anticorrosion performance. The possibility of stress corrosion can also be reduced by annealing to eliminate the internal stress of metal components.

9.2.2 The Use of Various Protective Layers

The essence of this method is to isolate the metal from the surrounding media in order to prevent the formation of corrosion batteries on the metal surface. The requirement is that the protective layer should have good continuity and compactness, and maintain a high degree of stability and robustness in the use of the medium. In production practice, all kinds of protective layers can be selected reasonably according to the use of metal parts. The commonly used protective layers are metal layer and non-metallic layer. The metal protective layer maintains the luster, conductivity, heat conduction and other characteristics of the metal. The cost of non-metallic protective layer is low and the process is relatively simple, but the characteristics of metal are buried. The anodizing of non-ferrous metal aluminum, the blackening, bluing and phosphating of ferrous metals and alloys can prevent the electrochemical corrosion of metals, but these methods are not suitable for machine parts which are prone to friction. In a word, we should consider the conditions for the use of metal parts and the requirements for protection before we decide what kind of protective layer is more appropriate.

The specific categories of protective layers commonly used are as follows:

- **Protective layer**
  - **Metal layer**
    - Package plating: Cu, brass, Ni, Al, *etc.*
    - Spray plating: Zn, Al, Pd, Cu, copper alloy, Fe, *etc.*
    - Penetration plating: Zn, Al, Si, Cr, *etc.*
    - Immersion plating: Zn, Al, Sn, *etc.*
    - Electroplating: Zn, Cd, Sn, Ni, Cu, Cr, Ag, gold alloy, *etc.*
    - Electroless plating: Ni, Cu
  - **Non-metallic layer**
    - Organics: Paint, resin, rubber, asphalt, *etc.*
    - Inorganics: Enamel, cement, acid-resistant materials, *etc.*
    - Oxidation treatment of non-ferrous metals
    - Passivation treatment of non-ferrous metals
    - Sodium iron oxidation treatment, phosphorus treatment

9.2.3 Corrosion Inhibitor Method

In the medium, a small amount of substances that can block or slow down the electrode process are added to prevent metal corrosion. This method is called corrosion inhibitor method, and the added substance is called corrosion inhibitor or inhibitor. The essence of corrosion inhibitor is to increase the resistance and electrode polarization, so as to reduce $I_{\text{corrosion reality}}$. The material that can increase anode passivation and polarization makes the $I_{\text{corrosion reality}}$ decrease is called anode corrosion inhibitor. The commonly used anode corrosion inhibitors are oxidizing substances, such as chromate, dichromate, nitrate, nitrous acid and so on, which make the passivation film formed in the anode to prevent metal corrosion. The material that can increase the cathodic resistance and polarize to reduce $I_{\text{corrosion reality}}$ is called cathodic corrosion inhibitor. Common cathodic inhibitors include zinc salt, calcium bicarbonate, heavy metal salts and organic amines, agar, dextrin, animal glue, *etc.* Some of these substances can form insoluble hydroxides or carbonates with OH^- near the cathode, covering the surface of the cathode, increasing the cathode resistance and cathode polarization, reducing $I_{\text{corrosion reality}}$ (oxygen absorption corrosion). The corrosion inhibition of organic amines is generally considered to be:

$$R_3N + H^+ \Longrightarrow (R_3NH)^+$$

The resulting $(R_3NH)^+$ is adsorbed on the metal surface, preventing the H^+ discharge and thus reducing anode corrosion.

In recent years, due to the increasingly complex structure of some equipment and instruments, it is generally difficult to fill all pores and gaps with corrosion inhibitors. Therefore, atmospheric corrosion can be avoided by starting research on volatile compounds and putting them in packaging materials or in closed spaces where protected products are stored. This compound is usually called gas phase corrosion inhibitor. For example, benzotriazole is a solid compound, because of its large vapor pressure, its vapor can saturate the space very quickly and be adsorbed on the metal surface. Therefore, even if the electrolyte solution is gathered on the metal surface, benzotriazole vapor can hinder the corrosion process.

9.2.4 Electrochemical Protection

The essence of electrochemical protection is to apply DC power (or protective screen) by making metal as cathode and cathodic polarization to protect it (called cathodic protection), or to connect metal to the positive pole of DC current for anodic polarization, so that the metal is passivated and the corrosion rate of metal is greatly reduced (called anodic protection).

(1) Cathodic protection

Cathodic protection is one of the effective methods to prevent metal corrosion, which is often used in the anticorrosion protection of underground pipelines, coolers, ships, seaplanes, undersea metal equipment, *etc.*, as shown in Fig. 9-4.

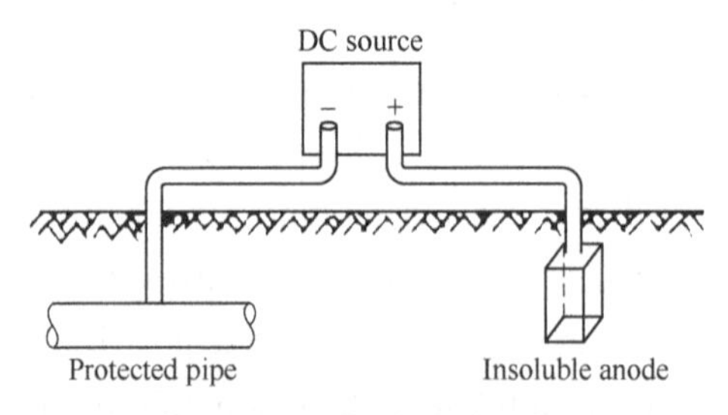

Fig.9-4 Cathodic protection

In cathodic protection law, a metal or alloy with negative potential can also be connected on metal equipment, and the protected metal is called cathodic protection through the current generated by the large potential difference between the two. The metal or alloy with negative potential is corroded as anode, so it is called sacrifice anode protection law.

(2) Anodic protection

Anodic protection is not as widely applied as cathodic protection, and it usually has certain conditions for the protected metal, that is, the metal is likely to produce stable passivation film under the condition of a given medium. The medium must have certain passivation ability and can protect passivation under a small anodic current density. For example, the corrosion rate of stainless steel in 9.3-15.1mol/kg H_2SO_4 can be drastically reduced at 18-50℃.

There are many methods to prevent metal corrosion, but which one should be chosen must be based on the properties of the metal, the conditions of use, the requirements for protection, economic accounting and so on. In addition several methods can be used at the same time to complement each other. Therefore, it is a necessary for engineers and technicians to correctly select corrosion-resistant metals to manufacture metal components, to design metal components reasonably according to the use conditions, and to select the method of protecting metals according to the causes of electrochemical corrosion.

Although metal corrosion brings great harm to production, it can also benefit the production. For example, in chemical cutting and printed circuit plate production, corrosion is used for processing. The following is a brief introduction to the principle of printed circuit plate making method.

One of the methods of printing circuit is to print the circuit on copper foil by photocopying on the copper plate (fiberglass insulation board with copper foil on a surface), and then corrode the copper, which is not protected by photosensitive adhesive, with iron trichloride solution, so that the printed circuit board with clear circuit line can be obtained. The reason why iron trichloride can corrode copper can be seen from the algebra value of electrode potential:

$$\varphi^{\ominus}(Fe^{3+}/Fe^{2+}) = +0.77V,\ \varphi^{\ominus}(Cu^{2+}/Cu) = +0.34V, \varphi^{\ominus}(Cu^{+}/Cu) = +0.521V$$

Because the electrode potential of copper is smaller than that of Fe^{3+}/Fe^{2+} electron pair, copper can be used as reductant and oxidant in iron trichloride solution. The reactions are as follows:

$$2FeCl_3 + Cu = 2FeCl_2 + CuCl_2$$

$$FeCl_3 + Cu = FeCl_2 + CuCl\downarrow$$

9.3 Protection of Polymer Materials

At present, the scale of the polymer synthesis industry in the world has reached an annual output of about 150 million tons, which exceeds the annual output of the steel industry. The annual output per capita in developed countries has reached 80 to 120kg, while the current annual output per capita in China is only about 12 kilograms. From the most common daily necessities to the most cutting-edge high-tech products, polymer materials are indispensable. Polymer materials are the fastest growing category in the field of materials.

In the process of using polymer materials, due to the combined effects of environmental factors such as heat, oxygen, water, light, microorganisms, chemical media, *etc.*, the chemical composition and structure of polymer materials will undergo a series of changes, and the physical properties will also change accordingly. These changes and phenomena are called aging. The essence of polymer material aging is the change of its physical structure or chemical structure. The following introduces the influence mechanism and anti-aging measures of environmental factors on the aging of polymer materials.

9.3.1 Influence Mechanism of Environmental Factors

The physical properties exhibited by polymer materials are closely related to their chemical structure and aggregated state structure. The chemical structure is a long-chain structure in which polymers are connected by covalent bonds. The aggregated state structure is that many large molecules are arranged and stacked by intermolecular forces. The spatial structure are crystalline state, amorphous, and crystalline-amorphous. The intermolecular forces that maintain the aggregated structure include ionic bonding force, metal bonding force, covalent bonding force, and van der Waals force. Environmental factors can cause changes in intermolecular force, even chain breakage, or the shedding of certain groups. It will eventually destroy the aggregate structure of the material and change the physical properties of the material.

(1) Influence of temperature

As the temperature increases, the movement of the polymer chain intensifies. Once the dissociation energy of the chemical bond is exceeded, it will cause thermal degradation or group shedding of the polymer chain; lower temperature will often affect the mechanical properties of the material, which is closely related to the criticality of mechanical properties. Temperature points include glass transition temperature T_g, viscous flow temperature T_f and melting point T_m. The physical state of the material can be divided into glass state, high elasticity state, viscous flow state. On both sides of the critical temperature, the aggregate state structure of polymer material long chains will produce significant changes, which will significantly change the physical properties of the material.

Rubber is a highly cross-linked amorphous polymer, and the use environment should ensure that it is in a high elastic state. The use temperature must be higher than the glass transition temperature, lower than the viscosity flow temperature and decomposition temperature. Fiber is a highly crystalline polymer material. The use temperature is required to be far below the melting point T_m for easy ironing. For crystalline plastics the glass transition temperature $T_g <$use temperature $<$ melting point T_m, but for amorphous plastics the use temperature must be lower than the glass transition temperature T_g of about 50-75℃.

In the extremely cold region, temperature has a great influence on the properties of plastics and rubber products. For crystalline plastics, if the ambient temperature is lower than the glass transition temperature of the material, the free movement of polymer segments will be hindered, which shows that the plastics become brittle, hard and easy to break; the cold environment has little effect on amorphous plastics. For rubber products, the performance of rubber products below glass transition temperature will be similar to that of crystalline plastics, and the properties of rubber will be lost. Cold environment has no effect on the physical properties of fiber materials.

(2) Influence of humidity

The effect of humidity on polymer materials can be attributed to the swelling and dissolution of materials by moisture, which changes the intermolecular force to maintain the aggregation structure of polymer materials, thus destroying the aggregation state of materials, especially for non-crosslinked amorphous polymers. It is very obvious that the swelling and even aggregation of polymer materials will cause damage to the properties of the materials; for the crystalline form of plastics or fibers, due to the existence of water permeability restrictions, the effect of humidity is not very obvious.

(3) Influence of oxygen

Oxygen is the main cause of aging of polymer materials. Because of the permeability of oxygen, crystalline polymers are more resistant to oxidation than amorphous polymers. Oxygen first attacks the weak links in the main chain of polymers, such as double bonds and hydroxyl. The hydrogen and other groups or atoms on the group and *tert*-carbon atom form the polymer peroxide radical or peroxide, and then cause the fracture of the main chain in this part. When the molecular weight of the polymer decreases significantly, the glass transition temperature decreases, and the polymer becomes sticky. In the presence of some initiators or transition metal elements which are easy to decompose into free radicals, the oxidation reaction tends to be intensified.

(4) Photoaging

Whether the polymer irradiated with light will cause the molecular chain to break, depending on the relative size of the light energy and the dissociation energy and the sensitivity of the polymer chemical structure to light waves. Due to the existence of the ozone layer and the atmosphere on the surface of the earth, the wavelength of sunlight that can reach the ground is between 290nm and 4300nm. The light wave energy is greater than the chemical bond dissociation energy. Only the ultraviolet region of the light wave will cause the breakage of polymer chemical bonds.

Table 9-2 is a comparison table of chemical bond energy and ultraviolet wavelength with similar energy. The ultraviolet wavelength is 300nm to 400nm, which can be absorbed by polymers containing carbonyl groups and double bonds, which breaks the macromolecular chain, changes the chemical structure, and changes the material properties. Polyethylene terephthalate has strong absorption of 280nm ultraviolet light, and the degradation products are mainly CO, H_2, CH_4. Polyolefins containing only C—C bonds have no absorption of ultraviolet light, but a small amount of impurities, such as carbonyl groups, unsaturated bonds, hydroperoxide groups, catalyst residues, aromatic hydrocarbons and transition metal elements, can promote the photooxidation of polyolefins.

Table 9-2 Chemical Bond Energy and Ultraviolet Wavelength with Similar Energy

Chemical bonds	Bond energy/(kJ/mol)	Wavelength/nm	Light energy/(kJ/mol)
C—H	413.6	290	418
C—F	441.2	272	446
C—O	351.6	340	356
C—C	347.9	342	354
C—N	290.9	400-410	303-297
C—Cl	328.6	350-364	346-333
N—H	389.3	300-306	404-397
O—H	463	259	468

(5) Effects of chemical media

Similar to the action of water and oxygen in the environment, the chemical medium can work only when it penetrates in to the polymer materials, including the influence on covalent bonds and secondary bonds. The former can lead to a series of irreversible chemical processes such as chain breaking, crosslinking and addition of polymer chain. The latter will change the aggregation state of the material and lead to the change of the corresponding physical and mechanical properties.

The physical changes such as environmental stress cracking, dissolution and plasticity are typical performance of chemical medium aging of molecular materials. When there is a small amount of non-solvent liquid medium on the surface of the bidirectionally loaded polymer, it may generate small cracks or crazing, which is called environmental stress cracking. This surface phenomenon is the result of the local surface stress exceeding its yield stress under the plastic of chemical medium and the concentration of material surface stress. In some cases, environmental stress cracking can be changed by means of polymer. Increasing the molecular weight and chain branch degree can reduce the crystallization of the polymer and improve its environmental stress cracking resistance. When a small amount of solvent is in contact with the stressed polymer, it can cause dissolution, which can occur in both amorphous and crystalline polymers. Morphological studies show that the polymer dissolution is actually the result of the relocation under the action of internal stress. The way to eliminate dissolution is to decrease the internal stress of the material. Annealing after forming and material processing are beneficial to eliminate the internal stress. In the case of continuous contact of liquid medium, the interaction between polymer and small molecule replaces the interaction between polymers, making the polymer more flexible. This indicates that the glass transition temperature decreases. The strength, hardness and elastic modulus of the alloy decreased, and the elongation at break increased.

(6) Biological aging

Polymer materials are in a certain environment for a long time. Due to the strong genetic variability of microorganisms, enzymes that can decompose and use these polymers will gradually evolve, so that they can be used as a carbon source or energy source for growth, although the degradation rate is extremely low. But this potential hazard does exist. For some polymer packaging, it is hoped that it can be quickly biodegraded after use.

Polymer materials add phenols and organic compounds containing copper, mercury or tin. Compounds can prevent their hydrolysis. For polymers that want to have bacteriolysis, natural polymer materials are used, after chemical or physical modification, to increase their strength as packaging. After the 1990s, natural polymer—starch, cellulose, chitin and their modified polymer compounds—have been widely used in various application fields as degradable plastics. Polysaccharides natural polymers and their modified compounds can be processed into degradable disposable films, sheets, containers and foamed products by blending with general plastics. The waste can be hydrolyzed into small molecular compounds step by step through the intervention of amylase and other polysaccharide natural polymer decomposing enzymes which widely exists in the natural environment, and finally decompose into pollution-free carbon dioxide and water, and return to the biosphere. Based on these advantages, natural polymer compounds of polysaccharides represented by starch are still an important part of degradable plastics.

9.3.2 Anti-aging Measures

(1) Temperature

For crystalline plastics and rubber, the temperature should be above the glass transition temperature. When the environmental temperature is lower than the glass transition temperature of the material, the physical properties will change and affect the performance. Plasticization is a special lubrication effect. The action mechanism of plasticizer includes molecular plasticity (including internal plasticity) and structural plasticity. At the molecular level, the mixing of plasticizer and polymer reduces the interaction between polymer chains and increases the flexibility of polymer. The internal structural plasticity is to weaken the interaction between polymers by changing the chemical composition of polymers, so as to achieve the purpose of plasticization.

The service temperature of amorphous plastics should be lower than the glass transition temperature, the service temperature of crystalline plastics and fibers should be much lower than the melting point, and the service temperature of rubber should be lower than the viscous flow temperature. If some polymer materials are used at high temperature for a long time, there is also a risk of aging. Increasing the rigidity of polymer chain, such as introducing benzene ring into the side chain, properly increasing the crystallinity, cross-linking degree and molecular weight of the material can improve the melting point or viscous flow temperature, but the machinability of the material may become difficult.

(2) Humidity

Polyester, polyacetal, polyamide and polysaccharide polymers can be hydrolyzed by water or acid under the catalysis of acid. In areas with severe air pollution and frequent acid rain, the use of such polymer materials will be restricted. If a waterproof film can be covered on the surface of such materials, the occurrence of hydrolytic aging can be reduced or even avoided.

(3) Oxygen

In the polymer processing, amine antioxidants, phenolic antioxidants, sulfur-containing organic compounds and phosphorus-containing compounds can react quickly with peroxy free radicals and terminate the chain reaction early. According to the mechanism of action, antioxidants are divided into free radical receptor type and free radical decomposition type. Free radical receptor type antioxidants such as certain amines and phenolic antioxidants, which can be combined with polymer radicals or peroxygen free radicals, react quickly to reduce their activity, and they also become. Free radicals with low activity that cannot continue the chain reaction. Free radical decomposing antioxidants such as sulfur-containing organic compounds and phosphorus-containing compounds can free polymer peroxides group into a stable hydroxy compound. However, for phenolic antioxidants, due to the tendency of hydroperoxide to decompose into free radicals, the best stabilizer system should be composed of antioxidants and hydroperoxide cracking inhibitors. If free radical receptor type antioxidants are used together with free radical decomposition type antioxidants, they often produce better synergistic effects. Since the presence of certain transition metal elements will aggravate the oxidative aging of polymer materials, metal chelating agents must be added during the forming process to form complexes with them and eliminate their catalytic effect.

(4) Photoaging

In the material processing, if the light stabilizer is added, the aging degradation of the material can be avoided. According to the action mechanism, this kind of light stabilizer includes light shielding agent, ultraviolet absorbent, quenching agent and free radical capture agent. The light shielding agent can reflect ultraviolet light, avoid its penetrating into the polymer, reduce the photoexcitation reaction. The stabilizer which acts as light shielding includes carbon black, titanium dioxide and so on. The ultraviolet absorber can absorb ultraviolet light and the absorber is in the excited state, and then emit fluorescence, phosphor or heat and return to the ground state. The free radical catcher can effectively capture the polymer free radicals and terminate the chain reaction.

With the widespread application of polymer materials, the aging of polymer materials has become an important factor restricting the application of polymers. According to the different application environments of materials, the research on the aging and anti-aging of materials should be synchronized with the production and processing of materials. The research on production and processing methods is relatively thorough. Due to the complexity of environmental factors, the research on aging and anti-aging is relatively lagging. Basic research and application research in this area should be strengthened. The research results guide the production and processing of materials, will make the application of polymer materials get greater development.

9.4 Electronic Packaging Materials

As the integration of integrated circuit increases rapidly, the calorific value of the chip increases sharply and the life of the chip decreases. It has been reported that for every 10℃ increase in temperature, the failure rate due to the shortened life of GaAs or Si semiconductor chips is three times as high. The reason is that in microelectronic integrated circuits and high-power rectifier devices, thermal fatigue caused by poor heat dissipation between materials and thermal stress caused by thermal expansion coefficient mismatch. The key to solve this problem is reasonable encapsulation. Electronic packaging materials mainly include substrate, wiring, frame, interlayer media and sealing materials. The earliest materials used for packaging are ceramics and metals. With the continuous improvement of circuit density and function, more and higher requirements are put forward for packaging technology, which also promotes the development of packaging materials.

9.4.1 Main Performance Requirements for Electronic Packaging Materials

The packaging material plays a role of supporting and protecting semiconductor chips and electronic circuits, and assists in dissipating the heat generated during the operation of the circuit. As an ideal electronic packaging material, it must meet the following basic requirements:

① low thermal expansion coefficient;

② good thermal conductivity;

③ good air tightness and ability to withstand the impact of high temperature, high humidity, corrosion and radiation and other harmful environments on electronic devices;

④ high strength and stiffness, which plays a supporting and protective role in the chip;

⑤ good machining and welding performance to facilitate machining into a variety of complex shapes;

⑥ the density requirements for electronic packaging materials used in the aerospace field and other portable electronic devices are as small as possible to reduce the mass of the device.

9.4.2 Commonly Used Electronic Packaging Materials

(1) Plastic packaging material

Plastic packaging has the advantages of low price, lightweight, good insulation and so on. Compared with other packaging materials, plastic packaging materials are the most important packaging materials that can realize miniaturization, lightweight and low cost of electronic products. The materials used in plastic packaging are mainly thermosetting plastics, including phenolic, polyester, epoxy and organosilicon, among which epoxy resin is the most widely used. However, plastic packaging materials such as epoxy materials have inferior air tightness, and most of them are sensitive to humidity in the process of reflow welding. The water absorbed by the sealing material is easily expanded by heat, which will lead to the burst of the plastic sealing device. The thermodynamic properties of epoxy resin are greatly affected by moisture. At high temperature, humidity will reduce the glass transition temperature, elastic modulus and strength of the material. Water and gas will also cause corrosion of the metal layer in the packaging device. The dielectric constant of plastic sealing material is changed, which seriously affects the reliability of packaging performance, so it cannot meet the very high reliability requirements of military and civil products. In addition, most of the plastic packaging contain lead, which is very toxic.

(2) Ceramic packaging material

Ceramic packages are hermetic packages, and ceramic package materials mainly include Al_2O_3, BeO and AlN, *etc.* The advantages of ceramic packages are good moisture resistance, high mechanical strength, small thermal expansion coefficient and high thermal conductivity. Al_2O_3 ceramic is currently the most mature ceramic encapsulation material. It is widely used for its low price, good thermal shock resistance and good electrical insulation, and mature manufacturing and processing technology. However, due to relatively low thermal conductivity of Al_2O_3 ceramics, it is impossible to use them in large amounts in high-power integrated circuits. BeO ceramic has high thermal conductivity, but its toxicity and high production cost limit its production and application. AlN ceramics have good thermal conductivity and lower coefficient of thermal expansion (CTE) that is more compatible with chip materials, and are considered to be the most promising packaging materials. However, the preparation process of AlN ceramics is complicated and costly, so large-scale production and application have not been possible so far.

(3) Metal packaging material

The metal packaging material has the advantages of high mechanical strength and excellent heat dissipation performance. The traditional metal packaging materials are Cu, Al, Mo, W, kovar, invar, and W/Cu and Mo/Cu alloys. Al has low density and high thermal conductivity, and it is easy to machining. The coefficient of thermal expansion, CTE, is up to $20\times10^{-6}K^{-1}$. Both the density and CTE of Cu are very high. The Fe-Co-Ni alloy and Fe-Ni alloy have low CTE (similar to silicon and nickel), but they have small thermal conductivity and high density and are not suitable for packaging avionics equipment. Although Mo and W have ideal CTE values, their thermal conductivity is not as good as that of Al and Cu. Their density is 3 to 4 times greater than that of Al, and the wettability with Si is not good. Molybdenum, tungsten, tungsten-copper, molybdenum-copper, copper-invar copper, and copper-molybdenum-copper alloys that have evolved are superior to kovar in terms of thermal conductivity, but their mass is greater than kovar.

(4) Metal based composite packaging material

Ceramic materials, such as Si and GaAs for chips and Al_2O_3 and BeO for substrates, have a coefficient of thermal expansion (CTE) of 4×10^{-6} to $7\times10^{-6}K^{-1}$, and have high thermal conductivity. The CTE of Al and Cu is as high as $20\times10^{-6}K^{-1}$. Such substrates and packaging materials will generate large thermal stress, which is the reason for brittle cracks in integrated circuits and substrates. It can be seen from the above that plastic packaging materials, ceramic packaging materials, and metal packaging materials all have these or those shortcomings, and they have been unable to meet the development of modern electronic packaging technology. Therefore, in recent years, many researchers have been committed to research and development of new electronic packaging materials. Metal-based composite packaging materials can obtain different properties through different combinations of matrix and reinforcement.

Al, Mg, Cu or their alloys are usually chosen as the matrix of metal matrix composites. These pure metals or alloys have good thermal and electrical properties, good machinability and weldability, and low densities (such as aluminum and magnesium). The reinforcement should have low CTE, high thermal conductivity, good chemical stability, low cost, and good wettability with the metal matrix. Metal-based electronic packaging composites have high thermal physical properties and good packaging properties. It has the following characteristics: (i) Different CTE matching packaging materials can be prepared by changing the type, volume fraction and arrangement of the reinforcement or by changing the heat treatment process of the composites; (ii) The CTE of composite material is low, which can match with the CTE of electronic device material, at the same time has high thermal conductivity and low density; (iii) The material preparation process is mature, and the emergence of net molding process reduces the subsequent processing of composite materials, so that the production cost is constantly reduced.

① Aluminum-based composite packaging materials Particle reinforced aluminum matrix composites are the most mature metal matrix composites. SiCP/Al was widely studied and applied in the past few decades. Its matrix may be pure aluminum, but most of them are all kinds of aluminum alloys. Alcoa prepared porous SiC/Al alloy composite material by vacuum die casting. The SiC/Al material prepared by this method has strong commercial competitiveness because of high SiC volume fraction. LANXIDE prepared reticular SiCP/Al composites by pressureless infiltration technology. Its CTE can be adjusted by the proportion of SiCP.

The high-silicon aluminum alloy refers to an aluminum-based matrix composites with Si content of 30% to 50%. When the Si content is greater than 50%, it is called an aluminum-silicon alloy. These alloy materials became research hot spot in the 1990s because of three outstanding advantages and excellent comprehensive performance. First, it's a lightweight alloy based on the lightest metals; Second, the physical properties can be designed by changing the alloy composition; Third, it is low cost. The preparation methods of high silicon aluminum alloy composites mainly include: (i) Pressure infiltration method; (ii) Pressureless infiltration method; (iii) Powder metallurgy method; (iv) Vacuum hot pressing method; (v) Jet deposition method.

Presently, silicon aluminum composites has become an electronic packaging material with broad application prospects. They have attracted more and more attention in the world, especially in the field of aerospace. The latest research progress is the silicon content exceeded 70% in the composites. This lightweight electronic packaging material will be used in the electronic system of aerospace vehicles.

However, there is no report on the preparation process of large-size plates. The silicon aluminum alloy material has become a potential electronic packaging material with broad application prospects, which has attracted more and more attentions, especially in the aerospace field. The Si content of the high-silicon aluminum alloy reported in China is only about 17% to 30%. Its research work mainly uses its low density, low expansion and high wear resistance to manufacture wear-resistant parts for automobile engines. Research on high-silicon aluminum alloy materials for electronic packaging is still in its infancy.

② Carbon fiber reinforced magnesium matrix composites   Compared with magnesium alloys, carbon fiber-reinforced magnesium-based composite materials have the advantages of low density, high thermal conductivity, high electrical conductivity, good vibration damping properties, excellent electromagnetic shielding properties and easy processing. Disadvantages of these materials are dimensional in stability, high coefficient of thermal expansion and low creep resistance. It has broad application prospects in aerospace, electronic packaging, automotive industry, military manufacturing and other high-precision instruments. Due to the low melting point around 650℃ of magnesium and its alloys, carbon / magnesium composite materials are usually prepared by liquid methods, such as vacuum pressure infiltration, die casting, vacuum suction casting and casting. In the preparation process, the biggest difficulty is that liquid magnesium cannot wet carbon fibers and cannot form a good interface bond. Therefore, the physical and chemical behavior of liquid magnesium and carbon fibers at the interface will greatly affect the performance of the composite.

9.4.3 Preparation Method of Metal-based Composite Electronic Packaging Materials

(1) Liquid state method

This method includes gas pressure infiltration casting method, squeeze casting method and pressureless infiltration casting method. The gas pressure infiltration casting method is to use the gas transmission pressure to infiltrate the molten metal into the preform to obtain a composite material. Prefabricated parts can be made by ordinary pressing, slurry casting and injection molding methods. This method is very effective for the production of electronic packaging materials, and it is possible to obtain composite materials with a volume percentage of added particles of 50% to 80%. However, it has the disadvantages of slower production process and lower applied pressure.

The squeeze casting method is to make the reinforcement into a preform, and the preform is put into a mold, then the molten metal penetrates into the reinforcement preform through the action of liquid pressure, although there will be some residual gas in the electronic packaging material produced. However, the quality of the materials is good, and at the same time it has the advantages of short production cycles and mass production. The disadvantage is that the production cost is relatively high, which has high requirements on penetration pressure and mold; the shape complexity of the parts is greatly limited.

The manufacturing process of the pressureless infiltration casting method is: putting the base alloy ingot on the preform, passing into a controlled atmosphere containing N_2, and heating until the alloy melts into the preform. The advantage is that the amount of SiCP can be changed according to needs, and a grid-shaped electronic packaging material with a complicated shape can be produced. The main disadvantage is that it must be carried out in a controlled atmosphere of N_2, some parts of the preform cannot be fully penetrated, there is a certain amount of pores in the product, and the production process takes a long time.

(2) Solid state method

The solid-state method includes a solid-state diffusion method and a powder metallurgy method. The solid-state diffusion method is one of the methods for manufacturing continuous fiber-reinforced metal matrix composite materials. This method is complicated in process, high in cost, and difficult. The powder metallurgy method consists of mixing powder (matrix and reinforcement) in a certain ratio, pressing, sintering under the protection of vacuum or inert gas, and then hot isostatic pressing or isostatic pressing rolling. The advantage of the powder metallurgy method is that the particles of the material are evenly distributed, the mechanical properties are relatively good, the range of the matrix and the reinforcement is optional, and the volume fraction of the reinforcement can reach 55%. It is very suitable for preparing various low-expansion particle reinforced aluminum matrix composites material. The disadvantage is that the cost of raw materials and equipment is high, the internal structure of the manufactured composite material is uneven, and the hole rate is large, so the composite material must be subjected to secondary plastic processing to improve its comprehensive mechanical properties. In addition, the powder-metallurgical French-German process is more complicated and must be carried out under a sealed vacuum or protective atmosphere. The equipment and production costs are high, and the structure and size of the parts are limited.

(3) Spray deposition

The use of spray deposition technology to prepare particle-reinforced metal matrix composites is an important direction for the development of this technology in recent years. However, most of the current domestic and foreign preparation techniques are to spray a certain amount of reinforcement phase particles into the atomizing cone during the spray deposition forming process, and co-deposit on the depositor after forced mixing with the metal droplets to obtain a composite material blank. The biggest disadvantage of this type of method is the low utilization rate of enhanced particles and the high cost of material preparation.

Nowadays, integrated circuits are developing towards miniaturization, high density assembly, low cost, high performance and high reliability, which puts forward higher requirements for substrate, wiring materials, sealing materials and interlayer dielectric materials, and requires the emergence of electronic packaging materials with good performance and low cost. This provides a huge space for the development of metal-based electronic packaging in line with the development of materials. By changing the shape, size and volume fraction of the reinforcements in metal matrix composites, we can find an electronic packaging material which not only matches the thermal properties of the substrate, but also has good mechanical properties. In addition the manufacturing method is also economical and applicable, which is the development direction of metal based electronic packaging composites.

Exercises

1. What is a corrosive battery? How many types of corrosive battery are there?

2. Try to find some practical examples of metal corrosion protection in daily life and then explain which type of method be used.

3. Where is the joint between a copper faucet and an iron water pipe vulnerable to corrosion? What is the difference in mechanism between this corrosion and that occurs with paper clips?

4. In an iron-corrode battery, if the only difference between two points on the iron block is the concentration of oxygen, where the partial pressure of oxygen is 100kPa at one point and 10.0kPa at the other, what is the potential difference of oxygen between the two points?

Chapter 10 Metal Coordination Compound

Teaching contents	Learning requirements
Basic concepts and nomenclature	Describe how ligands and metals form complexes, and explain the meaning of coordination number Distinguish between a monodentate and a bidentate ligand, and describe chelation Use nomenclature rules to assign the names of common transition metal complexes
Equilibria involving complex ions	Use the formation constant K_f of a complex ion to determine the concentrations of ions in solution Describe how to use precipitation, acid-base, redox, and complex-ion formation reactions in qualitative cation analysis
Chemical bond theory in coordination compound	Describe what happens to the metal *d*-orbitals in an octahedral ligand field, and explain what is meant by spectrochemical series Discuss how colors are displayed by coordination complexes
Applications of coordination chemistry	Discuss some important applications of coordination complexes

At the beginning of the 18th century, German paint manufacturer Diesbach synthesized a blue pigment for the first time with the composition $KCN \cdot Fe(CN)_2 \cdot Fe(CN)_3$ and named it Prussian Blue. After that, the French chemist Tassaert prepared cobalt trichloride hexaammine $CoCl_3 \cdot 6NH_3$ in 1798. As the chemical formula indicated, both compounds were complex compounds formed from simple compounds, and are now commonly referred to as coordination compounds. The chemical formulas of these two complexes are written as $KFe[Fe(CN)_6]$ and $[Co(NH_3)_6]Cl_3$, respectively. At the same time, Tassolt also keenly realized that the forming of this stable complex compounds certainly had new meanings that were not yet understood by chemists at that time. The successful preparation of $KFe[Fe(CN)_6]$ and $[Co(NH_3)_6]Cl_3$ aroused great interest of chemists to study similar systems, marking the real beginning of coordination chemistry.

It took another century or so before people really understood the new meaning that Tassolt realized. Our current understanding of the nature of metal complexes is based on the insights provided by the young Swiss chemist Werner (A. Werner 1866-1919) in 1893. His insights were later referred to as Werner coordination theory, and thus won the Nobel Prize in Chemistry in 1913. This theory has had a profound influence on the development of inorganic chemistry and chemical bond theory in the 20th century.

Today, coordination chemistry has developed into a specialized discipline of chemistry. Studies in modern biochemistry and molecular biology have found that coordination compounds play an important role in biological life activities. It is not only the central subject of modern inorganic chemistry, but it also has important significance for the research of analytical chemistry, catalytic kinetics, electrochemistry, quantum chemistry, *etc.*

10.1 Basic Concepts and Nomenclature

10.1.1 Coordinate Bonds and Coordination Compounds

Coordinate bond is a covalent chemical bond between two atoms that is produced when one atom shares a pair of electrons with another atom lacking such a pair. It is also called coordinate covalent bond. For example, AgCl dissolves in excess ammonia:

$$H_3N:+AgCl+:NH_3 = [H_3N:Ag:NH_3]^+ + Cl^-$$

The atoms that provide a pair of electrons in the reaction become the donor of the electron pair, such as the N atom in NH_3 molecule in the above example. An atom (or ion) that accepts a pair of electrons is called an electron pair acceptor, and it must have an empty orbital, as Ag^+ ion in the above example. The generated $[Ag(NH_3)_2]^+$ is called coordinate ion, and the positive coordinate ion and Cl^- ion make up the compound $[Ag(NH_3)_2]Cl$. It can also be a compound composed of a negative coordinate ion and another positive ion, such as $K_3[Fe(CN)_6]$. Compounds containing complex ions like these are called complex salts. There are also some molecules that are generated by the coordination of neutral molecules and neutral atoms. The uncharged molecules are called coordinate molecules, such as $[Ni(CO)_4]$. Coordinate salts and coordinate molecules are collectively called coordination compounds, or complexes for short.

10.1.2 Composition and Type of Complexes

The composition of the complex can be determined. For example: adding H_2O_2 to the ammonia solution of $CoCl_2$ can obtain a kind of orange-yellow crystal $CoCl_3 \cdot 6NH_3$. After dissolving this crystal in water, adding $AgNO_3$ solution will precipitate AgCl precipitation immediately, the amount of precipitation is equivalent to the total amount of chlorine in the compound:

$$CoCl_3 \cdot 6NH_3 + 3AgNO_3 = 3AgCl\downarrow + Co(NO_3)_3 \cdot 6NH_3$$

Obviously, the chloride ion in this compound is free and can independently display its chemical properties. Although the content of ammonia in this compound is high, its aqueous solution is neutral or weakly acidic. Adding a strong base at room temperature does not produce ammonia gas. Only when it is heated to boiling, ammonia gas is released and produces precipitation of cobalt oxide :

$$2\,(CoCl_3 \cdot 6NH_3) + 6KOH \xlongequal{\text{heat}} Co_2O_3\downarrow + 12NH_3 + 6KCl + 3H_2O$$

The aqueous solution of this compound was tested with carbonate or phosphate, and the presence of cobalt ions could not be detected. These tests prove that the Co^{3+} and NH_3 molecules in the compound have been combined to form a complex ion $[Co(NH_3)_6]^{3+}$, thus to a certain extent, the chemical properties of Co^{3+} and NH_3 when they exist independently are lost.

In the above complex, Co^{3+} is called the metal center (or central ion). It can be a metal ion or an atom, or a non-metallic atom (such as a Si atom), but they all accept electron pairs. The six coordinated NH_3 molecules are called ligands. The atoms in the ligand that are directly bonded to the central ion are called coordinate atoms, which are electron pair donors.

The heart of Werner's theory proposed in 1893 was that certain metal atoms, primarily those of transition metals, have two types of valence or bonding capacity. The primary valence is based on the number of electrons the atom loses in forming the metal ion. A secondary valence is responsible for the bonding of other groups called ligands to the central metal ion.

In modern usage, a complex is any species involving coordination of ligands to a metal center. The complex can be a cation, an anion, or a neutral molecule. Compounds that is complexes or contains complex ions are known as coordination compounds. Square brackets are traditionally used to enclose the formula of a complex ion or a neutral coordination compound. It is important to note that counterions in a coordination compound are ionic species that do not form covalent bonds with the transition metal; they simply balance the charge on the complex ion. Other species that can be written outside the square brackets include waters of hydration, water molecules that are trapped in the solid crystal lattice but are not directly bonded to the transition metal.

$[Co(NH_3)_6]^{3+}$	$[CoCl_4(NH_3)_2]^-$	$[CoCl_3(NH_3)_3]$	$K_4[Fe(CN)_6]$
Complex cation	Complex anion	Neutral complex	Coordination compound

The coordination number of a complex is the number of points around the metal center at which bonds to ligands can form. Coordination numbers ranging from 2 to 12 have been observed, although 6 is by far the most common number, followed by 4. Coordination number 2 is limited mostly to complexes of Cu(I), Ag(I), and Au(I). Coordination numbers greater than 6 are not often found in members of the first transition series, but are more common in those of the second and third series. Stable complexes with coordination numbers 3 and 5 are rare. The coordination number observed in a complex depends on a number of factors, such as the radius ratio of the central metal atom to the attached ligands.

Coordination numbers of some common ions are listed in Table 10-1. The four most commonly observed geometric shapes of complex ions are shown in Fig.10-1.

Table 10-1 Some Common Coordination Numbers of Metal Ions

+1 Metal ion		+2 Metal ion		+3 Metal ion	
Cu^+	2,4	Ca^{2+}	6	Al^{3+}	4,6
Ag^+	2	Mg^{2+}	6	Cr^{3+}	6

(Continued)

+1 Metal ion		+2 Metal ion		+3 Metal ion	
Au^+	2,4	Fe^{2+}	6	Fe^{3+}	6
		Co^{2+}	4,6	Co^{3+}	6
		Cu^{2+}	4,6	Au^{3+}	4
		Zn^{2+}	4,6		

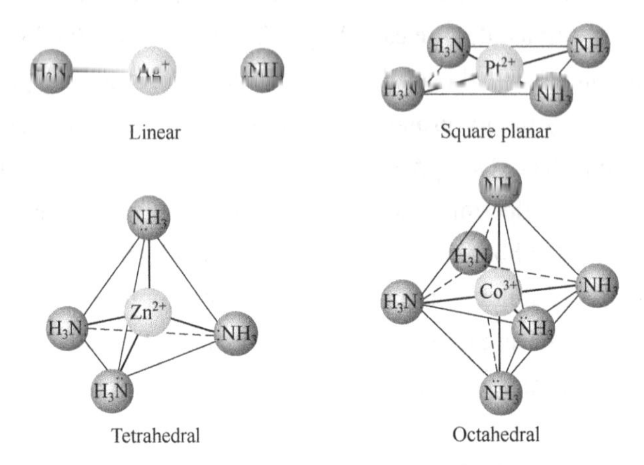

Fig.10-1 Structures of some complex ions

(Attachment of the NH_3 molecules occurs through the lone-pair electrons on the N atoms. In each complex, all the ligands are the same and hence no distortions of these shapes are observed)

10.1.3 Ligands

A common feature shared by the ligands in coordination complexes is the ability to donate electron pairs to central metal atoms or ions. A ligand that uses one pair of electrons to form one point of attachment to the central metal atom or ion is called a monodentate ligand, such as F^-, Cl^-, Br^-, I^-, OH^- (hydroxyl), H_2O, SCN^-, NH_3, CN^-, CO, *etc.* The ligands containing 2, 3 ··· coordinating atoms are called bidentate, tridentate, ··· ligands, collectively called polydentate ligands.

Ligands that have more than one potential attachment point are called multidentate ligands or chelating ligands. Chelating ligands are identified by the number of potential attachment points. For example, a bidentate ligand such as ethylenediamine ($H_2NCH_2CH_2NH_2$, abbreviated "en") has two attachment points, and a hexadentate ligand has six attachment points. Due to their complex formulas, many multidentate ligands are given abbreviations that are used in coordination compound formulas. Some examples include "en" (for ethylenediamine, $H_2NCH_2CH_2NH_2$), and "ox" (for the oxalate ion, $^-O_2CCO_2^-$). Some common multidentate ligands are shown in Table 10-2.

It should be noted that the number of ligands should not be regarded as the coordination number of the central ion in the polydentate ligand complex. In Fig.10-2, the coordination numbers of Cu^{2+} and Fe^{2+} are 4 and 6 respectively.

Table 10-2 Some Common Multidentate Ligands

Symbol	Name	Chemical formula	Coordinating atoms
en	ethylenediamine	$NH_2CH_2CH_2NH_2$	2
tm	trimethylene diamine	$NH_2CH_2CH_2CH_2NH_2$	2
dien	diethyltriamine	$NH_2CH_2CH_2NHCH_2CH_2NH_2$	3
gly	glycine ion	$NH_2CH_2COO^-$	2
ox	oxalate ion	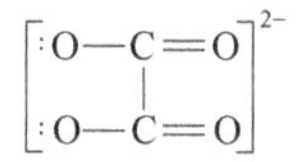 $\left[\begin{matrix} :O-C=O \\ \vert \\ :O-C=O \end{matrix}\right]^{2-}$	2
EDTA	ethylenediaminetetraacetate ion	 $(^-:OOCCH_2)_2\ddot{N}CH_2CH_2\ddot{N}(CH_2COO:^-)_2$	6
bipy	2,2′-bipyridine	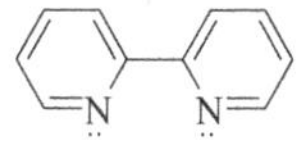 N N	2
phen	1,10-phenanthroline	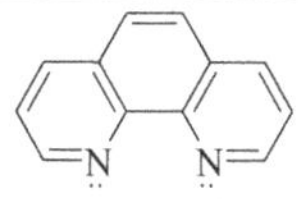 N N	2

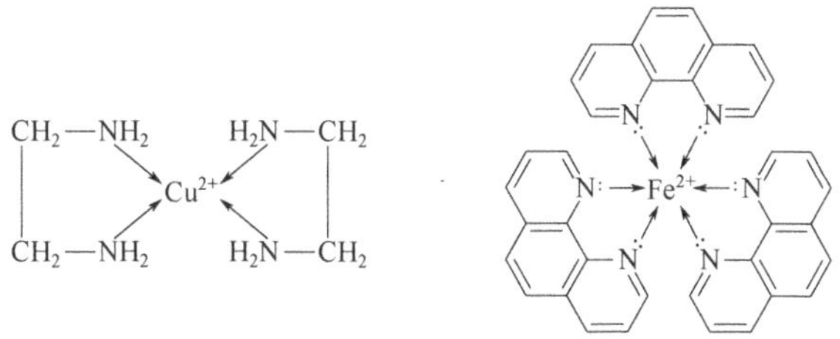

Fig.10-2 Polydentate ligand complexes

With the continuous development of the discipline of coordination chemistry, various complexes with novel configurations have been continuously discovered. In 1965, the first molecular nitrogen complex $[Ru(NH_3)N_2]Cl_2$ prepared by Canadian chemists A.D. Allen and C.V. Senoff inspired people to explore nitrogen fixation at normal temperature and pressure. The research on the structure and mechanism of action of nitrogenases that began later, led to the marginal discipline of chemical simulation of biological nitrogen fixation. Chinese scientist Lu Jiaxi and others have achieved world-class results in this regard. The first olefin complex Zeise salt $K[PtCl_3(C_2H_4)]\cdot H_2O$ shown in Fig.10-3(a) synthesized in 1927 and the ferrocene $[Fe(C_5H_5)_2]$ shown in Fig.10-3(b) prepared in 1951 made people aware hydrocarbons containing unsaturated bonds and delocalized electrons in π orbitals (such as ethylene, benzene C_6H_6, cyclopentadienyl C_5H_5, *etc.*) can also be used as electron donors to form complexes. Ferrocene's famous "sandwich" structure was quickly deduced from IR spectroscopy and then detailed structural data was measured by X-ray diffraction. The stability, structure and bonding status of ferrocene greatly stimulated the imagination of chemists, and promoted a series of synthesis, characterization and theoretical work, which led to the rapid development of d-block metal organic chemistry. Two fruitful chemists, E. Fischer (Germany) and G. Wilkinson (UK), won the Nobel Prize in 1973 for their outstanding contributions[1]. Similarly, in the late 1970s, a stable compound [Fig.10-3(d)] formed by pentamethylcyclopentadienyl (C_5Me_5) and elements in the f-block was successfully prepared, and soon ushered in the prosperous period of the development of metal organic chemistry in the f-block.

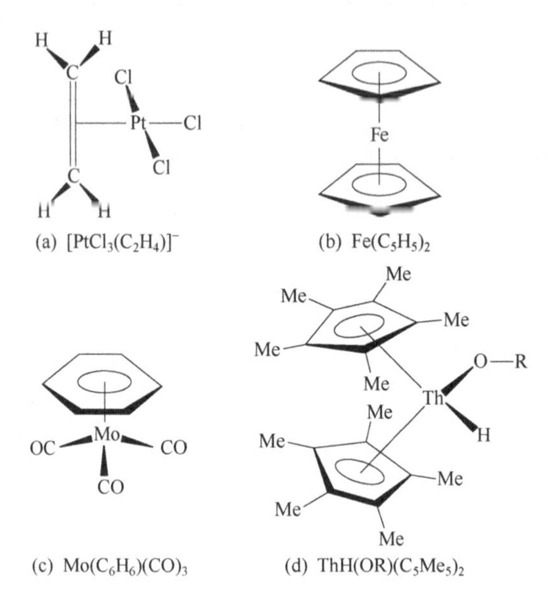

Fig.10-3 Diversity of complex structure

10.1.4 Naming Coordination Compounds

Coordination compounds are named according to an established system of rules.

① If the coordination compound contains a complex ion, name the cation first, followed by the anion name.

② In naming the complex ion or neutral coordination compound, name the ligands first and then name the transition metal.

③ Ligands names:

i. Name the ligands in alphabetical order. When using prefixes, do not alphabetize the prefix—alphabetize only the names of the ligands.

ii. If there is more than one particular ligand, use a prefix to indicate the number of ligands. Monodentate ligands use the prefixes di-, tri-, tetra-, penta-, and hexa- to indicate 2, 3, 4, 5, and 6 ligands, respectively. For complex ligand names or names that contain a prefix such as di- or tri-, place the ligand name in parentheses and use the prefixes bis-, tris-, tetrakis-, pentakis-, and hexakis- to indicate 2, 3, 4, 5, and 6 ligands, respectively.

iii. If the ligand is an anion, change the name to end in -o. (See Table 10-3.)

---

[1] *J. Organometal. Chem.*, **100**, 273(1975).

Table 10-3 Names of Some Common Ligands

Neutral ligands		Anionic ligands	
NH_3	Ammine	Cl^-	Chloro
H_2O	Aqua	Br^-	Bromo
CO	Carbonyl	CN^-	Cyano
NO	Nitrosyl	OH^-	Hydroxo
$H_2NCH_2CH_2NH_2$	Ethylenediamine	$C_2O_4^{2-}$	Oxalato
$P(C_6H_5)_3$	Triphenylphosphine	SCN^- (*S*-bonded)	Thiocyanato
		NCS^- (*N*-bonded)	Isothiocyanato
		NO_2^- (*N*-bonded)	Nitro
		ONO^- (*O*-bonded)	Nitrito

iv. If the ligand is neutral, name it using its common name. There are some exceptions to this rule, as shown in Table 10-3.

④ If the complex ion is an anion, the suffix -ate is added to the metal name. If the metal is Fe, Cu, Ag, or Au, use the Latin name:

Iron Ferrate Copper Cuprate Silver Argentate

Tin Stannate Lead Plumbate Gold Aurate

⑤ The oxidation state (in Roman numerals) of the metal is added after its name, and put in parentheses.

Consider the following examples that demonstrate how to name some simple coordination compounds.

$$[Co(NH_3)_6]Cl_3$$

This compound has six ammonia ligands (prefix = hexa-, ligand name = ammine) and a cobalt ion with a +3 oxidation state. The counterion is the chloride ion. The compound name is hexaamminecobalt(Ⅲ) chloride. Notice the space in the name between the cation (the complex ion) and the anion.

$$K_4[Fe(CN)_6]$$

The cation in this compound is the potassium ion, and the anion is a complex ion. The complex ion has six cyanide ligands (prefix = hexa-, ligand name = cyano), an iron ion with a +2 oxidation state, and an overall negative charge (metal name = ferrate). The compound name is potassium hexacyanoferrate(Ⅱ).

$$[Pt(en)Cl_2]$$

In this neutral coordination compound there are two chloride ligands (prefix = di-, ligand name = chloro), an ethylenediamine ligand (abbreviated "en" in the compound formula), and a platinum ion with a +2 oxidation state. The compound name is dichloroethylenediamine-platinum(Ⅱ).

$$[Cr(H_2O)_4Cl_2]Cl$$

The compound has four water ligands (tetraaqua), two chloride ion ligands (dichloro), a Cr^{3+} ion, and chloride ion counterions. The combination of the metal and ligands gives a complex ion with an overall +1 charge, so only one Cl^- counterion is needed to balance the charge on the complex ion. The compound name is tetraaquadichlorochromium(Ⅲ) chloride.

10.2 Complex Ion Equilibria

10.2.1 Dissociation Balance and Equilibrium Constant of Complex Ion

The central ion and the ligand form the inner coordination sphere of the complex. The inner coordination sphere (or coordination sphere) is the portion of the coordination compound where species are directly bonded to the central ion. The outer coordination sphere is the portion of the coordination compound that does not include the metal and any ligands covalently bonded to the metal. It includes counterions and waters of hydration.

The dissociation between the inner coordination sphere and the outer coordination sphere in solution is similar to the situation in strong electrolyte and the dissociation between the central ion and the ligand in the inner sphere is similar to that of weak electrolyte. For example: There are two dissociation methods in the dark blue solution $[Cu(NH_3)_4]SO_4$ formed by $CuSO_4$ and ammonia:

Completely dissociated $[Cu(NH_3)_4]SO_4 = [Cu(NH_3)_4]^{2+} + SO_4^{2-}$

Partial dissociated $[Cu(NH_3)_4]^{2+} = Cu^{2+} + 4NH_3$

The dissociation constant can be expressed as

$$K_d = \frac{m_r(Cu^{2+})m_r^4(NH_3)}{m_r\{[Cu(NH_3)_4]^{2+}\}}$$

The larger the value of the dissociation constant K_d, the greater the concentration of the central ion and ligands in the solution, that is, the greater the dissociation tendency of the complex ion. The stability of complex ions in solution can also be expressed by the constant of complex balance:

$$Cu^{2+} + 4NH_3 = [Cu(NH_3)_4]^{2+}$$

$$K_f = \frac{m_r\{[Cu(NH_3)_4]^{2+}\}}{m_r(Cu^{2+})m_r^4(NH_3)}$$

The formation constant, K_f, of a complex ion is the equilibrium constant describing the formation of a complex ion from a central ion and its ligands. For the same type of complexes, the larger the value of K_f, the more stable the complex. Obviously, K_f and K_d have a reciprocal relationship with each other:

$$K_f = K_d^{-1}$$

The formation constants of some complex ions are listed in Table 10-4. In fact, the formation of complex ions such as $[Cu(NH_3)_4]^{2+}$ with a coordination number greater than 1 in the solution is carried out in stages.

Table 10-4 The Formation Constants of Some Complex Ions

Complex ion	K_f	$\lg K_f$
$[Ag(CN)_2]^-$	1.26×10^{21}	21.2
$[Ag(NH_3)_2]^+$	1.12×10^{7}	7.05
$[Ag(S_2O_3)_2]^{3-}$	2.89×10^{13}	13.46
$[AgCl_2]^-$	1.10×10^{5}	5.04
$[AgBr_2]^-$	2.14×10^{7}	7.33
$[AgI_2]^-$	5.50×10^{11}	11.74
$[Ag(py)_2]^+$	1.0×10^{10}	10.0
$[Co(NH_3)_6]^{2+}$	1.29×10^{5}	5.11
$[Cu(CN)_2]^-$	1.00×10^{24}	24.0
$[Cu(SCN)_2]^-$	1.52×10^{5}	5.18
$[Cu(NH_3)_2]^+$	7.24×10^{10}	10.86
$[Cu(NH_3)_4]^{2+}$	2.09×10^{13}	13.32
$[Cu(P_2O_7)_2]^{6-}$	1.0×10^{9}	9.0
$[FeF_6]^{3-}$	2.04×10^{14}	14.31
$[Fe(CN)_6]^{3-}$	1.0×10^{42}	42
$[Hg(CN)_4]^{2-}$	2.51×10^{41}	41.4
$[HgI_4]^{2-}$	6.76×10^{29}	29.83
$[HgBr_4]^{2-}$	1.0×10^{21}	21.00
$[HgCl_4]^{2-}$	1.17×10^{15}	15.07
$[Ni(NH_3)_6]^{2+}$	5.50×10^{8}	8.74
$[Ni(en)_3]^{2+}$	2.14×10^{18}	18.33
$[Zn(CN)_4]^{2-}$	5.0×10^{16}	16.7
$[Zn(NH_3)_4]^{2+}$	2.87×10^{9}	9.46
$[Zn(en)_2]^{2+}$	6.76×10^{10}	10.83

Note: J A Dean Lange's Handbook of Chemistry. 13th ed. 1985. Tab.5-14, Tab.5-15. Temperature 20-25℃.

The following issues should be considered when comparing the stability of different complex ions.

① For the same type of complex ions, such as $[HgI_4]^{2-}$ and $[Cu(NH_3)_4]^{2+}$, the strength of stability can be obtained by comparing their formation constant, K_f. The larger K_f (or the smaller K_d), the more stable the complex ions is.

② For different types of complex ions (those with different numbers of ligands), the strength of stability cannot be obtained by simply comparing the value of K_f, and their stability needs to be determined through appropriate calculation or experiments.

Example 10-1 A solution in which 0.10mol of $CuSO_4$ is dissolved in 1.00kg of 6.0mol/kg ammonia. Try to find the concentration of each component in the mixed solution.

Analysis Because the concentration of ammonia is much greater than that of Cu^{2+}, it can be considered that Cu^{2+} is almost entirely coordinated by NH_3 . Then the solution should contain 0.1mol/kg of $[Cu(NH_3)_4]^{2+}$, the concentration of free ammonia is (6.0−4×0.1) = 5.6mol/kg.

Solution Let x be the concentration of Cu^{2+} ion, then

$$Cu^{2+} + 4NH_3 \rightleftharpoons [Cu(NH_3)_4]^{2+}$$

Equilibrium concentration/(mol/kg) $\quad x \quad 5.6+4x \quad 0.10-x$

Since K_f is quite large (2.1×10^{13}), $0.10-x \approx 0.1$, $5.6+4x \approx 5.6$

$$K_f = \frac{m_r\{[Cu(NH_3)_4]^{2+}\}}{m_r(Cu^{2+})m_r^4(NH_3)} = \frac{0.10}{x(5.6)^4} = 2.1\times10^{13}$$

$$x = \frac{0.10}{(5.6)^4\times2.1\times10^{13}} = 4.8\times10^{-18}(mol/kg)$$

Therefore, the concentration of each component in the solution is

$$m\{[Cu(NH_3)_4]^{2+}\} = 0.10mol/kg$$

$$m(NH_3) = 5.6mol/kg$$

$$m(SO_4^{2-}) = 0.10mol/kg$$

$$m(Cu^{2+}) = 4.8\times10^{-18}mol/kg$$

The stability of the corresponding complex ions can be compared by calculating the center ion concentration of the complex solution.

10.2.2 The Coordination Equilibria

As other equilibria, changing the balance conditions will move complex-ion equilibria. If the concentration of ligands or central ion is reduced, the complex-ion equilibria move toward the dissociation direction until a new equilibrium is established.

(1) Precipitation or dissolution of insoluble substances

After adding Na_2S solution in $[Ag(S_2O_3)_2]^{3-}$ solution, due to the formation of Ag_2S precipitate with very low solubility, the concentration of Ag^+ in the solution is reduced, and the equilibrium moved in the direction of dissociating the complex ion $[Ag(S_2O_3)_2]^{3-}$.

$$2[Ag(S_2O_3)_2]^{3-} \rightleftharpoons 2Ag^+ + 4S_2O_3^{2-}$$

$$+$$

$$Na_2S \rightleftharpoons S^{2-} + 2Na^+$$

$$\Vert$$

$$Ag_2S\downarrow$$

The corresponding ion equation is

$$2[Ag(S_2O_3)_2]^{3-} + S^{2-} = Ag_2S\downarrow + 4S_2O_3^{2-}$$

On the contrary, after adding AgBr(s) to $Na_2S_2O_3$ solution

$$\begin{array}{c} AgBr(s) = Ag^+ + Br^- \\ + \\ 2S_2O_3^{2-} \\ \| \\ [Ag(S_2O_3)_2]^{3-} \end{array}$$

Due to the formation of stable $[Ag(S_2O_3)_2]^{3-}$, the concentration of Ag^+ in the solution is reduced, $m(Ag^+)m(Br^-) < K_{sp}^{\ominus}(AgBr)$, and as a result AgBr(s) is dissolved. The ion equation is

$$AgBr(s) + 2S_2O_3^{2-} = [Ag(S_2O_3)_2]^{3-} + Br^-$$

As can be seen from the above two examples, there are two kinds of equilibrium in the system, complex-ion equilibria and dissolution equilibria. The precipitating agent (S^{2-} and Br^-) competes with the complexing agent ($S_2O_3^{2-}$) for metal ions. The ability to compete depends mainly on the stability of the complex ions and the solubility of insoluble substances. Generally speaking, when the K_f of the complex ion is relatively large and the K_{sp} of the insoluble substance is not very small, the insoluble substance is easily dissolved by the complexing agent. When the K_f of the complex ion and the K_{sp} of the insoluble matter are both small, the complex ion is easily destroyed by the precipitant. More accurate judgment can be made through calculation.

(2) Generating more stable complex ion

Adding sufficient amount of NaCN to the solution containing complex ion $[Ag(NH_3)_2]^+$, then $[Ag(NH_3)_2]^+$ can be almost completely dissociated and generate complex ion $[Ag(CN)_2]^-$. The reaction equation is

$$[Ag(NH_3)_2]^+ + 2CN^- = [Ag(CN)_2]^- + 2NH_3\uparrow$$

The reason why the above reaction can occur is that the stability of complex ion $[Ag(CN)_2]^-$ ($K_f = 1.26\times10^{21}$) is greater than the stability of complex ion $[Ag(NH_3)_2]^+$ ($K_f = 1.12\times10^7$). The essence of this type reaction is the competition result of the two complex ion equilibrium with different stability. The above reaction process is

$$\begin{array}{c} [Ag(NH_3)_2]^+ = Ag^+ + 2NH_3 \\ + \\ 2NaCN = 2CN^- + 2Na^+ \\ \| \\ [Ag(CN)_2]^- \end{array}$$

It is easy to know that the conversion between two complex ions contains the same central ion depends on the stability of two complex ions. Of course, this conversion is also related to the concentration of complexing agent and precipitating agent. When it is not easy to determine, it needs to be determined through calculation.

(3) Generating hard-to-dissociate substance

If the ligand generated by the dissociation of complex ions can react with other substances to produce a more hard-to-dissociate substance, the dissociation equilibrium of the ligand can be moved to the dissociation direction. For example, add an acid (such as HCl) to the solution containing complex ion $[Cu(NH_3)_4]^{2+}$, because the H^+ in the acid easily combines with the NH_3 molecule to form a more stable ion, NH_4^+, which reduces the concentration of NH_3 in the solution. The balance of complex ion $[Cu(NH_3)_4]^{2+}$ moves to the dissociation direction:

$$[Cu(NH_3)_4]^{2+} \rightleftharpoons Cu^{2+} + 4NH_3$$

$$+$$

$$4HCl \rightleftharpoons 4Cl^- + 4H^+$$

$$\Vert$$

$$4NH_4^+$$

The total reaction equation is

$$[Cu(NH_3)_4]^{2+} + 4H^+ \rightleftharpoons Cu^{2+} + 4NH_4^+$$

(4) Occurrence of redox reaction

When the ligand generated by the dissociation of the complex ion can undergo a redox reaction with a certain substance, the complex-ion equilibria will move toward the dissociation direction. For example, adding NaClO solution to $[Cu(CN)_4]^{2-}$ solution, because ClO^- is an oxidant, the concentration of CN^- ion generated by dissociation of complex ion $[Cu(CN)_4]^{2-}$ can be reduced, so complex-ion $[Cu(CN)_4]^{2-}$ equilibrium moves to the dissociation direction.

$$[Cu(CN)_4]^{2-} \rightleftharpoons Cu^{2+} + 4CN^-$$

$$+$$

$$4NaClO \rightleftharpoons 4Na^+ + 4ClO^-$$

$$\Vert$$

$$4CNO^- + 4Cl^-$$

The total reaction equation is

$$[Cu(CN)_4]^{2-} + 4ClO^- \rightleftharpoons Cu^{2+} + 4CNO^- + 4Cl^-$$

This reaction is the main chemical reaction for the treatment of cyanide-containing waste water by alkaline chlorination in environmental protection.

It should be noted that there are some compounds similar to the coordinate compound, such as alum, $KAl(SO_4)_2·12H_2O$. They are different from complex ions. After dissolving in water, they are all dissociated into simple hydrated ions as K^+, Al^{3+} and SO_4^{2-} , and only a tiny amount of $[Al(SO_4)_2]^-$. But there is no absolute boundary for this difference. For example, the complex ion $[Cu(NH_3)_4]^{2+}$ can still be dissociated into a small amount of Cu^{2+} and NH_3. Some complex salts such as $KCl·MgCl_2·6H_2O$ also have unstable complex ion $[MgCl_3]^-$ in solution. Therefore, these can also be regarded as a complex with an extremely unstable inner coordination sphere.

Many compounds exist in the form of complexes. Many common crystalline hydrates are complexes. For example, the structural formulas of salts such as magnesium chloride and aluminum chloride containing 6 molecules of crystal water are $[Mg(H_2O)_6]Cl_2$ and $[Al(H_2O)_6]Cl_3$, respectively. However, some of the crystalline water in some crystalline hydrates can also be located in the outer coordination sphere. For example, the structural formula of $CuSO_4·5H_2O$ is $[Cu(H_2O)_4]SO_4·H_2O$. The blue color alum and many copper salts solution are actually $[Cu(H_2O)_4]^{2+}$.

10.3 The Chemical Bond Theory in Coordination Compounds

The chemical bond theory in coordination compounds mainly discusses the interaction force between the metal ion and the individual atoms in the ligands. At present, there are three theories to discuss this force: valence bond theory, crystal field theory and molecular orbital theory (also called coordination field theory).

10.3.1 Valence Bond Theory

The hybridization of the orbitals in the same atom and the overlap of the orbitals between different atoms constitute the core argument of the valence bond theory. Pauling (Pauling L.C., 1901-1994, American chemist) first applied the valence bond theory to the complex, then modified and supplemented the theory later. According to valence bond theory of complex, the coordinate atoms in the ligand form a coordination bond with a lone pair of electrons "dropping into" the hybrid empty orbitals of the central metal ion.

(1) The empty orbital of the central metal ion accepts the lone pair of electrons to form a coordination bond

The d-block elements and elements in the periodic table close to the transition series, especially metal elements, the ions of them have empty valence orbital, which can be used to accept lone pair of electrons given by ligands and easily form complex ions. But the main group elements can also form complex ions, such as hydrated ions of Li^+, Na^+, F^-, Cl^- and others, and $[Sn(OH)_4]^{2-}$, $[SiF_6]^{2-}$, $[AlF_6]^{3-}$ *etc.*

Common monodentate ligands (ligands containing a coordination atom), such as F^-, Cl^-, Br^-, I^-, OH^-, H_2O, SCN^-, NH_3, CN^-, CO *etc.*, can provide lone pair of electrons. Electron pair donors are usually those with more electronegativity, such as F, O, S, N in the above ligands.

(2) The bonded hybrid orbital determines the configuration of complex ion

In the complex ions, the valence orbital of the central ions are usually $(n-1)$d, ns, np and nd orbitals with similar energies. They form hybrid orbitals under the action of ligands. Because the hybrid orbitals extend in specific directions, the complex ions also have a certain spatial configuration. Now take $[Cu(NH_3)_2]^+$ as an example.

The electron configuration of Cu^+ ion is

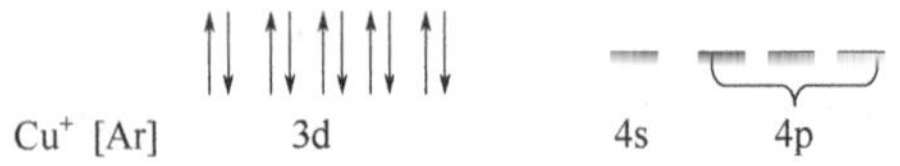

Cu^+ [Ar] 3d 4s 4p

The 3d subshell is full, while the 4s and 4p subshells are empty. Each empty orbital can accept one pair of electrons given by NH_3 and a total of 4 pairs of electrons form a complex ion with the highest coordination number of 4. But the coordination number of Cu^+ is usually 2, this because the nature of the ligand (such as charge, radius, *etc.*) also affects the coordination number. In complex ion $[Cu(NH_3)_2]^+$, Cu^+ uses sp hybrid orbitals to form coordination bonds with NH_3, and its electron configuration is

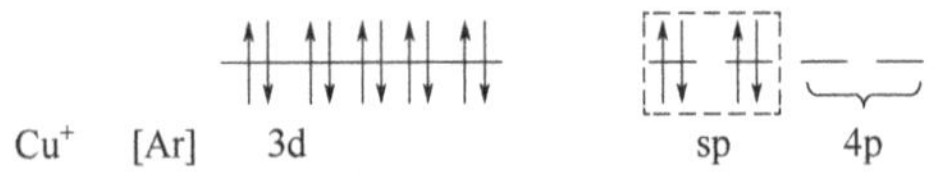

Cu^+ [Ar] 3d sp 4p

Since the two sp hybrid orbitals are linearly distributed, the spatial configuration of $[Cu(NH_3)_2]^+$ is linear. Several hybrid orbitals and spatial configurations of complex ions are listed in Table 10-5.

Table 10-5 Hybrid Orbitals and Configurations of Some Complex Ions

CN	Hybrid orbital	Configurations	Complex ions
2	sp	Linear	$[Cu(NH_3)_2]^+$ $[Ag(NH_3)_2]^+$
3	sp^2	Plane triangle	$[Cu(CN)_3]^{2-}$
4	sp^3	Tetrahedron	$[Zn(NH_3)_4]^{2+}$ $[Ni(NH_3)_4]^{2+}$ $[Cu(CN)_4]^{3-}$
4	dsp^2/spd^2	Square-planar	$[Ni(CN)_4]^{2-}$ $[Cu(NH_3)_4]^{2+}$
5	dsp^3	Triangular double cone	$[Ni(CN)_5]^{3-}$ $[Fe(CO)_5]$
6	d^2sp^3/sp^3d^2	Octahedron	$[Fe(CN)_6]^{3-}$ $[Fe(CN)_6]^{4-}$ $[FeF_6]^{3-}$ $[Co(NH_3)_6]^{3+}$

(3) The hybrid orbital of central ion determines the type of complex ion

When the orbitals of the central ion in secondary outer shell is not completely filled, there may be two cases when forming the complex ion.

For example, the electron configuration when forming complex ion $[Ni(NH_3)_4]^{2+}$ is

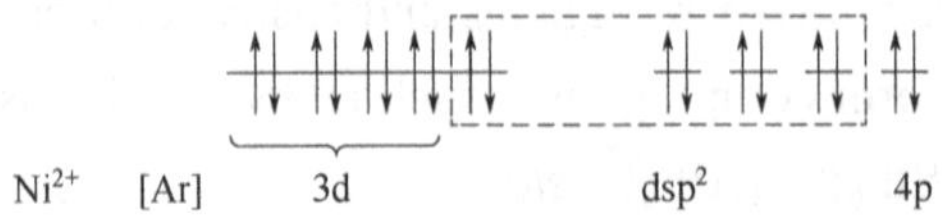

Ni^{2+} [Ar] 3d dsp^2 4p

The eight d electrons of Ni^{2+} have not been redistributed and still occupy five 3d orbitals, that is, there are only 2 unpaired electrons in the 3d subshell. The 4s and 4p orbitals form four sp^3 hybrid orbitals through hybridization. The spatial geometry of these hybrid orbital is regular tetrahedron. The central ion still maintains its original electronic configuration, and the lone pair of electrons of the ligand only enters the outer orbitals, and the corresponding complex ion is called the outer orbital complex.

Another case is complex ion $[Ni(CN)_4]^{2-}$.

Ni^{2+} [Ar] 3d sp^3

Under the influence of the ligand CN^-, Ni^{2+} redistributes two unpaired 3d electrons and one 3d orbital is vacated. This 3d orbital hybridize with one 4s and two 4p empty orbitals and generate four dsp^2 hybrid orbitals. The spatial geometry of these hybrid orbitals is a square-planar. The electron configuration of the central ion changes, unpaired electrons are paired, and the lone pair of electrons of the ligand enter the inner orbital, and the corresponding complex ion is called the inner orbital complex.

In the complex ions $[FeF_6]^{3-}$ and $[Fe(CN)_6]^{3-}$, the central ion Fe^{3+} is bonded to the ligand by sp^3d^2 and d^2sp^3 hybrid empty orbitals, respectively. The spatial geometry of such hybrid orbitals is an octahedron. However, $[FeF_6]^{3-}$ belongs to the outer orbital complex ion, and $[Fe(CN)_6]^{3-}$ belongs to the inner orbital complex ion. Some outer orbital complexes and inner orbital complexes are shown in Table 10-6. A simple identification method is: the hybrid types with s as the initial letter are outer orbital complexes; the hybrid types with d as the initial letter are outer orbital complexes.

Since the energy of the $(n-1)$d orbital is lower than nd orbital, the coordination bond in inner orbital complex is stronger than that in outer orbital complex. Generally, the inner orbital complex is more stable than the outer orbital complex ion when we compare the stability of the same type of complex ion formed by the same metal ion (K_f is shown in Table 10-4). In solution, the former is more difficult to dissociate than the latter. Moreover, the coordination bonds of inner-orbital complexes generally have covalent bond properties, while the coordination bonds of outer-orbital complexes have some ionic bond properties.

Table 10-6 Some Outer Orbital and Inner Orbital Complexes

Central metal	Complex	Hybrid types of central atom
Ag^+	$[Ag(NH_3)_2]^+$	sp
Cu^+	$[Cu(NH_3)_2]^+$	sp
Cu^+	$[Cu(CN)_4]^{3-}$	sp^3
Cu^{2+}	$[Cu(NH_3)_4]^{2+}$	dsp^2
Zn^{2+}	$[Zn(NH_3)_4]^{2+}$	sp^3
Cd^{2+}	$[Cd(CN)_4]^{2-}$	sp^3
Fe	$[Fe(CO)_5]$	dsp^3

(Continued)

Central metal	Complex	Hybrid types of central atom
Fe^{3+}	$[FeF_6]^{3-}$	sp^3d^2
Fe^{3+}	$[Fe(CN)_6]^{3-}$	d^2sp^3
Fe^{2+}	$[Fe(CN)_6]^{4-}$	d^2sp^3
Fe^{2+}	$[Fe(H_2O)_6]^{2+}$	sp^3d^2
Mn^{2+}	$[MnCl_4]^{2-}$	sp^3
Mn^{2+}	$[Mn(CN)_6]^{4-}$	d^2sp^3
Cr^{3+}	$[Cr(NH_3)_6]^{3+}$	d^2sp^3

Whether the complex is inner orbital type or outer orbital type is mainly determined by the charge and electron configuration of the central ion and the electronegativity of the coordination atom. Coordination atoms with large electronegativity (such as F^- ions, halogen ions and O atoms in H_2O molecules) often form outer orbital complexes; coordination atoms with low electronegativity (such as C atom in CO molecule and anion CN^-) often form inner orbital complexes. The electronegativity of the N atom in the NH_3 molecule is moderate, forming outer orbital complexes with some central atoms, and forming inner orbital complex with other central atoms.

The valence bond theory can give a good explanation of the formation, spatial configuration and coordination number of the central ion, but also has its limitations. For example, it cannot explain why the electrons of the central ion in the inner orbital complex should be redistributed. Moreover, due to the different number of d electrons, the stability of the complex is different. The theories of chemical bonding do not help much in explaining the characteristic colors and magnetic properties of complex ions. In transition metal ions, we need to focus on how the electrons in the d orbitals of a metal ion are affected when they are in a complex. A theory that provides that focus and an explanation of these properties is crystal field theory.

10.3.2 Crystal Field Theory (CFT)

In the crystal field theory, bonding in a complex ion is considered to be an electrostatic attraction between the positively charged nucleus of the central metal ion and electrons in the ligands. Repulsions also occur between the ligand electrons and electrons in the central ion. In particular, the crystal field theory focuses on the repulsions between ligand electrons and d electrons of the central ion.

(1) Splitting of d energy levels in octahedral field

All five of d orbitals are alike in energy when in an isolated atom or ion, but they are unlike in their spatial orientations. One of them d_{z^2} is directed along the z axis, and another $d_{x^2-y^2}$ has lobes along the x and y axes. The remaining three have lobes extending into regions between the perpendicular x, y, and z axes. In the presence of ligands, because repulsions exist between ligand electrons and d electrons, the d-orbital energy levels of the central metal ion are raised. As we will see soon, however, they are not all raised to the same extent.

Fig.10-4 depicts six anions (ligands) approaching a central metal ion along the x, y, and z axes. This direction of approach leads to an octahedral complex. Repulsions between ligand electrons and d-orbital electrons are strengthened in the direct head-to-head approach of ligands to the d_{z^2} orbitals and $d_{x^2-y^2}$ orbitals. These two orbitals have their energy raised with respect to an average d-orbital energy for a central metal ion in the field of the ligands. For the other three orbitals (d_{xy}, d_{xz} and d_{yz}) with ligands approach between the lobes of the orbitals, there is a gain in stability over the head-to-head approach; these orbital energies are lowered with respect to the average d-orbital energy. The difference in energy between the two groups of d orbitals is called crystal field splitting and is represented by the symbol Δ_o, with the subscript o emphasizing that the crystal field splitting shown in Fig.10-5 is for an octahedral complex.

$$\Delta_o = \text{(the energy of } d_{x^2-y^2},\ d_{z^2}) - \text{(the energy of } d_{xy},\ d_{xy},\ d_{yz})$$

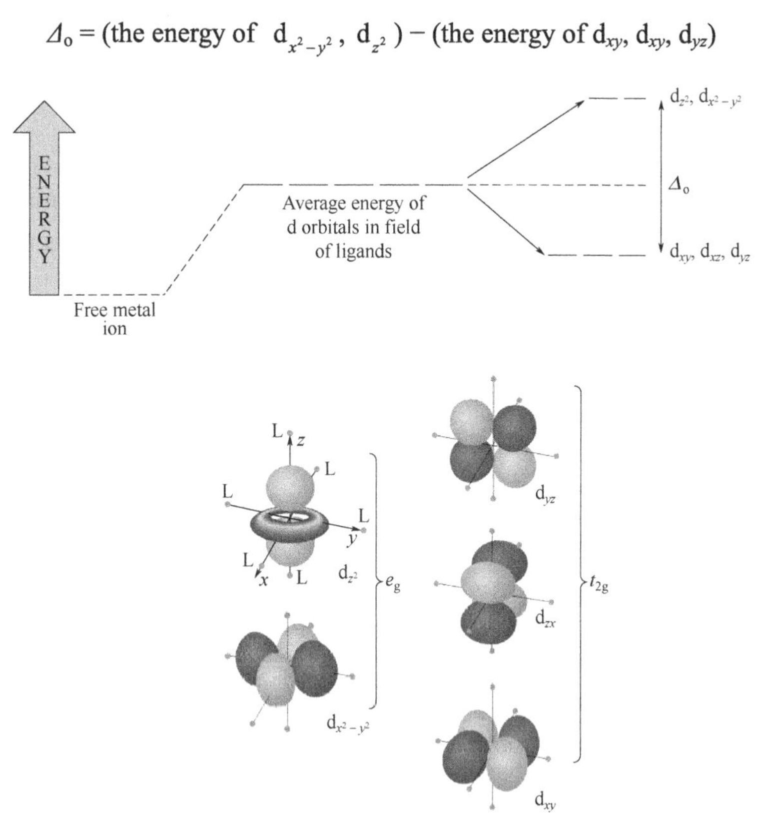

Fig.10-4 Splitting of d energy levels in the formation of an octahedral complex ion

(2) Weak field ligand and strong field ligand

The splitting energy Δ_o is numerically equivalent to the energy absorbed by an electron transitioning from t_{2g} orbital to e_g orbital, which can be measured by spectroscopic experiments. Different ligands can be arranged in order of their abilities to produce a splitting of the d energy levels. This arrangement is known as the spectrochemical series.

For different central atoms, the order may change slightly. The ligand at the front of the sequence (generally bounded by H_2O) is a weak-field ligand, and the ligand at the back of the sequence (generally bounded by NH_3) is a strong-field ligand.

Weak field

(small Δ_o)

$I^- < Br^- < Cl^- < \underline{S}CN^- < F^- < OH^- < ox^{2-} < ON\underline{O}^- < H_2O < SC\underline{N}^- < EDTA$

$< NH_3 \approx py < en < SO_3^{2-} < phen < \underline{N}O_2^- < \underline{C}N^- < CO$

(large Δ_o)

The underline indicates the donor atom strong field

(3) High spin complex and low spin complex

Consider these two complexes of Co(Ⅲ): $[CoF_6]^{3-}$ and $[Co(NH_3)_6]^{3+}$. Placing the electrons in the lower levels confers extra stability (lower energy) on the complex, but some of this stability is offset because it requires energy, called the pairing energy (*P*), to force an electron into an orbital that is already occupied by an electron. Alternatively, the electron could be assigned to either the $d_{x^2-y^2}$ or d_{z^2} orbital, avoiding the pairing energy. To pair or not to pair, that is the question.

Whether the electron enters the lowest level and becomes paired or, instead, enters the upper level with the same spin depends on the magnitude of Δ_o. If Δ_o is greater than the pairing energy *P*, greater stability is obtained if the electron is paired with one in the lower level. If Δ_o is less than the pairing energy, greater stability is obtained by keeping the electrons unpaired. The F^- ion is a weak-field ligand, whereas NH_3 is a strong-field ligand. Because the crystal field splitting for NH_3 is greater than the pairing energy for Co^{3+}, we have the following situations.

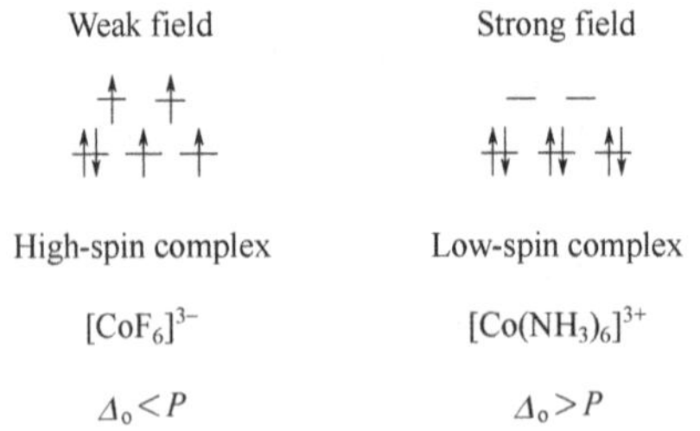

Weak field	Strong field
↑ ↑	— —
↑↓ ↑ ↑	↑↓ ↑↓ ↑↓
High-spin complex	Low-spin complex
$[CoF_6]^{3-}$	$[Co(NH_3)_6]^{3+}$
$\Delta_o < P$	$\Delta_o > P$

The two situations represent two electron spin states. Complexes with a large number of single electrons are called high-spin complexes. Complexes that does not have or contain a small number of single electrons are called low-spin complexes. By comparison, it is not difficult to find that the high-spin complex and the low-spin complex correspond to the outer-orbital complex and the inner-orbital complex in the valence bond theory, respectively.

Similar to the d^6 configuration, the transition metal ions in the d^4, d^5, and d^7 configurations will also face two options, forming corresponding high-spin complexes or low-spin complexes. There can only be one type of ion configuration for d^1 to d^3 and d^8 to d^{10}.

(4) Crystal field stabilization energy (CFSE)

Whether the formation of high spin or low spin, the complex is in the lowest energy state. There is no gain or loss of total energy during the energy level splitting process. The energy lost by one set of orbitals (t_{2g} orbitals) is obtained by another set of orbitals (e_g orbitals), which is:

$$2E(e_g) + 3\,E(t_{2g}) = 0$$

That is, the energy gained equals the energy lost. Also, the energy difference between the orbitals is

$$(\text{the energy of } d_{x^2-y^2},\ d_{z^2}) - (\text{the energy of } d_{xy},\ d_{xz},\ d_{yz} \text{ orbitals}) = \Delta_o$$

Then

$$E(e_g) = 0.6\Delta_o$$
$$E(t_{2g}) = -0.4\Delta_o$$

The splitting is as shown:

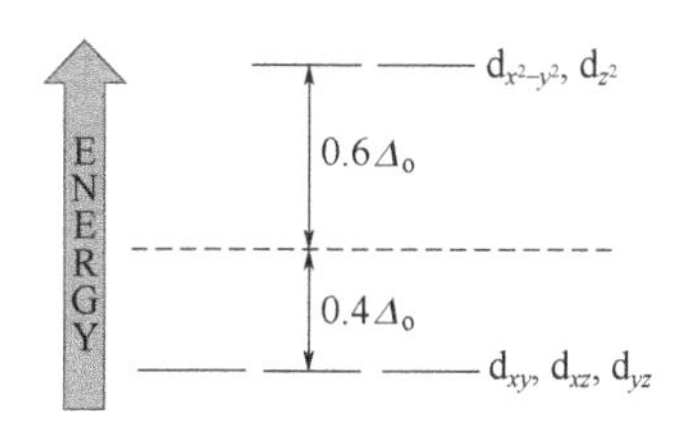

The splitting of the two groups of orbitals is not equal with respect to the average energy of the d orbitals. It means filling an electron in the $t_{2g}(d_{xy}, d_{xz}, d_{yz})$ orbital corresponds to an increase in the stable energy of $0.4\Delta_o$, and filling an electron in the $e_g(d_{x^2-y^2}, d_{z^2})$orbital corresponds to the decrease in the stable energy of $0.6\Delta_o$. This extra stabilization energy resulting from the d orbital splitting is called crystal field stabilization energy (CFSE). The CFSE for the four d configurations are:

d^3: CFSE $= -1.2\Delta_o$ $[3\times(-0.4\Delta_o)]$

d^8: CFSE $= -1.2\Delta_o+3P$ $[6\times(-0.4\Delta_o)+2\times(0.6\Delta_o)+3P]$

d^6(high-spin): CFSE $= -0.4\Delta_o+P$ $[4\times(-0.4\Delta_o)+2\times(0.6\Delta_o)+P]$

d^6(low-spin): CFSE $= -2.4\Delta_o+3P$ $[6\times(-0.4\Delta_o)+3P]$

Comparing the two arrangements of d^6 configuration, the CFSE of low spin is greater than that of high spin, and the complex should be more stable. Many well-known Co^{3+} complexes are low-spin complexes, and six d electrons are filled in pairs in t_{2g} orbitals. They are all very stable diamagnetic complexes. $[Co(NH_3)_6]Cl_3$ does not lose NH_3 at temperatures up to 200℃ and does not decompose when recrystallized in hot concentrated hydrochloric acid.

(5) The colors of transition metal complexes

The crystal field theory satisfactorily explained the color of the transition metal complex.Crystal field splitting of the d energy levels produces the energy difference, Δ, that accounts for the colors of complex ions. Promotion of an electron from a lower level to a higher d level results from the absorption of the appropriate components of white light; the transmitted light is colored. Subtracting one color from white light leaves the complementary color. A solution containing $[Cu(H_2O)_4]^{2+}$ absorbs most strongly in the yellow region of the spectrum (about 580nm). The wavelength components of the light transmitted combine to produce the color blue. Thus, aqueous solutions of copper(Ⅱ) compounds usually have a characteristic blue color. In the presence of high concentrations of Cl^-, copper(Ⅱ) forms the complex ion $[CuCl_4]^{2-}$. This species absorbs strongly in the blue region of the spectrum. The transmitted light is yellow, as is the color of the solution.

Another example is $[Ti(H_2O)_6]^{3+}$. It has a maximum absorption wavenumber of 20300cm^{-1}. When the visible light passes through the aqueous solution, the blue-green light (wavelength about 500×10^3pm) is absorbed. The visible purple transmitted and reflected light is the complementary color of blue-green light.

(6) The magnetic properties of transition metal complexes

The crystal field theory is also used to explain the magnetic properties of the complex. The contribution to the free ion magnetic moment comes from the orbital angular momentum and spin angular momentum of the electrons. In some cases, due to the interaction between the electrons and the environment, the orbital angular momentum of the metal ion in the complex is quenched, and it can be considered that only spin angular momentum works. The magnetic moment μ of a complex can be calculated theoretically. The relationship between the magnetic moment and the number of unpaired electrons n is as follows:

$$\mu = \sqrt{n(n+2)}\mu_B$$

μ_B is a Bohr magneton ($1\mu_B = 9.274\times10^{-24} A\cdot m^2$). The magnetic moment can also be determined experimentally by a magnetic balance. The agreement between the calculated results and the experimental results in Table 10-7 can illustrate the success of the crystal field theory.

Table 10-7 Magnetic Moments of Some Transition Metal Complexes

Central ion	Unpaired electron number, n	Calculated μ/μ_B	Experimental μ/μ_B
Ti^{3+}	1	1.73	1.7-1.8
V^{3+}	2	2.83	2.7-2.9
Cr^{3+}	3	3.87	3.8
Mn^{3+}	4	4.90	4.8-4.9
Fe^{3+}	5	5.92	5.9

Example 10-2 The magnetic moment of a Co^{2+} octahedral complex was measured to be 4.0μ_B, try to find the d electrons configuration of Co^{2+}.
Solution Co^{2+} in the complex is d^7 configuration, and two possible configuration are $t_{2g}^5 e_g^2$ and $t_{2g}^6 e_g^1$. The former is a high-spin arrangement with three unpaired electrons, and the latter is a low-spin arrangement with one unpaired electron. Since the experimental magnetic moment is close to three unpaired electrons, the theoretically calculated spin magnetic moment is 3.87μ_B, which is far from the calculated result of one electron 1.73μ_B, it is not difficult to infer that d electrons adopt a high spin arrangement.

10.4 Applications of Coordination Chemistry

10.4.1 Nitrogenase (Nitrogen Fixation)

Agricultural production requires a large amount of nitrogen fertilizer. The nitrogen in the air is a rich natural resource, but it cannot be directly absorbed by plants. It must be converted into ammonia or ammonium salts (so-called fixed nitrogen) to be absorbed by plants. At present, the industrial production of fixed nitrogen is under the severe conditions of high temperature and high pressure, and a catalyst is also needed to synthesize ammonia. There is no such enhanced reaction conditions in the biosphere, but a completely different and more complicated route to ammonia synthesis.

The most fundamental reason why various nitrogen-fixing microorganisms have peculiar nitrogen-fixing skills is that they all have nitrogen-fixing enzymes as biological catalysts for nitrogen-fixing reactions. Biological nitrogen fixation uses ATP (adenosine triphosphate) as the reducing agent, and the related reduction half reaction can be expressed as:

$$N_2 + 16MgATP + 8e^- + 8H^+ \longleftrightarrow 2NH_3 + 16MgADP + 16P_i + H_2$$

P_i in the formula represents inorganic phosphate. The attractive feature of this process is that it occurs at room temperature and pressure, and the rhizobia of various legumes (such as clover, alfalfa, kidney bean, pea, *etc.*), some other fungi and blue-green algae have the ability to convert N_2 to NH_3 at normal temperature and pressure.

The details of nitrogen fixation mechanism are still unknown, but nitrogenases arc known to be involved in ferrithioprotein and molybdenum ferrithioprotein. Biochemists have isolated metal containing coenzymes in the catalytic process, and bioinorganic chemists have prepared a variety of model compounds with active sites. An important breakthrough is the acquisition of the coenzyme $MoFe_7S_8$ [Fig.10-5(a)] and its associated “P” cluster compounds, which generally refers to a polyhedron with three or more atoms and formed by connecting a group of peripheral atoms or ligands. The core atoms in cluster compounds can be main group elements or transition elements. Fig.10-5(b) is the Fe_8 “P” cluster crystal determined by X-ray diffraction [M. K. Chan, J. Kim, and D. C. Ress, Science, 260, 792(1993)].

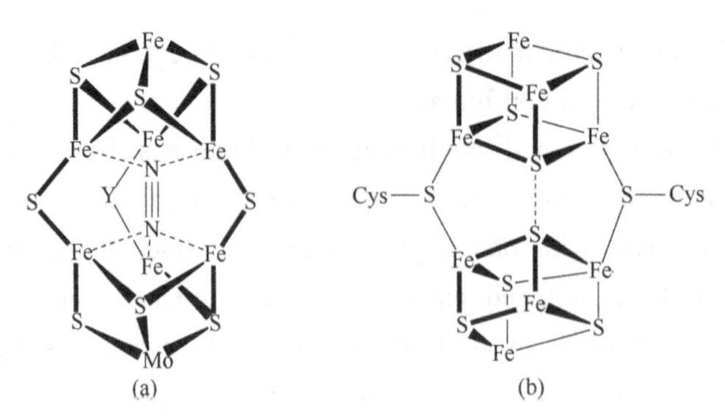

Fig.10-5 (a) $MoFe_7S_8$ in nitrogenase; (b) The Fe_8 "P" cluster related to (a)

10.4.2 Cisplatin: A Cancer-fighting Drug

Chemotherapy is a treatment used for some types of cancer. The treatment employs anticancer drugs to destroy cancer cells. An important cancer-fighting drug is cisplatin, which is commonly used to treat testicular, bladder, lung, esophagus, stomach, and ovarian cancers.

Cl, NH₃ / Pt / Cl, NH₃ — *cis*-$[PtCl_2(NH_3)_2]$ (cisplatin)

Cl, NH₃ / Pt / H₃N, Cl — *trans*-$[PtCl_2(NH_3)_2]$ (transplatin)

The compound *cis*-$[PtCl_2(NH_3)_2]$ was first described by Michele Peyrone in 1845, and known for a long time as Peyrone's chloride. The structure was deduced by Alfred Werner in 1893. In 1965, Barnett Rosenberg, Loretta van Camp *et al.* of Michigan State University discovered the anticancer activity of cisplatin. Not only did Rosenberg discover the anticancer activity of cisplatin, but he was also the first to report that the *trans*-isomer (transplatin) was ineffective in killing cancer cells.

Before examining the anticancer activity of cisplatin, let's first consider its synthesis. One method for making cisplatin starts from $K_2[PtCl_4]$,which is converted to $K_2[PtI_4]$ by treatment with an aqueous solution of KI:

$$K_2[PtCl_4] + 4KI = K_2[PtI_4] + 4KCl$$

In the next step, NH_3 is added, forming a yellow compound, *cis*-$[PtI_2(NH_3)_2]$. The formation of *cis*-$[PtI_2(NH_3)_2]$ is a key step, which occurs in two stages.

$$K_2[PtI_4] + NH_3 = [K]^+[PtI_3(NH_3)]^- + KI$$

$$[K]^+[PtI_3(NH_3)]^- + NH_3 = cis\text{-}[PtI_2(NH_3)_2] + KI$$

It might seem strange that only one isomer (the *cis* isomer) is obtained. One way to rationalize why this happens is to consider the ligands in $[PtI_3(NH_3)]^-$ in terms of their tendencies to direct an incoming ligand toward the trans position. Empirical studies indicate that NH_3 is a weaker *trans* director than I^-, and so the second molecule NH_3 is preferentially directed to a position that is *trans* to I^-, rather than being directed to a position that is *trans* to NH_3. Cl^- is a weaker *trans* director than I^-,so treatment of $K_2[PtCl_4]$ with NH_3 will give a lower yield of the *cis* isomer of $PtCl_2(NH_3)_2$. Consequently, the conversion of $K_2[PtCl_4]$ to $K_2[PtI_4]$ is an important step in the synthesis.

The remaining steps in preparing cisplatin are as follows. When *cis*-$[PtI_2(NH_3)_2]$ is treated with $AgNO_3$, insoluble AgI precipitates, leaving behind a solution of *cis*-$[Pt(NH_3)_2(H_2O)_2]^{2+}$. Finally, treatment with KCl gives a yellow precipitate *cis*-$[PtCl_2(NH_3)_2]$.

Cisplatin enters cancer cells mainly by diffusion. Once inside the cell, one of the chloride ions in cisplatin is replaced by a water molecule.

$$PtCl_2(NH_3)_2 + H_2O = [PtCl(H_2O)(NH_3)_2]^+ + Cl^-$$

Cisplatin

The anticancer activity of cisplatin is associated with the binding of $[PtCl(H_2O)(NH_3)_2]^+$ to a cellular DNA molecule. When $[PtCl(H_2O)(NH_3)_2]^+$ binds to a DNA molecule, structural deformations occur in the DNA molecule and these deformations, if not repaired by proteins in the cell, lead ultimately to cell death.

It may seem surprising that the *cis* and *trans* isomers of $PtCl_2(NH_3)_2$ exhibit a dramatic difference in anticancer activity, given their structural similarities. Overall, the *trans* isomer (transplatin) is more reactive and more potent than cisplatin but the higher reactivity leads ultimately to lower anticancer activity. Because of its increased reactivity, transplatin can undergo many side reactions before it reaches its target; thus transplatin is less effective in killing cancer cells. The discovery of the anticancer activity of cisplatin was nothing less than monumental. To date, thousands of platinum-containing compounds have been investigated as potential chemotherapy drugs.

10.4.3 Sequestering Metal Ions

Metal ions can act as unintended catalysts in promoting undesirable chemical reactions in a manufacturing process, or they may alter the properties of the material being manufactured. Thus, for many industrial purposes, it is imperative to remove mineral impurities from water. Often these impurities, such as Cu^{2+}, are present only in trace amounts, and precipitation of metal ions is feasible only if K_{sp} for the precipitate is very small. An alternative is to treat the water with a chelating agent. This reduces the free cation concentrations to the point at which the cations can no longer enter into objectionable reactions. The cations are sequestered. Among the chelating agents widely employed are the salts of ethylenediaminetetraacetic acid (H_4EDTA) usually as the sodium salt.

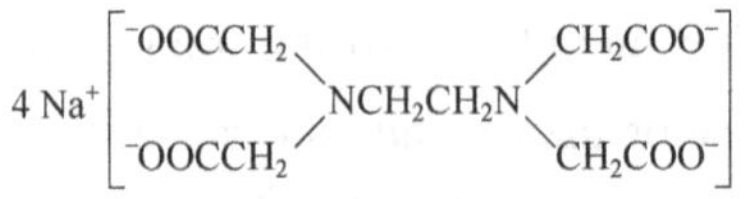

$$4\,Na^+\left[\begin{matrix} ^-OOCCH_2 & & CH_2COO^- \\ & NCH_2CH_2N & \\ ^-OOCCH_2 & & CH_2COO^- \end{matrix}\right]$$

A representative complex ion formed by a metal ion with the hexadentate anion is depicted in Fig.10-6. The high stability of such complexes can be attributed to the presence of five-member chelate rings. The stability of the complexes can also be attributed to the chelate effect.

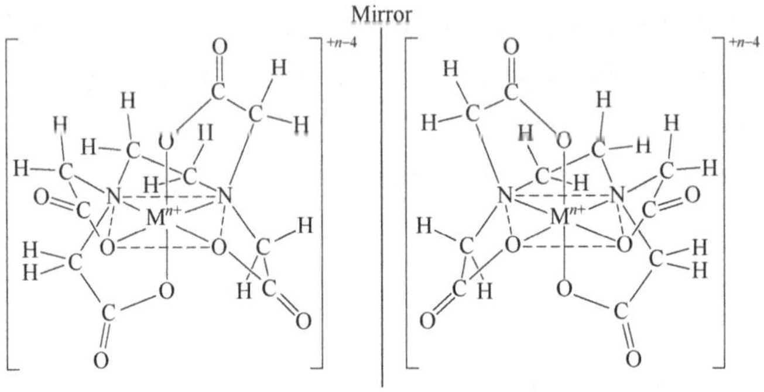

Fig.10-6　Structure of a metal-EDTA complex

In the presence of $EDTA^{4-}$(aq), Ca^{2+}, Mg^{2+} and Fe^{3+} in hard water are unable to form boiler scale or to precipitate as insoluble soaps. The cations are sequestered in the complex ions: $[Ca(EDTA)]^{2-}$, $[Mg(EDTA)]^{2-}$, and $[Fe(EDTA)]^{-}$, having K_f values of 4.0×10^{10}, 4.0×10^{8}, and 1.7×10^{24} respectively.

Chelation with EDTA can be used in treating some cases of metal poisoning. If a person with lead poisoning is fed $[Ca(EDTA)]^{2-}$, the following exchange occurs because $[Pb(EDTA)]^{2-}$ is even more stable than $[Ca(EDTA)]^{2-}$. The body excretes the lead complex, and the Ca^{2+} remains as a nutrient. A similar method can be used to rid the body of radioactive isotopes, as in the treatment of plutonium poisoning.

Some plant fertilizers contain EDTA chelates of metals such as Cu^{2+} as a soluble form of the metal ion for the plant to use. Metal ions can catalyze reactions that cause mayonnaise and salad dressings to spoil, and the addition of EDTA reduces the concentration of metal ions by chelation.

10.4.4 Photosynthesis

The reaction of converting H_2O and CO_2 into carbohydrates and O_2 using light as an energy source is one of the well-known redox reactions (Fig.10-7), but this reaction is difficult to occur thermodynamically.

The formation of carbohydrates includes the two processes of CO_2 reduction and oxidation of H_2O to O_2. The reaction has two photoreaction centers, photosystem I (PSI) and photosystem II (PSII), both of which occur in the cellular lipid of green leaves called chloroplasts.

PSI is based on chlorophyll a_1, and chlorophyll a_1 is a porphyrin complex containing metal Mg (Fig.10-8). Excited by sunlight PSI can be used as a reducing agent to reduce the Fe-S complex, and the electrons obtained by the complex are finally used to reduce CO_2. The oxidation capacity of the oxidized form PSI formed after the electron transfer is not enough to oxidize H_2O, but return to the reduced form. In the process of returning to the reduction state, two ADP molecules are converted into two ATP molecules by intermediate species including several iron-based redox couples and a quinone. The oxidation state of PSII is a strong oxidant sufficient to oxidize H_2O. The reaction can only be performed by a series of complex redox reactions catalyzed by manganese-based enzymes[1].

Fig.10-7 Photosynthesis

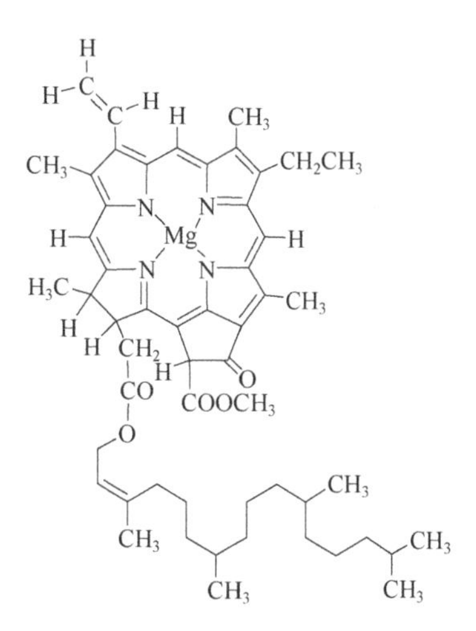

Fig.10-8 Chlorophyll a_1

10.4.5 Metalloprotein and Metalloenzyme

Amino acids are the basic structural units that make up proteins. Amino acids are combined with each other by peptide bonds to form peptide chains, and one or more peptide chains form protein molecules in various special ways. From the perspective of bioinorganic chemists, the peptide chains in amino acids and protein molecules are ligands for metal ions. The peptide chains of proteins belong to macromolecular ligands, also called biological ligands. Bioligands display a high degree of selectivity when they coordinate with metal ions. This selectivity means that metal macromolecules containing biological macromolecules have a high degree of functional specificity.

Heme (shown in Fig.10-9) is a cofactor consisting of an Fe^{2+} (ferrous) ion contained in the centre of a porphyrin, made up of four pyrrolic groups joined together by methine bridges.

---

[1] Yachandra VK, De Rose VJ, Latimer MJ, *et al*, *Science*, 260, 675 (1993). This enzyme seems to contain {2Mn(Ⅱ), 2Mn(Ⅳ)} or 4Mn(Ⅲ) mixed oxidation state cluster, which can transfer electrons to the photochemical active center.

Not all porphyrins contain iron, but a substantial fraction of porphyrin-containing metalloproteins have heme as their prosthetic group; these are known as hemoproteins. Hemes are most commonly recognized as components of hemoglobin, the red pigment in blood, but are also found in a number of other biologically important hemoproteins such as myoglobin, cytochrome, catalase, heme peroxidase, and endothelial nitric oxide synthase❶.

Hemoproteins have diverse biological functions including the transportation of diatomic gases, chemical catalysis, diatomic gas detection, and electron transfer. The heme iron serves as a source or sink of electrons during electron transfer or redox chemistry. In peroxidase reactions, the porphyrin molecule also serves as an electron source. In the transportation or detection of diatomic gases, the gas binds to the heme iron. During the detection of diatomic gases, the binding of the gas ligand to the heme iron induces conformational changes in the surrounding protein❷. In general, diatomic gases only bind to the reduced heme as ferrous Fe(Ⅱ), while most peroxidases cycle between Fe(Ⅲ) and Fe(Ⅳ) and hemoproteins involved in mitochondrial redox, oxidation-reduction, cycle between Fe(Ⅱ) and Fe(Ⅲ).

Hemoproteins achieve their remarkable functional diversity by modifying the environment of the heme macrocycle within the protein matrix❸. For example, the ability of hemoglobin to effectively deliver oxygen to tissues is due to specific amino acid residues located near the heme molecule❹. Hemoglobin reversibly binds to oxygen in the lungs when the pH is high, and the carbon dioxide concentration is low. When the situation is reversed (low pH and high carbon dioxide concentrations), hemoglobin will release oxygen into the tissues. This phenomenon, which states that hemoglobin's oxygen binding affinity is inversely proportional to both acidity and concentration of carbon dioxide, is known as the Bohr effect (shown in Fig.10-10)❺. The molecular mechanism behind this effect is the steric organization of the globin chain; a histidine residue, located adjacent to the heme group, becomes positively charged under acidic conditions (which are caused by dissolved CO_2 in working muscles, *etc.*), releasing oxygen from the heme group❻.

---

❶ Paoli M. Structure-function relationships in heme-proteins[J]. *DNA Cell Biol*, 2002, 21(4): 271-280. Alderton W K. Nitric oxide synthases: structure, function and inhibition[J]. *Biochem J.* 2001, 357(3): 593-615.

❷ Milani M. Structural bases for heme binding and diatomic ligand recognition in truncatedhemoglobins[J]. *J Inorg Biochem.* 2005, 99(1): 97-109.

❸ Poulos T. Heme enzyme structure and function[J]. *Chem Rev*, 2014, 114 (7): 3919-3962.

❹ Thom C S. Hemoglobin variants: biochemical properties and clinical correlates[J]. *Cold Spring Harb Perspect Med*, 2013, 3(3): a011858.

❺ Bohr C, Hasselbalch K, Krogh A. Concerning a Biologically Important Relationship—The Influence of the Carbon Dioxide Content of Blood on Its Oxygen Binding. From the Physiology Laboratory of the University of Copenhagen.

❻ Ackers G K, Holt J M. Asymmetriccooperativity in a symmetric tetramer: human hemoglobin[J]. *J Biol Chem*, 2006, 281 (17): 11441-3. [The most common type is heme B($C_{34}H_{32}O_4N_4Fe$); other important types include heme A($C_{49}H_{56}O_6N_4Fe$) and heme C($C_{34}H_{36}O_4N_4S_2Fe$)]

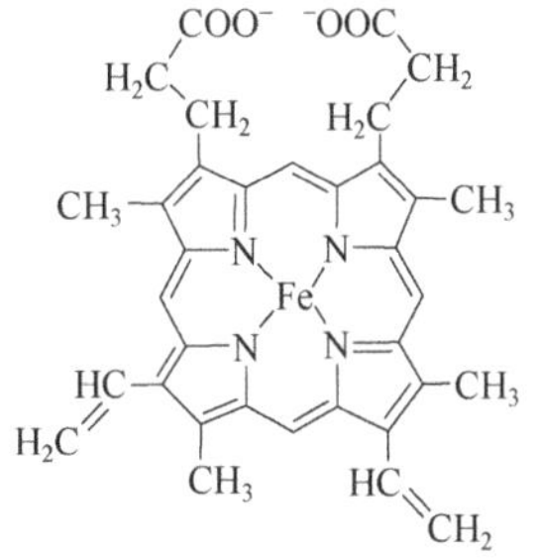

Fig.10-9 Heme B

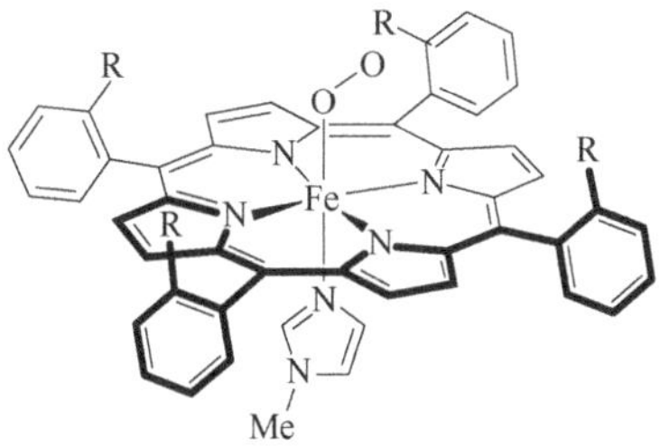

Fig.10-10 The coordination environment of Fe(Ⅱ) in myoglobin

Some biomacromolecules containing metal ions with different functions are listed in Table 10-8. The functions of hemoglobin and myoglobin are to transport and store oxygen, respectively. Metalloproteins with biocatalysis are called proteases, and many proteases are named after their functions. For example, oxygenases are used to catalyze oxygenation reactions, hydrogenases catalyze hydrogenation or hydrogen release reactions, and nitrogenases catalyze the conversion of N_2 to NH_3. Some other metalloenzymes are given in the Table 10-9, according to the metal involved.

Table 10-8 Some Biomacromolecules Containing Metal Ions

Protease (biocatalyst)	Protein for transmission and storage	Non-protein molecules
Oxygenase, Hydrogenase (Fe), Nitrogenase (Fe, Mo), Oxidase, Reductase, Superoxide dismutase, Hydroxylase (Fe, Mo, Cu), Superoxide dismutase (Mo, Cu, Zn), Carboxypeptide enzyme (Zn), Phosphatase (Zn, Cu, Mg), Aminopeptidase (Mg, Mn), Vitamin B_{12} coenzyme (Co)	Cytochrome, Iron-sulfur protein, Ferritin, Iron transfer protein, Myoglobin, Hemoglobin, Hemoglobin (Fe), Blue copper protein, Plasma ceruloplasmin, Hemocyanin (Cu)	Iron-containing cells (Fe), chlorophyll (Mg), bones (Ca, Si)

Note: the metal ions are given in parentheses.

Table 10-9 Some Other Metalloenzymes[1]

Ion	Examples of enzymes containing this ion
Magnesium	Glucose 6-phosphatase; Hexokinase; DNA polymerase
Manganese	Arginase; Oxygen-evolving complex
Iron	Catalase; Hydrogenase; IRE-BP; Aconitase
Cobalt	Nitrile hydratase; Methionyl aminopeptidase; Methylmalonyl-CoA mutase; Isobutyryl-CoA mutase
Nickel	Urease; Hydrogenase; Methyl-coenzyme M reductase (MCR)
Copper	Cytochrome oxidase; Laccase; Nitrous-oxide reductase; Nitrite reductase
Zinc	Carbonic anhydrase; Alcohol dehydrogenase; Carboxypeptidase; Aminopeptidase; Beta amyloid
Cadmium	Metallothionein; Thiolate proteins
Molybdenum	Nitrate reductase; Sulfite oxidase; Xanthine oxidase; DMSO reductase
Tungsten	Acetylene hydratase
various	Metallothionein; Phosphatase

[1] Nickel and Its Surprising Impact in Nature. *Metal Ions in Life Sciences.* **2** **(2008)**. *Wiley.* ISBN 978-0-470-01671-8; Metallomics and the Cell. *Metal Ions in Life Sciences.* **12** **(2013)**. Springer. ISBN 978-94-007-5561-1. ISSN 1868-0402; Cadmium: From Toxicology to Essentiality. *Metal Ions in Life Sciences.* **11.(2013)** Springer; The Metal-Driven Biogeochemistry of Gaseous Compounds in the Environment. *Metal Ions in Life Sciences.* **14.(2014)** Springer.

Exercises

1. Supply acceptable names for the following.

(a) $[Co(NH_3)_6]Cl_2$

(b) $[Cu(H_2O)_2(NH_3)_4]SO_4$

(c) $PtCl_2(en)$

(d) $[CrBr(H_2O)_5]^{2+}$

(e) $Rb[AgF_4]$

(f) $Na_2[Fe(CN)_5NO]$

(g) $K_2[Co(SCN)_4]$

2. Write appropriate formulas for the following.

(a) potassium hexacyanidoferrate(Ⅲ)

(b) bis(ethylenediamine)copper(Ⅱ) ion

(c) pentaaquahydroxidoaluminum(Ⅲ) chloride

(d) amminechloridobis(ethylenediamine) chromium(Ⅲ) sulfate

(e) tris(ethylenediamine)iron(Ⅲ) hexacyanidoferrate(Ⅱ)

3. Write the electronic configuration according to the magnetic moment of the following complexes.

(a) $[Mn(NCS)_6]^{4-}$ $(6.06\mu_B)$

(b) $[Cr(NH_3)_6]^{3+}$ $(3.9\mu_B)$

4. Explain the following observations in terms of complex-ion formation. (a) $Al(OH)_3(s)$ is soluble in NaOH(aq) but insoluble in $NH_3(aq)$; (b) $ZnCO_3(s)$ is soluble in $NH_3(aq)$, but ZnS(s) is not.

5. Explain the following observations in terms of complex-ion formation. (a) $CoCl_3$ is unstable in aqueous solution, being reduced to $CoCl_2$ and liberating $O_2(g)$. Yet, $[Co(NH_3)_6]Cl_3$ can be easily maintained in aqueous solution; (b) AgI is insoluble in water and in dilute $NH_3(aq)$, but AgI will dissolve in an aqueous solution of sodium thiosulfate.

Appendix

Ⅰ Some Basic Constants of Physics and Chemistry

(SI Base Units in 1986)

Quantity	Symbol	Value	Unit
Speed of light	c	299792458	m/s
Vacuum permeability	μ_0	$4\pi\times10^{-7}$	H/m
Vacuum permittivity [$=1/(\mu_0c^2)$]	ε_0	8.854188×10^{-12}	F/m
Gravitational constant	G	$6.67259(85)\times10^{-11}$	$m^3/(kg\cdot s^2)$
Planck constant	h	$6.6260755(40)\times10^{-34}$	J·s
$h/2\pi$	$\hbar$	$1.05457266(63)\times10^{-34}$	J·s
Elementary charge	e	$1.60217733(49)\times10^{-19}$	C
Mass of electron	m_e	$0.91093897(54)\times10^{-30}$	kg
Mass of proton	m_p	$1.6726231(10)\times10^{-27}$	kg
Proton-electron mass ratio	m_p/m_e	1836.152701(37)	—
Fine structure constant	α	$7.29735308(33)\times10^{-3}$	—
Rydberg constant	R_∞	10973731.534(13)	m^{-1}
Avogadro constant	L, N_A	$6.0221367(36)\times10^{23}$	mol^{-1}
Faraday constant	F	96485.309(29)	C/mol
Molar gas constant	R	8.314510(70)	J/(mol·K)
Boltzmann constant，R/L_A	k	$1.380658(12)\times10^{-23}$	J/K
Ev, (e/C) $J=\{e\}J$	eV	$1.60217733(49)\times10^{-19}$	J
Unified atomic mass unit	u	$1.6605402(10)\times10^{-27}$	kg

Ⅱ Thermodynamic Properties of Substances at 298.15K

(10^5Pa, roughly 1mol/L for aqueous solutions)

1. Elemental and inorganic

Substance	$\Delta_f H_m$/(kJ/mol)	$\Delta_f G_m$/(kJ/mol)	S_m/[J/(K·mol)]
Ag(s)	0	0	42.712
Ag_2CO_3(s)	−506.14	−437.09	167.36
Ag_2O(s)	−30.56	−10.82	121.71
$AgNO_3$(s)	−124.4	−33.41	140.9

(Continued)

Substance	$\Delta_f H_m$/(kJ/mol)	$\Delta_f G_m$/(kJ/mol)	S_m/[J/(K·mol)]
Al(s)	0	0	28.315
Al(g)	313.80	273.2	164.553
α-Al_2O_3	−1669.8	−2213.16	0.986
$Al_2(SO_4)_3$(s)	−3434.98	−3728.53	239.3
Br(g)	111.884	82.396	175.021
Br_2(g)	30.71	3.109	245.455
Br_2(l)	0	0	152.3
C(g)	718.384	672.942	158.101
C(Diamond)	1.896	2.866	2.439
C(Graphite)	0	0	5.694
CO(g)	−110.525	−137.285	198.016
CO_2(g)	−393.511	−394.38	213.76
Ca(s)	0	0	41.63
CaC_2(s)	−62.8	−67.8	70.2
$CaCO_3$(Calcite)	−1206.87	−1128.70	92.8
$CaCl_2$(s)	−795.0	−750.2	113.8
CaO(s)	−635.6	−604.2	39.7
$Ca(OH)_2$(s)	−986.5	−896.89	76.1
$CaSO_4$(Anhydrite)	−1432.68	−1320.24	106.7
Cl^-(aq)	−167.456	−131.168	55.10
Cl_2(g)	0	0	222.948
Cu(s)	0	0	33.32
CuO(s)	−155.2	−127.1	43.51
α-Cu_2O	−166.69	−146.33	100.8
F_2(g)	0	0	203.5
α-Fe	0	0	27.15
$FeCO_3$(s)	−747.68	−673.84	92.8
FeO(s)	−266.52	−244.3	54.0
Fe_2O_3(s)	−822.1	−741.0	90.0
Fe_3O_4(s)	−117.1	−1014.1	146.4
H(g)	217.94	203.122	114.724
H_2(g)	0	0	130.695
D_2(g)	0	0	144.884
HBr(g)	−36.24	−53.22	198.60
HBr(aq)	−120.92	−102.80	80.71
HCl(g)	−92.311	−95.265	186.786
HCl(aq)	−167.44	−131.17	55.10
H_2CO_3(aq)	−698.7	−623.37	191.2
HF(g)	-271.1	−273.2	173.8
HI(g)	−25.94	−1.32	206.42
H_2O(g)	−241.825	−228.577	188.823
H_2O(l)	−285.838	−237.142	69.940
H_2O(s)	−291.850	−234.03	39.4

(Continued)

Substance	$\Delta_f H_m$/(kJ/mol)	$\Delta_f G_m$/(kJ/mol)	S_m/[J/(K·mol)]
H_2O_2(l)	−187.61	−118.04	102.26
H_2S(g)	−20.146	−33.040	205.75
H_2SO_4(l)	−811.35	−866.4	156.85
H_2SO_4(aq)	−811.32	—	—
HSO_4(aq)	−885.75	−752.99	126.86
I_2(s)	0	0	116.7
I_2(g)	62.242	19.34	260.60
N_2(g)	0	0	191.598
NH_3(g)	−46.19	−16.603	192.61
NO(g)	89.860	90.37	210.309
NO_2(g)	33.85	51.86	240.57
N_2O(g)	81.55	103.62	220.10
N_2O_4(g)	9.660	98.39	304.42
N_2O_5(g)	2.51	110.5	342.4
O(g)	247.521	230.095	161.063
O_2(g)	0	0	205.138
O_3(g)	142.3	163.45	237.7
OH^-(aq)	−229.940	−157.297	−10.539
S(Monoclinic)	0.29	0.096	32.55
S(Rhombic)	0	0	31.9
S(g)	222.80	182.27	167.825
SO_2(g)	−296.90	−300.37	248.64
SO_3(g)	−395.18	−370.40	256.34
SO_4^{2-} (aq)	−907.51	−741.90	17.2

2. Organic compounds

Molecular formula	Status	$\Delta_f H_m$/(kJ/mol)	$\Delta_f G_m$/(kJ/mol)	S_m/[J/(K·mol)]
$CBrClF_2$	(g)	—	—	318.5
CBr_2Cl_2	(g)	—	—	347.8
CBr_2F_2	(g)	—	—	325.3
CCl_4	(l)	−128.2	—	—
	(g)	−95.8	—	—
CS_2	(l)	89.0	64.6	151.3
	(g)	116.6	67.1	237.8
$CHCl_3$	(l)	−134.5	−73.7	201.7
	(g)	−103.1	6.0	295.7
CHF_3	(g)	−695.4	—	259.7
$CHBr_3$	(l)	−28.5	−5.0	220.9
	(g)	17.0	8.0	330.9
CHI_3	(cr)	141.0	—	—
	(g)	—	—	356.2

(Continued)

Molecular formula	Status	$\Delta_f H_m$/(kJ/mol)	$\Delta_f G_m$/(kJ/mol)	S_m/[J/(K·mol)]
CH_2Cl_2	(l)	−124.1	—	177.8
	(g)	−95.6	—	270.2
CH_2N_2	(cr)	58.8	—	—
CH_2N_2	(g)	—	—	242.9
CH_2O	(g)	−108.6	−102.5	218.8
CH_2O_2	(l)	−424.7	−361.4	129.0
	(g)	−378.6	—	—
CH_3Br	(l)	−59.4	—	—
	(g)	−35.5	−26.3	246.4
CH_3Cl	(g)	−81.9	—	234.6
CH_3F	(g)	—	—	222.9
CH_3I	(l)	−12.3	—	163.2
	(g)	14.7	—	254.1
CH_4	(g)	−74.4	−50.3	186.3
CH_3NO	(l)	−254.0	—	—
CH_3NO_2	(l)	−113.1	−14.4	171.8
	(g)	−74.7	−6.8	275.0
CH_3NO_3	(l)	−159.0	−43.4	217.1
CH_4N_2O	(cr)	−333.6	—	—
CH_3OH	(l)	−239.1	−166.6	126.8
	(g)	−201.5	−162.6	239.8
CH_3SH	(l)	−46.4	−7.7	169.2
	(g)	−22.3	−9.3	255.2
CH_5N	(l)	−47.3	35.7	150.2
	(g)	−22.5	32.7	242.9
CH_6N_2	(l)	54.0	180.0	165.9
	(g)	94.3	187.0	278.8
C_2Cl_3N	(g)	—	—	—
C_2Cl_4	(l)	−50.6	3.0	266.9
C_2Cl_4O	(l)	−280.8	—	—
C_2Cl_6	(cr)	−202.8	—	237.3
C_2F_3N	(g)	−497.9	—	298.1
C_2F_4	(cr)	−820.5	—	—
	(g)	−658.9	—	300.1
C_2HCl	(g)	—	—	242.0
C_2HCl_3	(l)	−43.6	—	228.4
	(g)	−8.1	—	324.8
C_2HCl_3O	(l)	−236.2	—	—
	(g)	−196.6	—	—
$C_2HCl_3O_2$	(cr)	−503.3	—	—
C_2H_2	(g)	228.2	210.7	200.9

(Continued)

Molecular formula	Status	$\Delta_f H_m$/(kJ/mol)	$\Delta_f G_m$/(kJ/mol)	S_m/[J/(K·mol)]
C_2H_2O	(l)	−67.9	—	—
	(g)	−47.5	−48.3	247.6
$C_2H_2O_2$	(g)	−212.0	—	—
$C_2H_2O_4$	(cr)	−821.7	—	109.8
	(g)	−723.7	—	—
C_2H_3Br	(g)	79.2	81.8	275.8
C_2H_3Cl	(cr)	−94.1	—	—
	(l)	14.6	—	—
	(g)	37.3	53.6	264.0
C_2H_3IO	(l)	−162.5	—	—
$C_2H_3ClO_2$	(cr)	−510.5	—	—
C_2H_3N	(l)	31.4	77.2	149.6
	(g)	64.3	81.7	245.1
C_2H_4	(g)	52.5	68.4	219.6
C_2H_4O acetaldehyde	(l)	−191.8	−127.6	160.2
	(g)	−166.2	−132.8	263.7
C_2H_4O ethylene oxide	(l)	−77.8	−11.8	153.9
	(g)	−52.6	−13.0	242.5
C_2H_4O acetic acid	(l)	−484.5	−389.9	159.8
	(g)	−432.8	−374.5	282.5

Ⅲ Dissociation Equilibrium Constants of Weak Electrolytes in Water at 298.15K

Weak electrolyte	Molecular formula	K	pK
Arsenic acid	H_3AsO_4	$6.3\times10^{-3}(K_{a1})$ $1.0\times10^{-7}(K_{a2})$ $3.2\times10^{-12}(K_{a3})$	2.20 7.00 11.50
Arsenite	$HAsO_2$	$6.0\times10^{-10}(K_a)$	9.22
Boric acid	H_3BO_3	$5.8\times10^{-10}(K_a)$	9.24
Pyroboric acid	$H_2B_4O_7$	$1.0\times10^{-4}(K_{a1})$ $1.0\times10^{-9}(K_{a2})$	4 9
Carbonic acid	$H_2CO_3(CO_2+H_2O)$	$4.2\times10^{-7}(K_{a1})$ $5.6\times10^{-11}(K_{a2})$	6.38 10.25
Hydrocyanic acid	HCN	$6.2\times10^{-10}(K_a)$	9.21
Chromic acid	H_2CrO_4	$1.8\times10^{-1}(K_{a1})$ $3.2\times10^{-7}(K_{a2})$	0.74 6.50
Hydrofluoric acid	HF	$6.6\times10^{-4}(K_a)$	3.18
Nitrous acid	HNO_2	$5.1\times10^{-4}(K_a)$	3.29
Hydrogen peroxide	H_2O_2	$1.8\times10^{-12}(K_a)$	11.75
Phosphoric acid	H_3PO_4	$7.6\times10^{-3}(K_{a1})$ $6.3\times10^{-3}(K_{a2})$ $4.4\times10^{-13}(K_{a3})$	2.12 7.2 12.36

(Continued)

Weak electrolyte	Molecular formula	K	pK
Pyrophosphate	$H_4P_2O_7$	$3.0\times10^{-2}(K_{a1})$ $4.4\times10^{-3}(K_{a2})$ $2.5\times10^{-7}(K_{a3})$ $5.6\times10^{-10}(K_{a4})$	1.52 2.36 6.60 9.25
Phosphorous acid	H_3PO_3	$5.0\times10^{-2}(K_{a1})$ $2.5\times10^{-7}(K_{a2})$	1.30 6.60
Hydrogen sulfate	H_2S	$1.3\times10^{-7}(K_{a1})$ $7.1\times10^{-15}(K_{a2})$	6.88 14.15
Sulfuric acid	HSO_4^-	$1.0\times10^{-2}(K_a)$	1.99
Sulfurous acid	$H_3SO_3(SO_2+H_2O)$	$1.3\times10^{-2}(K_{a1})$ $6.3\times10^{-8}(K_{a2})$	1.90 7.20
Metasilicate	H_2SiO_3	$1.7\times10^{-10}(K_{a1})$ $1.6\times10^{-12}(K_{a2})$	9.77 11.8
Formic acid	HCOOH	$1.8\times10^{-4}(K_a)$	3.74
Acetic acid	CH_3COOH	$1.8\times10^{-5}(K_a)$	4.74
Monochloroacetic acid	$CH_2ClCOOH$	$1.4\times10^{-3}(K_a)$	2.86
Dichloroacetic acid	$CHCl_2COOH$	$5.0\times10^{-2}(K_a)$	1.30
Trichloroacetic acid	CCl_3COOH	$0.23(K_a)$	0.64
Glycine	NH_3CH_2COOH $NH_3CH_2COO^-$	$4.5\times10^{-3}(K_{a1})$ $2.5\times10^{-10}(K_{a2})$	2.35 9.60
Ascorbic acid	$CH_2(OH)-CH(OH)-CH$ (ring: O, C=O, C(OH)=C(OH))	$5.0\times10^{-5}(K_{a1})$ $1.5\times10^{-10}(K_{a2})$	4.30 9.82
Lactic acid	$CH_3CHOHCOOH$	$1.4\times10^{-4}(K_a)$	3.86
Benzoic acid	C_6H_5COOH	$6.2\times10^{-5}(K_a)$	4.21
Oxalic acid	$H_2C_2O_4$	$5.9\times10^{-2}(K_{a1})$ $6.4\times10^{-5}(K_{a2})$	1.22 4.19
d-Tartaric acid	$C_4H_6O_6$	$9.1\times10^{-4}(K_{a1})$ $4.3\times10^{-5}(K_{a2})$	3.04 4.37
Phthalate	$C_6H_4(COOH)_2$ (benzene ring with –COOH, –COOH)	$1.1\times10^{-3}(K_{a1})$ $3.9\times10^{-6}(K_{a2})$	2.95 5.41
Citric acid	CH_2COOH $C(OH)COOH$ CH_2COOH	$7.4\times10^{-4}(K_{a1})$ $1.7\times10^{-5}(K_{a2})$ $4.0\times10^{-7}(K_{a3})$	3.13 4.76 6.40
Phenol	C_6H_5OH	$1.1\times10^{-10}(K_a)$	9.95
Ethylenediaminetetraacetic acid	$H_6\text{-}EDTA^{2+}$ $H_5\text{-}EDTA^{+}$ $H_4\text{-}EDTA$ $H_3\text{-}EDTA^{-}$ $H_2\text{-}EDTA^{2-}$ $H\text{-}EDTA^{3-}$	$0.1(K_{a1})$ $3\times10^{-2}(K_{a2})$ $1\times10^{-2}(K_{a3})$ $2.1\times10^{-3}(K_{a4})$ $6.9\times10^{-7}(K_{a5})$ $5.5\times10^{-11}(K_{a6})$	0.9 1.6 2.0 2.67 6.17 10.26
Ammonia	NH_3	$1.8\times10^{-5}(K_b)$	4.74
Hydrazine	H_2NNH_2	$3.0\times10^{-6}(K_{b1})$ $1.7\times10^{-5}(K_{b2})$	5.52 14.12
Hydroxylamine	NH_2OH	$9.1\times10^{-6}(K_b)$	8.04
Methylamine	CH_3NH_2	$4.2\times10^{-4}(K_b)$	3.38
Ethylamine	$C_2H_5NH_2$	$5.6\times10^{-4}(K_b)$	3.25
Dimethylamine	$(CH_3)_2NH$	$1.2\times10^{-4}(K_b)$	3.93
Diethylamine	$(C_2H_5)_2NH$	$1.3\times10^{-3}(K_b)$	2.89

(Continued)

Weak electrolyte	Molecular formula	K	pK
Ethanolamine	$HOCH_2CH_2NH_2$	$3.2\times10^{-5}(K_b)$	4.50
Triethanolamine	$(HOCH_2CH_2)_3N$	$5.8\times10^{-7}(K_b)$	6.24
Hexamethylenetetramine	$(CH_2)_6N_4$	$1.4\times10^{-9}(K_b)$	8.85
Ethylenediamine	$H_2NH_2CCH_2NH_2$	$8.5\times10^{-5}(K_{b1})$ $7.1\times10^{-8}(K_{b2})$	4.07 7.15
Pyridine	N	$1.7\times10^{-5}(K_b)$	8.77

Ⅳ Solubility Product of Common Insoluble Electrolytes

Insoluble electrolyte	$K_{sp}^{\ominus}$	Insoluble electrolyte	$K_{sp}^{\ominus}$
AgAc	1.94×10^{-3}	$MnCO_3$	2.24×10^{-11}
AgBr	5.35×10^{-13}	$BaSO_4$	1.08×10^{-10}
AgCl	1.77×10^{-10}	BaS_2O_3	1.6×10^{-5}
Ag_2CO_3	8.46×10^{-12}	$Bi(OH)_3$	4.0×10^{-31}
$Ag_2C_2O_4$	5.40×10^{-12}	BiOCl	1.8×10^{-97}
Ag_2CrO_4	1.12×10^{-12}	Bi_2S_3	1×10^{-9}
$Ag_2Cr_2O_7$	2.0×10^{-7}	$CaCO_3$	3.36×10^{-9}
AgI	8.52×10^{-17}	$CaC_2O_4\cdot H_2O$	2.32×10^{-4}
$AgIO_3$	3.17×10^{-8}	CaC_rO_4	7.1×10^{-4}
$AgNO_2$	6.0×10^{-4}	CaF_2	3.45×10^{-11}
AgOH	2.0×10^{-8}	$CaHPO_4$	1.0×10^{-6}
Ag_3PO_4	8.89×10^{-17}	$Ca(OH)_2$	5.02×10^{-33}
Ag_2SO_4	1.20×10^{-5}	$Ca_3(PO_4)_2$	2.07×10^{-5}
$Ag_2S(\alpha)$	6.3×10^{-50}	$CaSO_4$	4.93×10^{-7}
$Ag_2S(\beta)$	1.09×10^{-49}	$CaSO_3\cdot0.5H_2O$	3.1×10^{-12}
$Al(OH)_3$	1.3×10^{-33}	$CdCO_3$	1.0×10^{-8}
AuCl	2.0×10^{-13}	$CdC_2O_4\cdot3H_2O$	1.42×10^{-14}
$AuCl_3$	3.2×10^{-25}	$Cd(OH)_2$(fresh)	2.5×10^{-27}
$Au(OH)_3$	5.5×10^{-46}	CdS	8.0×10^{-13}
$BaCO_3$	2.58×10^{-9}	$CoCO_3$	1.4×10^{-15}
BaC_2O_4	1.6×10^{-7}	$Co(OH)_2$(pink)	1.6×10^{15}
BaC_rO_4	1.17×10^{-10}	$Co(OH)_2$(blue)	5.92×10^{-44}
BaF_2	1.84×10^{-7}	$Co(OH)_3$	1.6×10^{-21}
$Ba_3(PO_4)_2$	3.4×10^{-23}	α-CoS(fresh)	4.0×10^{-25}
$BaSO_3$	5.0×10^{-10}	β-CoS(deposited)	2.0×10^{-21}
CuBr	6.27×10^{-9}	$Cr(OH)_3$	6.3×10^{-25}
CuCN	3.47×10^{-20}	$Mn(OH)_2$	1.9×10^{-13}
$CuCO_3$	1.4×10^{-10}	MnS(amorphous)	2.5×10^{-10}
CuCl	1.72×10^{-7}	MnS(crystallization)	2.5×10^{-13}
$CuCrO_4$	3.6×10^{-6}	Na_3AlF_6	4.0×10^{-10}

(Continued)

Insoluble electrolyte	$K_{sp}^{\ominus}$	Insoluble electrolyte	$K_{sp}^{\ominus}$
CuI	1.27×10^{-12}	$NiCO_3$	1.42×10^{-7}
CuOH	1.0×10^{-14}	$Ni(OH)_2$(fresh)	2×10^{-15}
$Cu(OH)_2$	2.2×10^{-20}	α-NiS	3.2×10^{-19}
$Cu_3(PO_4)_2$	1.40×10^{-37}	β-NiS	1.0×10^{-24}
$Cu_2P_2O_7$	8.3×10^{-16}	γ-NiS	2.0×10^{-26}
CuS	6.3×10^{-36}	$PbBr_2$	6.60×10^{-6}
Cu_2S	2.5×10^{-48}	$PbCl_2$	1.7×10^{-5}
$FeCO_3$	3.2×10^{-11}	$PbCO_3$	7.4×10^{-14}
$FeC_2O_4\cdot2H_2O$	3.2×10^{-7}	PbC_2O_4	4.8×10^{-10}
$Fe(OH)_2$	4.87×10^{-17}	$PbCrO_4$	2.8×10^{-13}
$Fe(OH)_3$	2.79×10^{-39}	PbF_2	7.12×10^{-7}
FeS	6.3×10^{-18}	PbI_2	9.8×10^{-9}
Hg_2Cl_2	1.43×10^{-18}	$Pb(OH)_2$	1.43×10^{-20}
Hg_2I_2	5.2×10^{-29}	$PbMoO_4$	1.0×10^{-13}
$Hg(OH)_2$	3.0×10^{-26}	PbS	8×10^{-28}
Hg_2S	1.0×10^{-47}	$PbSO_4$	2.53×10^{-8}
HgS (red)	4.0×10^{-53}	$Sn(OH)_2$	5.45×10^{-27}
HgS (black)	1.6×10^{-52}	$Sn(OH)_4$	1×10^{-56}
Hg_2SO_4	6.5×10^{-7}	SnS	1.0×10^{-25}
KIO_4	3.71×10^{-4}	$SrCO_3$	5.60×10^{-10}
$K_2[PtCl_6]$	7.48×10^{-6}	$SrC_2O_4\cdot H_2O$	1.6×10^{-7}
$K_2[SiF_6]$	8.7×10^{-7}	$SrCrO_4$	2.2×10^{-5}
Li_2CO_3	8.15×10^{-4}	$SrSO_4$	3.44×10^{-7}
LiF	1.84×10^{-3}	$ZnCO_3$	1.46×10^{-10}
$MgNH_4PO_4$	2.5×10^{-13}	$ZnC_2O_4\cdot2H_2O$	1.38×10^{-9}
$MgCO_3$	6.82×10^{-6}	$Zn(OH)_2$	3.0×10^{-17}
MgF_2	5.16×10^{-11}	α-ZnS	1.6×10^{-24}
$Mg(OH)_2$	5.61×10^{-12}	β-ZnS	2.5×10^{-22}

V Standard Electrode (Reduction) Potentials at 298.15K

1. In acidic solution

Electrode reaction	$E^{\ominus}/V$	Electrode reaction	$E^{\ominus}/V$
$Ag^++e^-=Ag$	0.7996	$V^{2+}+2e^-=V$	−1.175
$Ag^{2+}+e^-=Ag^+$	1.980	$Cd^{2+}+2e^-=Cd(Hg)$	−0.3521
$AgAc+e^-=Ag+Ac^-$	0.643	$Ce^{3+}+3e^-=Ce$	−2.483
$AgBr+e^-=Ag+Br^-$	0.07133	$Cl_2(g)+2e^-=2Cl^-$	1.35827
$Ag_2BrO_3+2e^-=2Ag+BrO_3^{2-}$	0.546	$HClO+H^++e^-=\frac{1}{2}Cl_2+H_2O$	1.611
$Ag_2C_2O_4+2e^-=2Ag+C_2O_4^{2-}$	0.4647	$HClO+H^++2e^-=Cl^-+H_2O$	1.482

(Continued)

Electrode reaction	$E^\ominus$/V	Electrode reaction	$E^\ominus$/V
$AgCl+e^-=Ag+Cl^-$	0.22233	$ClO_2+H^++e^-=HClO_2$	1.277
$Ag_2CO_3+2e^-=2Ag+CO_3^{2-}$	0.47	$HClO_2+2H^++2e^-=HClO+H_2O$	1.645
$Ag_2CrO_4+2e^-=2Ag+CrO_4^{2-}$	0.4470	$HClO_2+3H^++3e^-=\frac{1}{2}Cl_2+2H_2O$	1.628
$AgF+e^-=Ag+F^-$	0.779	$HClO_2+3H^++4e^-=Cl^-+2H_2O$	1.570
$AgI+e^-=Ag+I^-$	−0.15224	$ClO_3^-+2H^++e^-=ClO_2+H_2O$	1.152
$Ag_2S+2H^++2e^-=2Ag+H_2S$	−0.0366	$ClO_3^-+3H^++2e^-=HClO_2+H_2O$	1.214
$AgSCN+e^-=Ag+SCN^-$	0.08951	$ClO_3^-+6H^++5e^-=\frac{1}{2}Cl_2+3H_2O$	1.47
$Ag_2SO_4+2e^-=2Ag+SO_4^{2-}$	0.654	$ClO_3^-+6H^++6e^-=Cl^-+3H_2O$	1.451
$Al^{3+}+3e^-=Al$	−1.662	$ClO_4^-+2H^++2e^-=ClO_3^-+H_2O$	1.189
$AlF_6^{3-}+3e^-=Al+6F^-$	−2.069	$ClO_4^-+8H^++7e^-=\frac{1}{2}Cl_2+4H_2O$	1.39
$As_2O_3+6H^++6e^-=2As+3H_2O$	0.234	$ClO_4^-+8H^++8e^-=Cl^-+4H_2O$	1.389
$HAsO_2+3H^++3e^-=As+2H_2O$	0.248	$Co^{2+}+2e^-=Co$	−0.28
$H_3AsO_4+2H^++2e^-=HAsO_2+2H_2O$	0.560	$Co^{3+}+e^-=Co^{2+}$ (2mol/L H_2SO_4)	1.83
$Au^++e^-=Au$	1.692	$CO_2+2H^++2e^-=HCOOH$	−0.199
$Au^{3+}+3e^-=Au$	1.498	$Cr^{2+}+2e^-=Cr$	−0.913
$AuCl_4^-+3e^-=Au+4Cl^-$	1.002	$Cr^{3+}+e^-=Cr^{2+}$	−0.407
$Au^{3+}+2e^-=Au^+$	1.401	$Cr^{3+}+3e^-=Cr$	−0.744
$H_3BO_3+3H^++3e^-=B+3H_2O$	−0.8698	$Cr_2O_7^{2-}+14H^++6e^-=2Cr^{3+}+7H_2O$	1.232
$Ba^{2+}+2e^-=Ba$	−2.912	$HCrO_4^-+7H^++3e^-=Cr^{3+}+4H_2O$	1.350
$Ba^{2+}+2e^-=Ba(Hg)$	−1.570	$Cu^++e^-=Cu$	0.521
$Be^{2+}+2e^-=Be$	−1.847	$Cu^{2+}+e^-=Cu^+$	0.153
$BiCl_4^-+3e^-=Bi+4Cl^-$	0.16	$Cu^{2+}+2e^-=Cu$	0.3419
$Bi_2O_4+4H^++2e^-=2BiO^++2H_2O$	1.593	$CuCl+e^-=Cu+Cl^-$	0.124
$BiO^++2H^++3e^-=Bi+H_2O$	0.320	$F_2+2H^++2e^-=2HF$	3.053
$BiOCl+2H^++3e^-=Bi+Cl^-+H_2O$	0.1583	$F_2+2e^-=2F^-$	2.866
$Br_2(aq)+2e^-=2Br^-$	1.0873	$Fe^{2+}+2e^-=Fe$	−0.447
$Br_2(l)+2e^-=2Br^-$	1.066	$Fe^{3+}+3e^-=Fe$	−0.037
$HBrO+H^++2e^-=Br^-+H_2O$	1.331	$Fe^{3+}+e^-=Fe^{2+}$	0.771
$HBrO+H^++e^-=\frac{1}{2}Br_2(aq)+H_2O$	1.574	$[Fe(CN)_6]^{3-}+e^-=[Fe(CN)_6]^{4-}$	0.358
$HBrO+H^++e^-=\frac{1}{2}Br_2(l)+H_2O$	1.596	$FeO_4^{2-}+8H^++3e^-=Fe^{3+}+4H_2O$	2.20
$BrO_3^-+6H^++5e^-=\frac{1}{2}Br_2+3H_2O$	1.482	$Ga^{3+}+3e^-=Ga$	−0.560
$BrO_3^-+6H^++6e^-=Br^-+3H_2O$	1.423	$2H^++2e^-=H_2$	0.00000
$Ca^{2+}+2e^-=Ca$	−2.868	$H_2(g)+2e^-=2H^-$	−2.23
$Cd^{2+}+2e^-=Cd$	−0.4030	$HO_2+H^++e^-=H_2O_2$	1.495
$CdSO_4+2e^-=Cd+SO_4^{2-}$	−0.246	$O_2+4H^++4e^-=2H_2O$	1.229
$H_2O_2+2H^++2e^-=2H_2O$	1.776	$O(g)+2H^++2e^-=H_2O$	2.421
$Hg^{2+}+2e^-=Hg$	0.851	$O_3+2H^++2e^-=O_2+H_2O$	2.076

(Continued)

Electrode reaction	$E^{\ominus}$ / V	Electrode reaction	$E^{\ominus}$ / V
$2Hg^{2+}+2e^-=Hg_2^{2+}$	0.920	$P(red)+3H^++3e^-=PH_3(g)$	−0.111
$Hg_2^{2+}+2e^-=2Hg$	0.7973	$P(white)+3H^++3e^-=PH_3(g)$	−0.063
$Hg_2Br_2+2e^-=2Hg+2Br^-$	0.13923	$H_3PO_2+H^++e^-=P+2H_2O$	−0.508
$Hg_2Cl_2+2e^-=2Hg+2Cl^-$	0.26808	$H_3PO_3+2H^++2e^-=H_3PO_2+H_2O$	−0.499
$Hg_2I_2+2e^-=2Hg+2I^-$	−0.0405	$H_3PO_3+3H^++3e^-=P+3H_2O$	−0.454
$Hg_2SO_4+2e^-=2Hg+SO_4^{2-}$	0.6125	$H_3PO_4+2H^++2e^-=H_3PO_3+H_2O$	−0.276
$I_2+2e^-=2I^-$	0.5355	$Pb^{2+}+2e^-=Pb$	−0.126
$I_3^-+2e^-=3I^-$	0.536	$PbBr_2+2e^-=Pb+2Br^-$	−0.284
$H_5IO_6+H^++2e^-=IO_3^-+3H_2O$	1.601	$PbCl_2+2e^-=Pb+2Cl^-$	−0.267
$2HIO+2H^++2e^-=I_2+2H_2O$	1.439	$PbF_2+2e^-=Pb+2F^-$	−0.344
$HIO+H^++2e^-=I^-+H_2O$	0.987	$PbI_2+2e^-=Pb+2I^-$	−0.365
$2IO_3^-+12H^++10e^-=I_2+6H_2O$	1.195	$PbO_2+4H^++2e^-=Pb^{2+}+2H_2O$	1.455
$IO_3^-+6H^++6e^-=I^-+3H_2O$	1.085	$PbO_2+SO_4^{2-}+4H^++2e^-=PbSO_4+2H_2O$	1.6913
$In^{3+}+2e^-=In^+$	−0.443	$PbSO_4+2e^-=Pb+SO_4^{2-}$	−0.358
$In^{3+}+3e^-=In$	−0.3382	$Pd^{2+}+2e^-=Pd$	0.951
$Ir^{3+}+3e^-=Ir$	1.159	$PdCl_4^{2-}+2e^-=Pd+4Cl^-$	0.591
$K^++e^-=K$	−2.931	$Pt^{2+}+2e^-=Pt$	1.118
$La^{3+}+3e^-=La$	−2.522	$Rb^++e^-=Rb$	−2.98
$Li^++e^-=Li$	−3.0401	$Re^{3+}+3e^-=Re$	0.300
$Mg^{2+}+2e^-=Mg$	−2.372	$S+2H^++2e^-=H_2S(aq)$	0.142
$Mn^{2+}+2e^-=Mn$	−1.185	$S_2O_6^{2-}+4H^++2e^-=2H_2SO_3$	0.564
$Mn^{3+}+e^-=Mn^{2+}$	1.5415	$S_2O_8^{2-}+2e^-=2SO_4^{2-}$	2.010
$MnO_2+4H^++2e^-=Mn^{2+}+2H_2O$	1.224	$S_2O_8^{2-}+2H^++2e^-=2HSO_4^-$	2.123
$MnO_4^{2-}+e^-=MnO_4^{3-}$	0.558	$H_2SO_3+4H^++4e^-=S+3H_2O$	0.449
$MnO_4^-+4H^++3e^-=MnO_2+2H_2O$	1.679	$SO_4^{2-}+4H^++2e^-=H_2SO_3+H_2O$	0.172
$MnO_4^-+8H^++5e^-=Mn^{2+}+4H_2O$	1.507	$2SO_4^{2-}+4H^++2e^-=S_2O_6^{2-}+2H_2O$	−0.22
$Mo^{3+}+3e^-=Mo$	−0.200	$Sb+3H^++3e^-=SbH_3$	−0.510
$N_2+2H_2O+6H^++6e^-=2NH_4OH$	0.092	$Sb_2O_3+6H^++6e^-=2Sb+3H_2O$	0.152
$N_2+6H^++6e^-=2NH_3(aq)$	−3.09	$Sb_2O_5+6H^++4e^-=2SbO^++3H_2O$	0.581
$N_2O+2H^++2e^-=N_2+H_2O$	1.766	$SbO^++2H^++3e^-=Sb+H_2O$	0.212
$N_2O_4+2e^-=2NO_2^-$	0.867	$Sc^{3+}+3e^-=Sc$	−2.077
$N_2O_4+2H^++2e^-=2HNO_2$	1.065	$Se+2H^++2e^-=H_2Se(aq)$	−0.399
$N_2O_4+4H^++4e^-=2NO+2H_2O$	1.035	$H_2SeO_3+4H^++4e^-=Se+3H_2O$	0.74
$2NO+2H^++2e^-=N_2O+H_2O$	1.591	$SeO_4^{2-}+4H^++2e^-=H_2SeO_3+H_2O$	1.151
$HNO_2+H^++e^-=NO+H_2O$	0.983	$SiF_6^{2-}+4e^-=Si+6F^-$	−1.24
$2HNO_2+4H^++4e^-=N_2O+3H_2O$	1.297	$SiO_2(quartz)+4H^++4e^-=Si+2H_2O$	0.857
$NO_3^-+3H^++2e^-=HNO_2+H_2O$	0.934	$Sn^{2+}+2e^-=Sn$	−0.1375
$NO_3^-+4H^++3e^-=NO+2H_2O$	0.957	$Sn^{4+}+2e^-=Sn^{2+}$	0.151
$2NO_3^-+4H^++2e^-=N_2O_4+2H_2O$	0.803	$Sr^++e^-=Sr$	−4.10
$Na^++e^-=Na$	−2.71	$Sr^{2+}+2e^-=Sr$	−2.89

(Continued)

Electrode reaction	$E^{\ominus}$/V	Electrode reaction	$E^{\ominus}$/V
$Nb^{3+}+3e^-=Nb$	−1.1	$Sr^{2+}+2e^-=Sr(Hg)$	−1.793
$Ni^{2+}+2e^-=Ni$	−0.257	$Te+2H^++2e^-=H_2Te$	−0.793
$NiO_2+4H^++2e^-=Ni^{2+}+2H_2O$	1.678	$V^{3+}+e^-=V^{2+}$	−0.255
$O_2+2H^++2e^-=H_2O_2$	0.695	$VO^{2+}+2H^++e^-=V^{3+}+H_2O$	0.337
$Te^{4+}+4e^-=Te$	0.568	$VO_2^++2H^++e^-=VO^{2+}+H_2O$	0.991
$TeO_2+4H^++4e^-=Te+2H_2O$	0.593	$V(OH)_4^++2H^++e^-=VO^{2+}+3H_2O$	1.00
$TeO_4^-+8H^++7e^-=Te+4H_2O$	0.472	$V(OH)_4^++4H^++5e^-=V+4H_2O$	−0.254
$H_6TeO_6+2H^++2e^-=TeO_2+4H_2O$	1.02	$W_2O_5+2H^++2e^-=2WO_2+H_2O$	−0.031
$Th^{4+}+4e^-=Th$	−1.899	$WO_2+4H^++4e^-=W+2H_2O$	−0.119
$Ti^{2+}+2e^-=Ti$	−1.630	$WO_3+6H^++6e^-=W+3H_2O$	−0.090
$Ti^{3+}+e^-=Ti^{2+}$	−0.368	$2WO_3+2H^++2e^-=W_2O_5+H_2O$	−0.029
$TiO^{2+}+2H^++e^-=Ti^{3+}+H_2O$	0.099	$Y^{3+}+3e^-=Y$	−2.37
$TiO_2+4H^++2e^-=Ti^{2+}+2H_2O$	−0.502	$Zn^{2+}+2e^-=Zn$	−0.7618
$Tl^++e^-=Tl$	−0.336		

2. In basic solution

Electrode reaction	$E^{\ominus}$/V	Electrode reaction	$E^{\ominus}$/V
$AgCN+e^-=Ag+CN^-$	−0.017	$S_4O_6^{2-}+2e^-=2S_2O_3^{2-}$	0.08
$[Ag(CN)_2]^-+e^-=Ag+2CN^-$	−0.31	$2SO_3^{2-}+2H_2O+2e^-=S_2O_4^{2-}+4OH^-$	−1.12
$Ag_2O+H_2O+2e^-=2Ag+2OH^-$	0.342	$Cu(OH)_2+2e^-=Cu+2OH^-$	−0.222
$2AgO+H_2O+2e^-=Ag_2O+2OH^-$	0.607	$2Cu(OH)_2+2e^-=Cu_2O+2OH^-+H_2O$	−0.080
$Ag_2S+2e^-=2Ag+S^{2-}$	−0.691	$[Fe(CN)_6]^{3-}+e^-=[Fe(CN)_6]^{4-}$	0.358
$H_2AlO_3^-+H_2O+3e^-=Al+4OH^-$	−2.33	$Fe(OH)_3+e^-=Fe(OH)_2+OH^-$	−0.56
$AsO_2^-+2H_2O+3e^-=As+4OH^-$	−0.68	$H_2GaO_3^-+H_2O+3e^-=Ga+4OH^-$	−1.219
$AsO_4^{3-}+2H_2O+2e^-=AsO_2^-+4OH^-$	−0.71	$2H_2O+2e^-=H_2+2OH^-$	−0.8277
$H_2BO_3^-+5H_2O+8e^-=BH_4^-+8OH^-$	−1.24	$Hg_2O+H_2O+2e^-=2Hg+2OH^-$	0.123
$H_2BO_3^-+H_2O+3e^-=B+4OH^-$	−1.79	$HgO+H_2O+2e^-=Hg+2OH^-$	0.0977
$Ba(OH)_2+2e^-=Ba+2OH^-$	−2.99	$2IO^-+2H_2O+2e^-=I_2+4OH^-$	0.42
$Be_2O_3^{2-}+3H_2O+4e^-=2Be+6OH^-$	−2.63	$IO^-+H_2O+2e^-=I^-+2OH^-$	0.485
$Bi_2O_3+3H_2O+6e^-=2Bi+6OH^-$	−0.46	$IO_3^-+2H_2O+4e^-=IO^-+4OH^-$	0.15
$BrO^-+H_2O+2e^-=Br^-+2OH^-$	0.761	$IO_3^-+3H_2O+6e^-=I^-+6OH^-$	0.26
$BrO_3^-+3H_2O+6e^-=Br^-+6OH^-$	0.61	$Ir_2O_3+3H_2O+6e^-=2Ir+6OH$	0.098
$Ca(OH)_2+2e^-=Ca+2OH^-$	−3.02	$La(OH)_3+3e^-=La+3OH^-$	−2.90
$Ca(OH)_2+2e^-=Ca(Hg)+2OH^-$	−0.809	$Mg(OH)_2+2e^-=Mg+2OH^-$	−2.690
$ClO^-+H_2O+2e^-=Cl^-+2OH^-$	0.81	$MnO_4^-+2H_2O+3e^-=MnO_2+4OH^-$	0.595
$ClO_2^-+H_2O+2e^-=ClO^-+2OH^-$	0.66	$MnO_4^{2-}+2H_2O+2e^-=MnO_2+4OH^-$	0.60
$ClO_2^-+2H_2O+4e^-=Cl^-+4OH^-$	0.76	$Mn(OH)_2+2e^-=Mn+2OH^-$	−1.56
$ClO_3^-+H_2O+2e^-=ClO_2^-+2OH^-$	0.33	$Mn(OH)_3+e^-=Mn(OH)_2+OH^-$	0.15
$ClO_3^-+3H_2O+6e^-=Cl^-+6OH^-$	0.62	$2NO+H_2O+2e^-=N_2O+2OH^-$	0.76
$ClO_4^-+H_2O+2e^-=ClO_3^-+2OH^-$	0.36	$SO_3^{2-}+3H_2O+4e^-=S+6OH^-$	−0.66
$[Co(NH_3)_6]^{3+}+e^-=[Co(NH_3)_6]^{2+}$	0.108	$2NO_2^-+3H_2O+4e^-=N_2O+6OH^-$	0.15

(Continued)

Electrode reaction	$E^{\ominus}$ / V	Electrode reaction	$E^{\ominus}$ / V
$Co(OH)_2+2e^-=Co+2OH^-$	−0.73	$NO_3^-+H_2O+2e^-=NO_2^-+2OH^-$	0.01
$Co(OH)_3+e^-=Co(OH)_2+OH^-$	0.17	$2NO_3^-+2H_2O+2e^-=N_2O_4+4OH^-$	−0.85
$CrO_2^-+2H_2O+3e^-=Cr+4OH^-$	−1.2	$Ni(OH)_2+2e^-=Ni+2OH^-$	−0.72
$CrO_4^{2-}+4H_2O+3e^-=Cr(OH)_3+5OH^-$	−0.13	$NiO_2+2H_2O+2e^-=Ni(OH)_2+2OH^-$	−0.490
$Cr(OH)_3+3e^-=Cr+3OH^-$	−1.48	$O_2+2H_2O+2e^-=H_2O_2+2OH^-$	−0.146
$Cu^{2+}+2CN^-+e^-=[Cu(CN)_2]^-$	1.103	$O_2+2H_2O+4e^-=4OH^-$	0.401
$[Cu(CN)_2]^-+e^-=Cu+2CN^-$	−0.429	$O_3+H_2O+2e^-=O_2+2OH^-$	1.24
$Cu_2O+H_2O+2e^-=2Cu+2OH^-$	−0.360	$2SO_3^{2-}+3H_2O+4e^-=S_2O_3^{2-}+6OH^-$	−0.571
$P+3H_2O+3e^-=PH_3(g)+3OH^-$	−0.87	$SO_4^{2-}+H_2O+2e^-=SO_3^{2-}+2OH^-$	−0.93
$H_2PO_2^-+e^-=P+2OH^-$	−1.82	$SbO_2^-+2H_2O+3e^-=Sb+4OH^-$	−0.66
$HPO_3^{2-}+2H_2O+2e^-=H_2PO_2^-+3OH^-$	−1.65	$SbO_3^-+H_2O+2e^-=SbO_2^-+2OH^-$	−0.59
$HPO_3^{2-}+2H_2O+3e^-=P+5OH^-$	−1.71	$SeO_3^{2-}+3H_2O+4e^-=Se+6OH^-$	−0.366
$PO_4^{3-}+2H_2O+2e^-=HPO_3^{2-}+3OH^-$	−1.05	$SeO_4^{2-}+H_2O+2e^-=SeO_3^{2-}+2OH^-$	0.05
$PbO+H_2O+2e^-=Pb+2OH^-$	−0.580	$SiO_3^{2-}+3H_2O+4e^-=Si+6OH^-$	−1.697
$HPbO_2^-+H_2O+2e^-=Pb+3OH^-$	−0.537	$HSnO_2^-+H_2O+2e^-=Sn+3OH^-$	−0.909
$PbO_2+H_2O+2e^-=PbO+2OH^-$	0.247	$Sr(OH)+2e^-=Sr+2OH^-$	−2.88
$Pd(OH)_2+2e^-=Pd+2OH^-$	0.07	$Te+2e^-=Te^{2-}$	−1.143
$Pt(OH)_2+2e^-=Pt+2OH^-$	0.14	$TeO_3^{2-}+3H_2O+4e^-=Te+6OH^-$	−0.57
$ReO_4^-+4H_2O+7e^-=Re+8OH^-$	−0.584	$Th(OH)_4+4e^-=Th+4OH^-$	−2.48
$S+2e^-=S^{2-}$	−0.47627	$Tl_2O_3+3H_2O+4e^-=2Tl^++6OH^-$	0.02
$S+H_2O+2e^-=HS^-+OH^-$	−0.478	$ZnO_2^{2-}+2H_2O+2e^-=Zn+4OH^-$	−1.215
$2S+2e^-=S_2^{2-}$	−0.42836		

Weast R C. Handbook of Chemistry and Physics,D-151.70th ed.1989-1990.

Ⅵ Saturated Vapor Pressure of Water at Different Temperatures

T/℃	*P*×10³/Pa	*T*/℃	*P*×10³/Pa	*T*/℃	*P*×10³/Pa
0	0.61129	14	1.5988	28	3.7818
1	0.65716	15	1.7056	29	4.0078
2	0.70605	16	1.8185	30	4.2455
3	0.75813	17	1.938	31	4.4953
4	0.81359	18	2.0644	32	4.7578
5	0.8726	19	2.1978	33	5.0335
6	0.93537	20	2.3388	34	5.3229
7	1.0021	21	2.4877	35	5.6267
8	1.073	22	2.6447	36	5.9453
9	1.1482	23	2.8104	37	6.2795
10	1.2281	24	2.985	38	6.6298
11	1.3129	25	3.169	39	6.9969
12	1.4027	26	3.3629	40	7.3814
13	1.4979	27	3.567	41	7.784

(Continued)

T/℃	P×10^3/Pa	T/℃	P×10^3/Pa	T/℃	P×10^3/Pa
42	8.2054	85	57.815	128	254.25
43	8.6463	86	60.119	129	262.04
44	9.1075	87	62.499	130	270.02
45	9.5898	88	64.958	131	278.2
46	10.094	89	67.496	132	286.57
47	10.62	90	70.117	133	295.15
48	11.171	91	72.823	134	303.93
49	11.745	92	75.614	135	312.93
50	12.344	93	78.494	136	322.14
51	12.97	94	81.465	137	331.57
52	13.623	95	84.529	138	341.22
53	14.303	96	87.688	139	351.09
54	15.012	97	90.945	140	361.19
55	15.752	98	94.301	141	371.53
56	16.522	99	97.759	142	382.11
57	17.324	100	101.32	143	392.92
58	18.159	101	104.99	144	403.98
59	19.028	102	108.77	145	415.29
60	19.932	103	112.66	146	426.85
61	20.873	104	116.67	147	438.67
62	21.851	105	120.79	148	450.75
63	22.868	106	125.03	149	463.1
64	23.925	107	129.39	150	475.72
65	25.022	108	133.88	151	488.61
66	26.163	109	138.5	152	501.78
67	27.347	110	143.24	153	515.23
68	28.576	111	148.12	154	528.96
69	29.852	112	153.13	155	542.99
70	31.176	113	158.29	156	557.32
71	32.549	114	163.58	157	571.94
72	33.972	115	169.02	158	586.87
73	35.448	116	174.61	159	602.11
74	36.978	117	180.34	160	617.66
75	38.563	118	186.23	161	633.53
76	40.205	119	192.28	162	649.73
77	41.905	120	198.48	163	666.25
78	43.665	121	204.85	164	683.1
79	45.487	122	211.38	165	700.29
80	47.373	123	218.09	166	717.83
81	49.324	124	224.96	167	735.7
82	51.342	125	232.01	168	753.94
83	53.428	126	239.24	169	772.52
84	55.585	127	246.66	170	791.47

(Continued)

T/℃	$P\times10^3$/Pa	T/℃	$P\times10^3$/Pa	T/℃	$P\times10^3$/Pa
171	810.78	214	2063.4	257	4465.1
172	830.47	215	2104.2	258	4539
173	850.53	216	2145.7	259	4613.7
174	870.98	217	2187.8	260	4689.4
175	891.8	218	2230.5	261	4766.1
176	913.03	219	2273.8	262	4843.7
177	934.64	220	2317.8	263	4922.3
178	956.66	221	2362.5	264	5001.8
179	979.09	222	2407.8	265	5082.3
180	1001.9	223	2453.8	266	5163.8
181	1025.2	224	2500.5	267	5246.3
182	1048.9	225	2547.9	268	5329.8
183	1073	226	2595.9	269	5414.3
184	1097.5	227	2644.6	270	5499.9
185	1122.5	228	2694.1	271	5586.4
186	1147.9	229	2744.2	272	5674
187	1173.8	230	2795.1	273	5762.7
188	1200.1	231	2846.7	274	5852.4
189	1226.1	232	2899	275	5943.1
190	1254.2	233	2952.1	276	6035
191	1281.9	234	3005.9	277	6127.9
192	1310.1	235	3060.4	278	6221.9
193	1338.8	236	3115.7	279	6317.2
194	1368	237	3171.8	280	6413.2
195	1397.6	238	3288.6	281	6510.5
196	1427.8	239	3286.3	282	6608.9
197	1458.5	240	3344.7	283	6708.5
198	1489.7	241	3403.9	284	6809.2
199	1521.4	242	3463.9	285	6911.1
200	1553.6	243	3524.7	286	7014.1
201	1568.4	244	3586.3	287	7118.3
202	1619.7	245	3648.8	288	7223.7
203	1653.6	246	3712.1	289	7330.2
204	1688	247	3776.2	290	7438
205	1722.9	248	3841.2	291	7547
206	1758.4	249	3907	292	7657.2
207	1794.5	250	3973.6	293	7768.6
208	1831.1	251	4041.2	294	7881.3
209	1868.4	252	4109.6	295	7995.2
210	1906.2	253	4178.9	296	8110.3
211	1944.6	254	4249.1	297	8226.8
212	1983.6	255	4320.2	298	8344.5
213	2023.2	256	4392.2	299	8463.5

(Continued)

T/℃	P×10³/Pa	T/℃	P×10³/Pa	T/℃	P×10³/Pa
300	8583.8	325	12046	350	16521
301	8705.4	326	12204	351	16825
302	8828.3	327	12364	352	16932
303	8952.6	328	12525	353	17138
304	9078.2	329	12688	354	17348
305	9205.1	330	12852	355	17561
306	9333.4	331	13019	356	17775
307	9463.1	332	13187	357	17992
308	9594.2	333	13357	358	18211
309	9726.7	334	13528	359	18432
310	9860.5	335	13701	360	18655
311	9995.8	336	13876	361	18881
312	10133	337	14053	362	19110
313	10271	338	14232	363	19340
314	10410	339	14412	364	19574
315	10551	340	14594	365	19809
316	10694	341	14778	366	20048
317	10838	342	14964	367	20289
318	10984	343	15152	368	20533
319	11131	344	15342	369	20780
320	11279	345	15533	370	21030
321	11429	346	15727	371	21286
322	11581	347	15922	372	21539
323	11734	348	16120	373	21803
324	11889	349	16320	—	—

Ⅶ Ground-state Electron Configurations

1 H hydrogen : $1s^1$	15 P phosphorus : [Ne] $3s^2\ 3p^3$
2 He helium : $1s^2$	16 S sulfur : [Ne] $3s^2\ 3p^4$
3 Li lithium : [He] $2s^1$	17 Cl chlorine : [Ne] $3s^2\ 3p^5$
4 Be beryllium : [He] $2s^2$	18 Ar argon : [Ne] $3s^2\ 3p^6$
5 B boron : [He] $2s^2\ 2p^1$	19 K potassium : [Ar] $4s^1$
6 C carbon : [He] $2s^2\ 2p^2$	20 Ca calcium : [Ar] $4s^2$
7 N nitrogen : [He] $2s^2\ 2p^3$	21 Sc scandium : [Ar] $3d^1\ 4s^2$
8 O oxygen : [He] $2s^2\ 2p^4$	22 Ti titanium : [Ar] $3d^2\ 4s^2$
9 F fluorine : [He] $2s^2\ 2p^5$	23 V vanadium : [Ar] $3d^3\ 4s^2$
10 Ne neon : [He] $2s^2\ 2p^6$	24 Cr chromium : [Ar] $3d^5\ 4s^1$
11 Na sodium : [Ne] $3s^1$	25 Mn manganese : [Ar] $3d^5\ 4s^2$
12 Mg magnesium : [Ne] $3s^2$	26 Fe iron : [Ar] $3d^6\ 4s^2$
13 Al aluminium : [Ne] $3s^2\ 3p^1$	27 Co cobalt : [Ar] $3d^7\ 4s^2$
14 Si silicon : [Ne] $3s^2\ 3p^2$	28 Ni nickel : [Ar] $3d^8\ 4s^2$

(Continued)

29 Cu copper : [Ar] $3d^{10}\ 4s^1$	74 W tungsten : [Xe] $4f^{14}\ 5d^4\ 6s^2$
30 Zn zinc : [Ar] $3d^{10}\ 4s^2$	75 Re rhenium : [Xe] $4f^{14}\ 5d^5\ 6s^2$
31 Ga gallium : [Ar] $3d^{10}\ 4s^2\ 4p^1$	76 Os osmium : [Xe] $4f^{14}\ 5d^6\ 6s^2$
32 Ge germanium : [Ar] $3d^{10}\ 4s^2\ 4p^2$	77 Ir iridium : [Xe] $4f^{14}\ 5d^7\ 6s^2$
33 As arsenic : [Ar] $3d^{10}\ 4s^2\ 4p^3$	78 Pt platinum : [Xe] $4f^{14}\ 5d^9\ 6s^1$
34 Se selenium : [Ar] $3d^{10}\ 4s^2\ 4p^4$	79 Au gold : [Xe] $4f^{14}\ 5d^{10}\ 6s^1$
35 Br bromine : [Ar] $3d^{10}\ 4s^2\ 4p^5$	80 Hg mercury : [Xe] $4f^{14}\ 5d^{10}\ 6s^2$
36 Kr krypton : [Ar] $3d^{10}\ 4s^2\ 4p^6$	81 Tl thallium : [Xe] $4f^{14}\ 5d^{10}\ 6s^2\ 6p^1$
37 Rb rubidium : [Kr] $5s^1$	82 Pb lead : [Xe] $4f^{14}\ 5d^{10}\ 6s^2\ 6p^2$
38 Sr strontium : [Kr] $5s^2$	83 Bi bismuth : [Xe] $4f^{14}\ 5d^{10}\ 6s^2\ 6p^3$
39 Y yttrium : [Kr] $4d^1\ 5s^2$	84 Po polonium : [Xe] $4f^{14}\ 5d^{10}\ 6s^2\ 6p^4$
40 Zr zirconium : [Kr] $4d^2\ 5s^2$	85 At astatine : [Xe] $4f^{14}\ 5d^{10}\ 6s^2\ 6p^5$
41 Nb niobium : [Kr] $4d^4\ 5s^1$	86 Rn radon : [Xe] $4f^{14}\ 5d^{10}\ 6s^2\ 6p^6$
42 Mo molybdenum : [Kr] $4d^5\ 5s^1$	87 Fr francium : [Rn] $7s^1$
43 Tc technetium : [Kr] $4d^5\ 5s^2$	88 Ra radium : [Rn] $7s^2$
44 Ru ruthenium : [Kr] $4d^7\ 5s^1$	89 Ac actinium : [Rn] $6d^1\ 7s^2$
45 Rh rhodium : [Kr] $4d^8\ 5s^1$	90 Th thorium : [Rn] $6d^2\ 7s^2$
46 Pd palladium : [Kr] $4d^{10}$	91 Pa protactinium : [Rn] $5f^2\ 6d^1\ 7s^2$
47 Ag silver : [Kr] $4d^{10}\ 5s^1$	92 U uranium : [Rn] $5f^3\ 6d^1\ 7s^2$
48 Cd cadmium : [Kr] $4d^{10}\ 5s^2$	93 Np neptunium : [Rn] $5f^4\ 6d^1\ 7s^2$
49 In indium : [Kr] $4d^{10}\ 5s^2\ 5p^1$	94 Pu plutonium : [Rn] $5f^6\ 7s^2$
50 Sn tin : [Kr] $4d^{10}\ 5s^2\ 5p^2$	95 Am americium : [Rn] $5f^7\ 7s^2$
51 Sb antimony : [Kr] $4d^{10}\ 5s^2\ 5p^3$	96 Cm curium : [Rn] $5f^7\ 6d^1\ 7s^2$
52 Te tellurium : [Kr] $4d^{10}\ 5s^2\ 5p^4$	97 Bk berkelium : [Rn] $5f^9\ 7s^2$
53 I iodine : [Kr] $4d^{10}\ 5s^2\ 5p^5$	98 Cf californium : [Rn] $5f^{10}\ 7s^2$
54 Xe xenon : [Kr] $4d^{10}\ 5s^2\ 5p^6$	99 Es einsteinium : [Rn] $5f^{11}\ 7s^2$
55 Cs caesium : [Xe] $6s^1$	100 Fm fermium : [Rn] $5f^{12}\ 7s^2$
56 Ba barium : [Xe] $6s^2$	101 Md mendelevium : [Rn] $5f^{13}\ 7s^2$
57 La lanthanum : [Xe] $5d^1\ 6s^2$	102 No nobelium : [Rn] $5f^{14}\ 7s^2$
58 Ce cerium : [Xe] $4f^1\ 5d^1\ 6s^2$	103 Lr lawrencium : [Rn] $5f^{14}\ 7s^2\ 7p^1$
59 Pr praseodymium : [Xe] $4f^3\ 6s^2$	104 Rf rutherfordium : [Rn] $5f^{14}\ 6d^2\ 7s^2$
60 Nd neodymium : [Xe] $4f^4\ 6s^2$	105 Db dubnium : [Rn] $5f^{14}\ 6d^3\ 7s^2$
61 Pm promethium : [Xe] $4f^5\ 6s^2$	106 Sg seaborgium : [Rn] $5f^{14}\ 6d^4\ 7s^2$
62 Sm samarium : [Xe] $4f^6\ 6s^2$	107 Bh bohrium : [Rn] $5f^{14}\ 6d^5\ 7s^2$
63 Eu europium : [Xe] $4f^7\ 6s^2$	108 Hs hassium : [Rn] $5f^{14}\ 6d^6\ 7s^2$
64 Gd gadolinium : [Xe] $4f^7\ 5d^1\ 6s^2$	109 Mt meitnerium : [Rn] $5f^{14}\ 6d^7\ 7s^2$
65 Tb terbium : [Xe] $4f^9\ 6s^2$	110 Ds darmstadtium : [Rn] $5f^{14}\ 6d^8\ 7s^2$
66 Dy dysprosium : [Xe] $4f^{10}\ 6s^2$	111 Rg roentgenium : [Rn] $5f^{14}\ 6d^9\ 7s^2$
67 Ho holmium : [Xe] $4f^{11}\ 6s^2$	112 Cn copernicium : [Rn] $5f^{14}\ 6d^{10}\ 7s^2$
68 Er erbium : [Xe] $4f^{12}\ 6s^2$	113 Nh nihonium : [Rn] $5f^{14}\ 6d^{10}\ 7s^2\ 7p^1$
69 Tm thulium : [Xe] $4f^{13}\ 6s^2$	114 Fl flerovium : [Rn] $5f^{14}\ 6d^{10}\ 7s^2\ 7p^2$
70 Yb ytterbium : [Xe] $4f^{14}\ 6s^2$	115 Mc moscovium : [Rn] $5f^{14}\ 6d^{10}\ 7s^2\ 7p^3$
71 Lu lutetium : [Xe] $4f^{14}\ 5d^1\ 6s^2$	116 Lv livermorium : [Rn] $5f^{14}\ 6d^{10}\ 7s^2\ 7p^4$
72 Hf hafnium : [Xe] $4f^{14}\ 5d^2\ 6s^2$	117 Ts tennessine : [Rn] $5f^{14}\ 6d^{10}\ 7s^2\ 7p^5$
73 Ta tantalum : [Xe] $4f^{14}\ 5d^3\ 6s^2$	118 Og oganesson : [Rn] $5f^{14}\ 6d^{10}\ 7s^2\ 7p^6$

References

[1] 阿特金斯 P, 普瓦拉 J, 基勒 J. 物理化学: 第 11 版[M]. 侯文华, 译. 北京: 高等教育出版社, 2021.

[2] 胡小玲, 苏克和. 物理化学简明教程[M]. 北京: 科学出版社, 2012.

[3] 耿旺昌, 西北工业大学基础化学教研组. 工程化学基础[M]. 西安: 西北工业大学出版社, 2017.

[4] 耿旺昌, 颜静, 陈芳, 等. 普通化学实验英文教程[M]. 北京: 电子工业出版社, 2019.

[5] 史启祯. 无机化学与化学分析[M]. 3 版. 北京: 高等教育出版社, 2011.

[6] 威勒, 奥弗顿, 洛克, 等. 无机化学: 第 6 版[M]. 李珺, 雷依波, 刘斌, 等译. 北京: 高等教育出版社, 2018.

[7] 韦廉臣. 格物探原[M]. 广州: 南方日报出版社, 2018.

[8] 周公度. 化学是什么[M]. 2 版. 北京: 北京大学出版社, 2019.

[9] Atkins P. Chemistry: A Very Short Introduction[M]. Oxford: Oxford University Press, 2015.

[10] Brock W H. The History of Chemistry: A Very Short Introduction [M]. Oxford: Oxford University Press, 2016: 2-3.

[11] Brown T E, Bursten B E, LeMay H E. Chemistry: The Central Science[M]. 14th ed. New York: Pearson, 2018.

[12] Ebbing D D, Gammon S D. General Chemistry[M]. 11th ed. Boston: Cengage Learning, 2017.

[13] Forrest S R. The path to ubiquitous and low-cost organic electronic appliances on plastic[J]. *Nature*, 2004, 428(6986): 911-918.

[14] Pauling L. General Chemistry[M]. New York: Dover, 1970.

[15] Petrucci R H, Herring F G, Madura J D, *et al*. General Chemistry: Principles and Modern Applications[M]. 11th ed. Toronto: Pearson, 2017.

[16] Principe L. In retrospect: the sceptical chymist[J]. *Nature*, 2011, 469(6): 30-31.

[17] Rose S. The Chemistry of Life[M]. 4th ed. London: Penguin, 1999.

Periodic Table of the Elements

Period \ Group	1	2	3	4	5	6	7	8	9	10	11	12	13	14	15	16	17	18
1	1 H 1.008																	2 He 4.00
2	3 Li 6.94	4 Be 9.01											5 B 10.81	6 C 12.01	7 N 14.01	8 O 16.00	9 F 19.00	10 Ne 20.18
3	11 Na 22.99	12 Mg 24.30											13 Al 26.98	14 Si 28.08	15 P 30.97	16 S 32.06	17 Cl 35.45	18 Ar 39.95
4	19 K 39.10	20 Ca 40.08	21 Sc 44.96	22 Ti 47.87	23 V 50.94	24 Cr 52.00	25 Mn 54.94	26 Fe 55.84	27 Co 58.93	28 Ni 58.69	29 Cu 63.55	30 Zn 65.38	31 Ga 69.72	32 Ge 72.63	33 As 74.92	34 Se 78.97	35 Br 79.90	36 Kr 83.80
5	37 Rb 85.47	38 Sr 87.62	39 Y 88.91	40 Zr 91.22	41 Nb 92.91	42 Mo 95.95	43 Tc 98	44 Ru 101.07	45 Rh 102.91	46 Pd 106.42	47 Ag 107.87	48 Cd 112.41	49 In 114.82	50 Sn 118.71	51 Sb 121.76	52 Te 127.60	53 I 126.90	54 Xe 131.29
6	55 Cs 132.91	56 Ba 137.33	57-71* La-Lu	72 Hf 178.49	73 Ta 180.95	74 W 183.84	75 Re 186.21	76 Os 190.23	77 Ir 192.22	78 Pt 195.08	79 Au 196.97	80 Hg 200.59	81 Tl 204.38	82 Pb 207.2	83 Bi 208.98	84 Po 208.98	85 At 209.99	86 Rn 222.02
7	87 Fr 223.02	88 Ra 226.03	89-103** Ac-Lr	104 Rf 267.12	105 Db 270.13	106 Sg 269.13	107 Bh 270.13	108 Hs 269.13	109 Mt 278.16	110 Ds 281.17	111 Rg 281.17	112 Cn 285.18	113 Nh 286.18	114 Fl 289.19	115 Mc 289.20	116 Lv 293.20	117 Ts 293.21	118 Og 294.21

*Lanthanoids	57 La 138.91	58 Ce 140.12	59 Pr 140.91	60 Nd 144.24	61 Pm 144.91	62 Sm 150.36	63 Eu 151.96	64 Gd 157.25	65 Tb 158.93	66 Dy 162.50	67 Ho 164.93	68 Er 167.26	69 Tm 168.93	70 Yb 173.05	71 Lu 174.97
**Actinoids	89 Ac 227.03	90 Th 232.04	91 Pa 231.04	92 U 238.03	93 Np 237.05	94 Pu 244.06	95 Am 243.06	96 Cm 247.07	97 Bk 247.07	98 Cf 251.08	99 Es 252.08	100 Fm 257.10	101 Md 258.10	102 No 259.10	103 Lr 262.11